T0299074

Geometric Methods in Elastic Theory of Membranes in Liquid Crystal Phases

Second Edition

Peking University–World Scientific Advanced Physics Series

ISSN: 2382-5960

Series Editors: Enge Wang *(Peking University, China)*
Jian-Bai Xia *(Chinese Academy of Sciences, China)*

Peking University-World Scientific Advanced Physics Series

Vol
2

Geometric Methods in Elastic Theory of Membranes in Liquid Crystal Phases

Second Edition

Zhanchun Tu
Beijing Normal University, China

Zhongcan Ou-Yang
Chinese Academy of Sciences, China

Jixing Liu
Chinese Academy of Sciences, China

Yuzhang Xie
Tsinghua University, China

World Scientific

NEW JERSEY • LONDON • SINGAPORE • BEIJING • SHANGHAI • HONG KONG • TAIPEI • CHENNAI

Published by

World Scientific Publishing Co. Pte. Ltd.

5 Toh Tuck Link, Singapore 596224

USA office: 27 Warren Street, Suite 401-402, Hackensack, NJ 07601

UK office: 57 Shelton Street, Covent Garden, London WC2H 9HE

Library of Congress Cataloging-in-Publication Data
Names: Tu, Zhanchun, author. | Ou-Yang, Zhong-Can, author. | Liu, Ji-Xing, author. |
    Xie, Yu-Zhang, author.
Title: Geometric methods in elastic theory of membranes in liquid crystal phases / Zhanchun Tu
    (Beijing Normal University, China), Zhongcan Ou-Yang (Chinese Academy of Sciences, China),
    Jixing Liu (Chinese Academy of Sciences, China), Yuzhang Xie (Tsinghua University, China).
Other titles: Peking University-World Scientific advance physics series ; v. 2.
Description: Second edition. | Singapore ; Hackensack, NJ : World Scientific, [2017] |
    Series: Peking University-World Scientific advanced physics series, ISSN 2382-5960 ; vol. 2 |
    Includes bibliographical references and index.
Identifiers: LCCN 2017032700| ISBN 9789813227729 (hardcover ; alk. paper) |
    ISBN 9813227729 (hardcover ; alk. paper)
Subjects: LCSH: Liquid crystal films. | Polymer liquid crystals. | Liquid membranes.
Classification: LCC QC173.4.L55 T89 2017 | DDC 530.4/29--dc23
LC record available at https://lccn.loc.gov/2017032700

British Library Cataloguing-in-Publication Data
A catalogue record for this book is available from the British Library.

The work is originally published by Peking University Press in 2014.
This edition is published by World Scientific Publishing Company Pte Ltd by arrangement with
Peking University Press, Beijing, China.
All rights reserved. No reproduction and distribution without permission.

B&R Book Program

Printed in Singapore

Preface

Since the publication of the first edition of this book 20 years ago, a lot of theoretical results on geometric theory of membrane elasticity have been achieved. In particular, the shape equation and boundary conditions for open lipid membranes were obtained. The main changes in this second edition are as follows: We add a new chapter (Chapter 4) to explain how to calculate variational problem on a surface with a free edge by using a new mathematical tool — moving frame method and exterior differential forms, and how to derive the shape equation and boundary conditions for open lipid membranes through this new method. In addition, we include the recent concise work on chiral lipid membranes as a section in Chapter 5. In Chapter 6, we mention some topics that we have not fully investigated, but they are also important to geometric theory of membrane elasticity.

We owe our sincere gratitude to our colleagues, Profs. Udo Seifert, Jemal Guven, Ivailo Mladenov, Qiang Du, and others since many supplemental results in this edition stem from their work. We also thank Pan Yang and Yang Wang for their patience and help in typing this manuscript. Finally, Zhan-Chun Tu, Zhong-Can Ou-Yang and Ji-Xing Liu would like to dedicate this edition to Prof. Yu-Zhang Xie who passed away on May 29, 2011.

Preface for the First Edition

Liquid crystal state was discovered by F. Reinitzer in 1888. A review article written by G. Friedel in 1921 summarizes the general properties of liquid crystals known up to that time. In this article, the correlation between liquid crystal and differential geometry was noticed for the first time. The focal conic texture of smectic liquid crystals was shown to have its geometrical origin from Dupin cyclides. However, for quite some time, geometrical studies on liquid crystals were in a stagnant state. Based on the structural similarity between smectic liquid crystal and fluid membrane, in 1973, W. Helfrich offered an elastic theory of membranes in analogy with the curvature elasticity theory of liquid crystals. Since then, the study on the correlation between liquid crystals, biomembranes and differential geometry has attracted great attention to both physicists and mathematicians. A review on the aspect of differential geometry in the study of biomembranes was given by mathematician J. C. C. Nitsch in 1993 [Q. Appl. Math. **51** (1993) 363]. This book is intended to serve as an introduction to those who are interested in this subject.

This book gives a comprehensive treatment on the conditions of mechanical equilibrium and the deformation of membranes as a surface problem in differential geometry. It is aimed at readers engaging in the field of investigation of the shape formation of membranes in liquid crystalline state with differential geometry. It does not offer a compiled survey of all the results obtained in this field. Since the pioneer publications by D. P. B. Canham and W. Helfrich on the geometrical form of membrane elasticity in the early 1970s, a lot of works on this interdisciplinary subject in physics, biology and mathematics have been published. The literatures concerning this field increase continuously. Many important results in which the shape problem of membrane vesicles is treated numerically have been published by authors such as W. Helfrich, H. Deuling, M. A. Peterson, S. Svetina, M. Wortis, R. Lipowski, U. Seifert, and others. However, in the personal interest of the present authors, the material chosen in this book is mainly limited to analytical results. They apologize to those authors whose valuable results are not included here.

One of us (Z.-C. Ou-Yang) wishes to express his grateful thanks to Prof. W. Helfrich for his kind guidance in leading him into this interesting field. We thank many colleagues for their collaborating research in this field. Special thanks are due to Profs. Huan-wu Peng, Bai-Lin Hao, Lu Yu and Zhao-Bin Su for their great support to our working in this interdisciplinary field of theoretical physics and biology. Also, we owe our sincere gratitude to many of our collaborators, Profs. Wei-mou Zheng, H. Naito, S. Komura, to name a few, with whom many results presented in this book were obtained. We thank Dr. Haijun Zhou for his patience and help in typing this manuscript. We also thank World Scientific Publishing for waiting patiently for the long delayed submission of the final edition of the manuscript.

Institute of Theoretical Physics   Zhong-Can Ou-Yang
Beijing, China   Ji-Xing Liu
December, 1998   Yu-Zhang Xie

Contents

1

Introduction to Liquid Crystal Biomembranes

In this chapter, we first give a general description of liquid crystalline states, which are different from the commonly known three states of matter. The essential characteristic of liquid crystal, unlike ordinary liquid state, is that the constituent molecules have no positional order but possess orientational order. Next, the two-component system in lyotropic liquid crystal is introduced, followed by discussions on the amphiphile–water system, especially the liquid crystalline state of biomembranes, mainly bilayers. A detailed discussion on the phase transitions in biomembranes is introduced next. The part relating to biochemistry of biomembrane on the molecular factors affecting the phase transitions in lipid bilayers is also included in our discussion. Finally, a comprehensive introduction of the methods of preparing artificial vesicles is given, which may be helpful for experimentalists in investigating the shape transitions of vesicles. The subject matter of this chapter is mainly based on several well-known books,[1] especially, many schematic figures are based on the book written by Datta.

1.1 Liquid Crystals

1.1.1 Mysterious Matter

Since the 1960s, there have been several valuable books which have helped us learn about a new state of matter — liquid crystal.[2,3] In his famous book,[2] Nobel Prize Laureate de Gennes remarked that liquid crystals are beautiful and mysterious. This state shows a different feature from our common knowledge that matter exists in three states: solid, liquid, and gas. We are all confused by the term liquid crystal when we encounter it

1

for the first time: How can something be in liquid state and in crystalline state at the same time?

To answer the question, let us start with a bit of history. The liquid crystalline state of matter was discovered by botanist Reinitzer in 1888 who observed that cholesteryl benzoate has "two melting points". Crystals of cholesteryl benzoate melt at 145°C to become a turbid liquid and turn clear at 179°C. This turbid state in a pure substance was unknown to people at that time. At first, Reinitzer thought it could be caused by the presence of impurities. However, even under high purification, the persistence of "two constant melting points" makes the "impurity point of view" very suspicious. In consideration of Reinitzer's suggestion, Lehmann, a physicist, collaborated with several chemists to study this systemically and found a large number of substances with "two constant melting points". They confirmed that the turbid liquids, like crystals, are optically birefringent. Since then, the state between the "two melting points" has been known variously as "anisotropic liquid", "paracrystal", "mesomorphous state", and by now commonly accepted the term "liquid crystal", a name first used by Lehmann. The relation of the new state with the common states of matter is illustrated in Fig. 1.1.

Experimental studies have revealed that not all molecules can achieve the liquid crystalline state.[2,3] The liquid crystal phase can be observed in certain organic compounds composed of elongated molecules with an axial ratio around 4-8:1. But one should note that some discotic molecules can also form liquid crystals.[3] Basically, the anisotropic geometry of the molecular shape is the origin of the formation of the liquid crystalline state. A liquid crystal can flow like an ordinary liquid but its other properties are strongly anisotropic. Among them, the optical anisotropy, such as birefringence and optical activity, is reminiscent of the crystal phase.

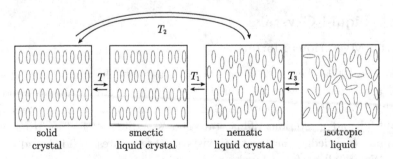

Fig. 1.1.   Two liquid crystal states and their relation with solid and liquid states.

1.1.2　Orientational Order

In order to understand the nature of the liquid crystalline state, let us recall the lattice structure of a crystal: the molecules are located on 3D periodic lattice sites, a state with long-range positional order. It is easy to recognize that X-ray diffraction patterns in crystalline state show point-like Bragg reflections. In contrast, molecules in isotropic liquid have only short-range positional order. Their X-ray diffraction patterns show only broad and diffuse Bragg reflections. In liquid crystals, the molecular long-range positional order disappears, while their orientational order is a long-range one (i.e., on a macroscopic scale, the directions of the long axes of the molecules are the same in thermal average). Thereby, the X-ray diffraction patterns in liquid crystals appear to be different from both crystals (without point-structure) and isotropic liquids (with some ring-like structure coming from cylindrical symmetry).

To specify the orientational order in liquid crystalline phase, we need two quantities: the local preferred direction, $n(r)$, and the amount of ordering, $S(r)$. $n(r)$ is a unit vector called the *director* and $S(r)$ is called the *order parameter*.[2] Here, r is the position vector in the liquid crystal. The existence of both the director-field and the order parameter-field reflects the long-range orientational order in the liquid crystalline state. The order parameter S is a measure of the degree of alignment of the long axes of the molecules and is usually defined by

$$S = 2\pi \int_0^\pi P_2(\cos\theta) f(\theta) \sin(\theta) \mathrm{d}\theta, \qquad (1.1)$$

where θ is the angle between the temporary direction of the molecular long axis and the director n, $f(\theta)$ is a distribution function of the molecular orientation corresponding to the local temperature, and $P_2(\cos\theta) = (3/2)\cos^2\theta - 1/2$. Equation (1.1) is simply the average value of $P_2(\cos\theta)$ over the orientation of all molecules, since $2\pi f(\theta)\sin\theta\mathrm{d}\theta$ is just the fraction of molecules in a cone making an angle between θ and $\theta + \mathrm{d}\theta$ with n. Here, we should remind our readers that in discotics the director n may be taken as the average direction of their short axes (normal to the surface of the molecules).

In the isotropic phase, the distribution of the long axes of the molecules is at random, i.e., $f(\theta) = 1/4\pi$, and one finds from Eq. (1.1) that $S = 0$. In the case of perfectly aligned molecules, $f(\theta) = \delta(\cos\theta - 1)/2\pi$ with $\delta(x)$ being the Dirac function of x, and $S = 1$. From the two extreme cases, one sees the advantage of choosing the average value of $P_2(\cos\theta)$ as the order parameter. Practically, S varies from 0.3 to 0.4 at T_c, the temperature of phase transition from liquid crystal state to isotropic liquid state, to about 0.8 at much lower temperatures. In the phase transition theory, the change of the order parameter reflects the nature of the transition. The order parameter can be determined experimentally such as by optical or magnetic NMR measurements.

The existence of orientational order also means that liquid crystal has cylindrical symmetry with director **n** as the axis of revolutionary symmetry. So far, it has been assumed implicitly that a liquid crystal molecule has mirror symmetry, i.e., $D_{\infty h}$ symmetry in point group theory. However, some molecules do have chiral asymmetry, i.e., D_{∞} symmetry. In this latter case, the liquid crystals show optical *dichroism*, i.e., they absorb one component of polarized light more than the other. They are also optically active, i.e., they can rotate the direction of the polarized light. To distinguish these two different symmetries in liquid crystals, it is necessary to introduce an additional order parameter, the chirality parameter, in the phase transition theory of liquid crystals. Macroscopically, this is reflected by the pitch of the bulk helical structure in the medium. Later on, we shall discuss both biomembranes with $D_{\infty h}$ and those with D_{∞} symmetry. They will be referred to as *fluid* membranes and *tilted chiral* membranes, respectively. However, we only need the macroscopic chiral order parameter (the pitch) in our discussion without introducing the microscopic chirality parameter.

1.1.3   Classification of Thermotropics

Roughly speaking, the liquid crystalline states are classified into two types, the *thermotropic* and the *lyotropic*. The thermotropic state belongs to a single component system while the lyotropic state has more than one component. In biomembranes, the lipid/water system is a two-component lyotropic liquid crystal. However, one should note that such a classification is not rigid, e.g., the lyotropic state may undergo a change under temperature change. So far, practically, our knowledge on the phase transition of liquid crystalline states comes mainly from studies on the thermotropic state. Therefore, in the following discussions, we shall give more details on thermotropic liquid crystals.

Following the classification introduced by Friedel,[4] thermotropic liquid crystals are classified into three substates: the *nematic* (N), the *cholesteric* (*Ch*), and the *smectic* (S). They are distinguished by the arrangement of the liquid crystal molecules with different heat contents as shown in Fig. 1.2.

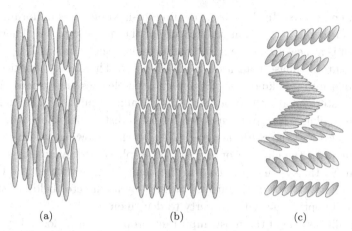

(a)   (b)   (c)

Fig. 1.2. Thermotropic liquid crystalline states: (a) nematic, (b) smectic, and (c) cholesteric.

One sees a helical arrangement of the cholesteric molecules, thereby this state is also named *twisted-nematic* (*TN*).

The names of the three different thermotropic liquid crystal phases have their structural meanings. "Nematic" means "thready" which comes from the thread-like disclinations appearing in the bulk of the nematic phase. One can recognize this phase by its thready texture. The nematic liquid crystal has a certain amount of long-range orientational order, but no long-range translational order. N-phase has D_∞ symmetry and the local axis of symmetry, the director, $\mathbf{n(r)}$, can change slowly and smoothly from point to point in the medium. The change of the director field is mainly affected by the boundary condition, the distribution of the disclinations, and the existence of external fields such as electric field or magnetic field. Without these affecting factors, in an ideal nematic liquid crystal, the directors are uniform in the whole sample.

If the constituent molecules are chiral or the nematic is doped with chiral molecules, at lower temperature, the liquid crystal can change from nematic into cholesteric phase in which there is a spatial variation of the director $\mathbf{n(r)}$ leading to a periodic helical structure with long-range orientational order. If the helical axis is along the z-direction, the helical structure can be described as

$$\mathbf{n(r)} = \left(\cos \frac{2\pi z}{p}, \, \sin \frac{2\pi z}{p}, \, 0 \right), \tag{1.2}$$

where p is the pitch of the helical structure and is temperature-dependent. When the temperature raises to a critical value, T_{ChN}, the pitch becomes infinity, i.e., $p = \infty$. In that case, the director-field becomes uniform, i.e., $\mathbf{n}$ is parallel to the z-direction, the feature of the nematic phase. Therefore, the pitch is a measure of the additional ordering, the chirality. T_{ChN} is the temperature of phase transition from Ch to N. Therefore, nematics may be divided into two kinds: one has no chiral molecules and cannot change into cholesteric phase, the other contains chiral molecules and can be in the cholesteric phase. Hence, we use two terms to distinguish them: the usual nematic (N) and the chiral nematic (N^*). In the following, we often use a star to express the chiral property. In the N-phase, the symmetry is $D_{\infty h}$ while in N^* the symmetry is D_{∞}. In the nematic state without the helical structure, one cannot distinguish N and N^* by its director field and has to measure its optical activity property to decide on it.

Chirality is one of the most important properties of biological systems. Hence, we would like to know more about the cholesteric state. Usually, in the cholesteric phase, there is a uniform helical axis. However, in some cholesteric liquid crystals, it is not possible topologically to twist about a single axis, but one can twist about two axes over a short distance to make a 3D lattice of a twisted structure with overall cubic symmetry. This sub-cholesteric phase often occurs within a half-degree temperature range between the cholesteric phase and the isotropic phase and forms a helical structure of very short pitch of the order of $p \simeq 300\,\mathrm{nm}$. This sub-cholesteric phase is called the blue phase on account of its blue color, a result of Bragg reflection from the twisted lattice. The lattice parameter is of the same order as the pitch of the cholesteric phase.

In the smectic phase, the molecules are arranged in layers of one-molecule thickness. Friedel named this phase "smectic", implying a parallelism with soap[4] which is also stratified. It differs from soap film due to the fact that the molecules in each layer are longitudinally parallel to each other. Hence, for each layer, one can define a director $\mathbf{n(r)}$ just as in nematic and cholesteric phases but with a 1D periodic arrangement. If the director $\mathbf{n}$ is normal to the layer surface, the substate is called the *smectic-A* (S_A) phase. If $\mathbf{n}$ is tilted to the layer surface, we have another substate, the *smectic-C* (S_C) phase. If in addition to the tilted feature the molecules are chiral, then the director-field can rotate about the normal direction of the layer in each layer to form a helical structure, and lead to another sub-phase, the tilted chiral smectic *smectic-C** (S_{C^*}). So far, more than six substates in smectics have been found. However, for our interest in liquid crystalline biomembranes, we introduce only the above three sub-smectic phases and schematize them as shown in Fig. 1.3.

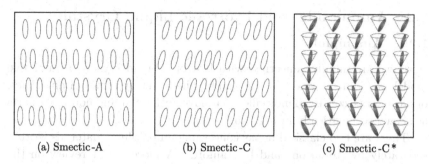

(a) Smectic-A   (b) Smectic-C   (c) Smectic-C*

Fig. 1.3.   Some substates of smectic liquid crystals.

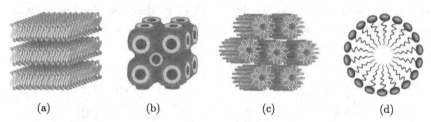

(a)   (b)   (c)   (d)

Fig. 1.4.   Lyotropic mesomorphism: (a) lamellar phase, (b) cubic phase, (c) hexagonal phase, (d) single spherical micellar phase.

1.1.4   Classification of Lyotropics

In thermotropic liquid crystals, the only factor that affects their phase transitions is the heat content. However, in multiple-component lyotropic liquid crystals more factors will affect their phase transitions. Even for a two-component system like the amphiphile-water system which is the simplest model of biomembranes, its phase transition phenomenon is very complicated. One would say that up to now, our knowledge on the phase transitions of lyotropic are not in true parallelism with that in the thermotropic case. Nevertheless, the following rough classification of lyotropics is now commonly accepted: *lamellar (L), cubic (C), hexagonal (H)*, and *micellar* phase. They are shown schematically in Fig. 1.4.

The details of these phases will be given in the next section. Here, we only want to point out that the phase transitions between the different phases shown in Fig. 1.4 depend mainly on the change of the water content. Under slow change of water content, the assemblage of the amphiphiles leads to dramatic changes of the phases and demonstrates the so-called *lyotropic mesomorphism*.

1.2   Amphiphiles and Lyotropic Liquid Crystals

1.2.1 Amphiphile

Here, we are mainly concerned with lyotropic liquid crystals as a simple model of biomembranes — the amphiphile-water system. *Amphiphiles*, or *surfactants*, consist of molecules that have both polar and non-polar parts as shown schematically in Fig. 1.5.

There are two classical books concerning amphiphilic materials written separately by Adamson[5] and by Tanford.[6] A more recent review on the physics of amphiphiles is the discussion at Les Houches in 1987, edited by Meunier, Langivan, and Boccara.[7] However, the term amphiphipathy was proposed a long time ago by Hartley[8] to describe the tendency to be adsorbed at interfaces or to form micelles in the solution for certain

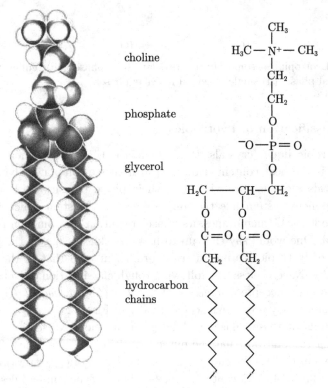

Fig. 1.5.   Schematic structure of an amphiphile molecule with two hydrocarbon chains (the non-polar part) attached to a single polar head group.

types of molecules. It expresses the natural ability of the molecules to form homogeneous solutions with water or other polar solvents, as well as with oils or non-polar solvents. Sodium hexadecylsulfate

$$C_{16}H_{33}SO_3^- N_a^+$$

may be considered as the prototype of all amphiphiles. Here, the non-polar hexadecyl aliphatic chain is *lipophilic* (or *hydrophobic*) and the polar sodium sulfate ionic group is hydrophilic. The two constituent parts of an amphiphile have very different solubilities in water or other solvents. In the present case, the sodium sulfate ionic group is practically infinitely soluble in water while the hexadecyl aliphatic or hydrocarbon part is much less soluble in water than in most non-aqueous solvents. This unique property leads to the presence of different aggregates of amphiphiles in the solution that will be discussed in the next several sub-sections.

1.2.2 Monolayer

If the concentration of amphiphiles in water is very low, a certain fraction of the molecules will form an amphiphilic monolayer serving as a partition between the surface and the bulk of water. The molecules in this mono-layer have their non-polar tails pointed away from water on account of their hydrophobic property. Gibbs had treated this phenomenon as a 2D effect in chemical physics over a century ago.[9] The water-amphiphile inter-face has lower energy or surface tension than pure water surface. Thus, in a Gedanken experiment as shown in Fig. 1.6, the float on the surface will move automatically to the right in order to lower the total surface energy.

In the above experimental set-up, to keep the float stationary, one has to apply a force proportional to the length of the line of contact between the float and the surface. The force can be interpreted as a 2D pressure. Typical pressure-area isotherms, according to Langmuir,[10] can be expressed by a 2D van der Waals equation analogous to that of the 3D real gas equation:

$$\left(\Pi + \frac{a}{A_0^2} \right) (A_0 - b) = k_B T, \tag{1.3}$$

where Π is the applied lateral pressure, A_0 is the mean area occu-pied per amphiphilic molecule, k_B is the Boltzmann constant, T is the absolute temperature, and a and b are constants taking care of the

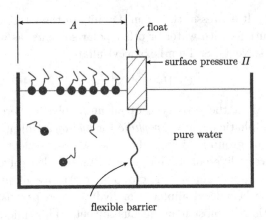

Fig. 1.6.   Gedanken experiment of monolayer surface pressure.

attractive intermolecular forces (van der Waals forces) and the finite size
of the amphiphile. In fact, this is an empirical formula. Some fine structures of $\Pi(A_0, T)$ have been explained as the effect of molecular orientation order.[11,12] In other words, the nature of liquid crystalline state
in monolayer is affected both by its heat content (T) and its geometrical content (A_0). If we refer to the amphiphilic monolayer as one of the
lyotropic liquid crystalline phases, the orientational order takes place in the
director in thermotropic liquid crystals. Thus, amphiphiles can be either
lyotropic or thermotropic. This is a unique feature of amphiphile-water
systems.

1.2.3 Micelle

For a monolayer in equilibrium with the solution, Gibbs showed that Π is
a function of the ambient pressure P, the temperature T, and the chemical
potential μ of the amphiphiles in the solution. The general principle of
physical chemistry states that

$$\mu = \mu_0 + k_B T \ln X, \tag{1.4}$$

where X is the activity of the amphiphiles in the bulk of the solution and
μ_0 characterizes the amphiphilic property of the molecules. For low concentration X is proportional to the concentration C of the amphiphiles in the
bulk. Equations (1.3) and (1.4) together indicate that Π increases with C.
However, experiments show that Π increases monotonically with C only for

C less than some *critical micelle concentration* (CMC). At CMC, the chemical potential of the amphiphiles in the monolayer becomes saturated and is equal to the chemical potential of some aggregated phase. Here, both μ and Π become saturated and fixed. The amphiphiles go into another phase to form micelles in which only the soluble (or polar) part of the molecule is exposed to the solvent and the insoluble part is buried in the depth of the aggregates (see Fig. 1.4).

In the micellar phase, there are several interesting questions: what are the physical origins that determine the size and the shape of the micelles? The reader may find some detailed answers in a book written by Israelachivili.[13] But the shape of the micelle is still an open problem. We would like to offer some new approaches to this problem. Now, let us put forward some more detailed descriptions on the rough classification of the micelle phase as shown in Fig. 1.4. At very low water/amphiphile ratio, stacked bilayer lamellae are formed with the polar part of the amphiphiles of adjacent lamellae facing each other as shown in Fig. 1.4(a). This is just the neat soap phase known in the detergent industry. It is quite similar to smectic liquid crystal (see Fig. 1.2). This explains why Friedel[4] gave the term "smectic" ($\sigma\mu\eta\gamma\mu\alpha$, soap) to smectic liquid crystals.

At slightly higher water/amphiphile ratio, still less than 50% of water, the lamellae phase changes into cubic phase in which the spheroidal aggregates have a cubic disposition (see Fig. 1.4(b)). With intermediate water/amphiphile ratio, the amphiphiles go into the hexagonal phase forming hexagonally arranged amphiphile-cylinders (Fig. 1.4(c)). At relatively high water/amphiphile ratio, the isolated spheroidal micelles are randomly distributed in water (Fig. 1.4(d)) and this has been named the *micellar* phase.

Diluting the micellar phase to the point of CMC, we will have a homogeneous solution of amphiphiles in bulk water with the formation of an amphiphilic monolayer at the air-water interface.

1.2.4   Phase Diagram

The phase transitions in lyotropic liquid crystals are affected mainly by both the temperature and the dilution. So far, most experimental studies on lyotropics have constructed the phase diagram for a given amphiphile in water in which the phase transition is described as a function of temperature and water/amphiphile ratio. A typical example is the phase diagram of soap obtained by Skoulius.[14] Figure 1.7 is a redrawing of Skoulius' results.

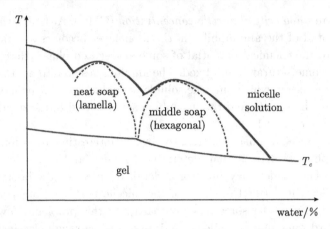

Fig. 1.7. Schematic redrawing of the phase diagram of a simple soap–water system obtained by Skoulius.[14]

From the phase diagram, one finds that besides the dilution effect in each sub-micellar phase, there exists a critical temperature T_c above which the system is in the lyotropic liquid crystal state, i.e., the hydrocarbon tails of the amphiphilic molecules are relatively flexible or fluid-like. In contrast, when $T < T_c$, the hydrocarbon chains are relatively rigid or solid-like and the amphiphile aggregates go into the *gel* phase. It is important that to keep the biomembrane (assumed as amphiphile–water system) in its functional state (assumed as liquid crystalline state), it is necessary to keep it in water within a proper temperature range.

1.3   Phase Transitions in Biomembranes

1.3.1   Fluid Mosaic Model

Now, let us pay our attention to the lyotropic liquid crystalline state in biomembranes. The accepted picture of biomembrane is the fluid mosaic model proposed by Singer and Nicholson in 1972.[15] As illustrated schematically in Fig. 1.8, the biomembrane is considered simply as a bilayer of amphiphilic lipids in which the lipid molecules can move freely on the membrane surface just like ordinary liquid molecules, while the protein molecules and the enzyme molecules are embedded in the lipid bilayer. Some of the protein molecules and the enzyme molecules may traverse in the entire lipid bilayer and communicate with both the inside of the cell and its environment. Some of them may be involved in the transport process.

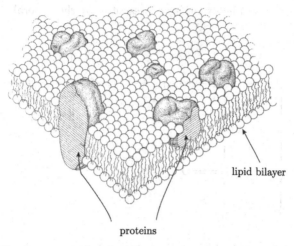

lipid bilayer

proteins

Fig. 1.8. Fluid mosaic model of biomembranes proposed by Singer and Nicholson.[15]

Other protein molecules and enzyme molecules are partially embedded in the bilayer and serve other biological functions. The fluid mosaic model allows the embedded proteins and enzymes to have considerable mobility.

1.3.2 Lipid Bilayer

The lipid molecules of the model biomembrane must organize themselves into a bilayer in which the hydrophilic polar heads shield the hydrophobic tails (the hydrocarbon chains) from the water surrounding the membrane. One should notice that the term "lipid" used by biochemists refers to all organic molecules of biological origin that are highly soluble in organic solvents and only weakly soluble in water. However, the lipids in biomembrane refer only to *phospholipids, glycolipids,* and cholesterol. Phospholipids and glycolipids are weakly soluble in water, while cholesterol is insoluble in water. Moreover, in biomembrane, the amount of cholesterol in the phospholipid bilayer may reach the ratio of one cholesterol molecule to one phospholipid molecule. Hence, the biomembrane seen as a bilayer of mixture of phospholipid (amphiphile) and cholesterol (non-amphiphile) is not quite like the amphiphilic bilayer discussed before. One can imagine that within certain limitations, the biomembrane viewed as liquid crystal film will show lyotropic mesomorphism. Most of the biolipids contain two hydrocarbon chains attached to a single polar head group, among which a large portion contains phosphoglycerides represented by the general formula[6]

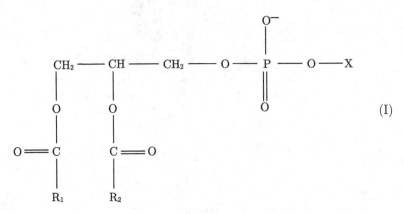

$$(I)$$

where R_1 and R_2 are hydrocarbon chains and X represents a polar head group. Such lipid molecules form small micelles only when their hydrocarbon chains are very short. Those with normal length hydrocarbon chains (about the length of 10 carbon atoms) mostly form bilayer structure in water. Actually, the consideration of biomembranes as lipid bilayers comes mainly from the fact that their constituent phospholipid molecule has two hydrocarbon chains. Theoretically, why double-chain phospholipids can form bilayer easily is still an open question.

1.3.3 Phase Transitions in Bilayer

Experimentally, when water is added to pure phospholipids, the sample swells to form different kinds of liquid crystal phases. Thereby, the lipids are called swelling amphiphiles. Using low-angle X-ray diffraction technique, Luzzati and his co-workers have studied the liquid crystalline structures of lipid-water systems extensively.[16] The basic feature is similar to what we have discussed in the section on phase transitions in micellar phase, but with some special aspects.

Addition of more water into the swelling lipid system tends to disperse the lipids without disrupting its ordered aggregated structure. This may be the reason for the formation of onion-like closed multiple bilayer structures, which cannot be sub-divided further without the rupture of the individual layers. Practically, one may use the ultrasonic technique (for example) to break the multilayered structure into small soluble particles to form single-wall (or single bilayer) vesicles. The particles so formed are just closed vesicles, i.e., closed structures of single bilayers.[17] Details of the preparation of vesicles will be given in the next section. Experimental studies show that thermodynamically the vesicles are the favored state of aggregated phospholipids. For given lipids and solvents, the vesicles have nearly uniform size and shape. These features make it possible for us to do a physical and geometrical study on biomembrane. This is just what we are trying to do.

The different liquid crystalline phases of amphiphile-water system are related directly to the water concentration of the system and the heat content. Bilayers consisting of phospholipids, glycolipids, and other amphiphiles undergo an endothermic pretransition from gel (or solid) phase to liquid crystalline state at a certain critical *lipid phase transition temperature* T_t. The lipid phase transition temperature depends upon the length of the acyl chain of the lipid molecule. As an example, for phospholipids composed of the same acyl chain at both positions, with each and every increase of the chain length by 2-methylene units, T_t increases from about 14°C to 17°C. The basic feature of the differential scanning calorimetric measurement done by Chapman and coworkers[18] is illustrated schematically in Fig. 1.9.

In Fig. 1.9, we notice that for pure lipid bilayer there is an endothermic pretransition at T_t and another main phase transition near T_t. However, with increasing cholesterol doping in the bilayer, both transitions disappear

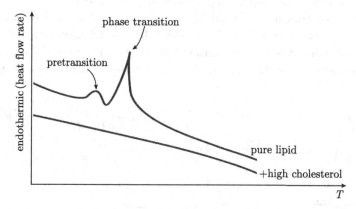

Fig. 1.9.   Schematic illustration of differential scanning calorimetric measurements of endothermic pretransition and main phase transition at T_t of a pure hpid bilayer (above). With increased cholesterol doping in the lipid bilayer, the two-phase transitions disappear (below).

gradually. This reveals again that real biomembranes are not in an ideal lyotropic liquid crystalline state.

1.3.4   Classification of Lipid Bilayer Phase

The liquid crystalline state of a lipid bilayer is specifically named the L_α phase, where L stands for the lamellar phase of the lyotropic liquid crystals. In this state, the two hydrocarbon chains of the lipid molecule are very *flexible* and their orientation on average is normal to the bilayer surface (see Fig. 1.10). If we treat the orientational direction of the chain as the director, the L_α phase is simply the smectic-A phase in thermotropic liquid crystals.

The solid state of lipid bilayer is called the L_β phase, in which the lipid molecules are arranged in quasi-hexagonal arrays in the bilayer plane with their two hydrocarbon chains more *rigid*, fully *extended*, and all in a *transversal* configuration (Fig. 1.10). This phase corresponds to the smectic-B phase in thermotropic liquid crystals. The subscripts "α" and "β" refer to the schematic drawing of the hydrocarbon chains in their corresponding phases.

There is another solid phase of the lipid bilayer, in which the lipid molecules are tilted with respect to the normal of the bilayer plane (Fig. 1.10). This special phase is called $L_{\beta'}$ and is often found in phosphatidylcholine. The tilt of the molecule comes from its characteristic head group orientation as proposed by Janiak and co-workers.[19]

The phase transition of lipid bilayer is more involved. For example, in Fig. 1.9, the small endothermic transition before the main order-disorder transition represents a pretransitional phase, the $P_{\beta'}$ phase. Experimental measurements, such as X-ray diffraction measurement, show that this

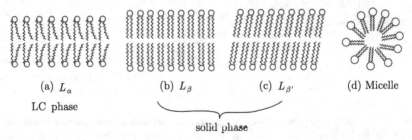

(a) L_α (b) L_β (c) $L_{\beta'}$ (d) Micelle

LC phase

solid phase

Fig. 1.10.   Schematic view of lipid bilayer in different phases: (a) fluid bilayer (L_α), (b) solid bilayer (L_β), (c) tilted solid bilayer $(L_{\beta'})$ and (d) micelle.

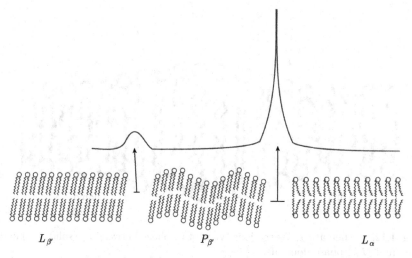

$L_{\beta'}$ $P_{\beta'}$ L_{α}

Fig. 1.11. Rippled lipid bilayer $(P_{\beta'})$ in pretransition phase.

phase has a ripple structure with periodicity of about 120–160 Å, corresponding to the parking of 12–20 molecules.[19] On account of the parking geometry, this phase is also called the *ripple* phase (Fig. 1.11). So far, the mechanism of the formation of the long-range 1D order in $P_{\beta'}$ phase is not quite clear. It seems that in pure lipid bilayer it comes from a cooperative change in the orientation of the lipid–lipid head group. In the presence of other lipids, presumably, they interfere with the long-range order distribution. One should notice that the pretransition $(P_{\beta'})$ phase seems to be suppressed by increasing the cholesterol content. The theory of the formation of ripple structure in bilayer still offers a great challenge in physics of biomembrane.

Besides L_{α} , L_{β}, $L_{\beta'}$, and $P_{\beta'}$ phases, lipid bilayer may exist in another more complex state, the *interfacial lipid* (IL) *phase.*[20] This phase exists between the solid domain and the liquid crystal domain (Fig. 1.12). It is difficult to relate the IL phase to any phase known in thermotropic liquid crystals. In our opinion, this phase may correspond to defects in thermotropic liquid crystals. Thereby, it is completely an open question in the theory of lipid bilayer physics.

In Fig. 1.12, the drawing of the molecules in the fluid phase domain is lighter than that in the solid phase domain. This is based on the property of the configurations of the hydrocarbon chains. More quantitative discussions on this aspect will be given in Sec. 1.4.

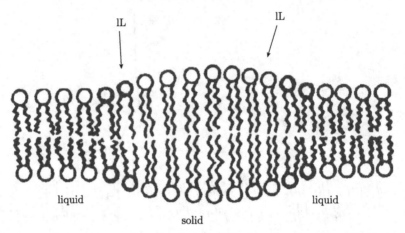

Fig. 1.12.   Schematic cross-sectional view of IL phase between the solid (L_β) and the fluid (L_α) phase domains.

1.4   Simple Biochemistry of LC Biomembranes

1.4.1   Effect of Chain Length

We do not intend to give a complete survey on all aspects of membrane biochemistry. We will discuss only some molecular characteristics relevant to the theory of membrane elasticity that will affect the behavior of phase transitions in biomembrane.

First of all, it must be emphasized that most biological lipids have two hydrocarbon chains, only very few of them have a single hydrocarbon chain. Moreover, the hydrocarbon chains of biological lipids are very long and few have less than 16 carbon atoms. Usually, the hydrocarbon chain of biological lipids has about 20 carbon atoms. The chain length of a saturated hydrocarbon chain with n carbon atoms may be estimated by the Tanford formula[6]:

$$l \leq l_{\max} = (1.54 + 1.265n) \text{ Å}. \tag{1.5}$$

From the Tanford formula, the estimated length of a lipid molecule with hydrocarbon chains of 24 carbon atoms is about 70 Å. This gives the thickness of the bilayer in the L_β phase (Fig. 1.12) and the thickness in the fluid phase should be less than 70 Å. Measurements on real bilayers of egg yolk phosphatidylcholine give the thickness of the bilayer as $d = 2l - 50$ Å, a value well within the range given by the Tanford inequality (1.5). It has also been found that in the same lipid bilayer, the surface area per head group is about $60 \, \text{Å}^2$ or $30 \, \text{Å}^2$ per hydrocarbon chain. This value of area per head group is greater than $21 \, \text{Å}^2$, the head group area of close-packing chains.[6] From these experimental data and with theory in chemistry, one can carry out a check on the axial ratio of the hydrocarbon chain, $(50/2) : 2\sqrt{2/\pi}$, which is about 4.8. It is very interesting to notice that the value of the axial ratio of biological lipid is just in the same range of the axial ratio of the molecules which can achieve liquid crystal states as pointed out in Sec. 1.1. This is one of the reasons to believe that biomembranes can exist in liquid crystalline phase.

1.4.2 Double Bond Effect

For lipids in biomembrane, an almost universal feature is the presence of double bonds, i.e., *unsaturation* in their hydrocarbon chains. In phospho-glycerides, with molecular formula (I) given in Sec. 1.3.2, it is found that chain R_1 is often a *saturated* (i.e., single bonds only) fatty acyl group and R_2 an unsaturated one (i.e., including some double bonds).[21] This unsaturated double bond usually leads to a *cis*-packing of the hydrocarbon chain (see Fig. 1.13). The *cis* double bond forms a permanent link that makes the practical area per hydrocarbon chain in the lipid bilayer larger than that of the close-packed chain, as shown in the data given in the last sub-section. It also keeps the membrane fluid remaining in the fluid state until a lower temperature T_t than that of its saturated analog is reached.

Experiments show that the more the number of double bonds there are, the lower the lipid phase transition temperature T_t is. This unique property allows some kinds of fish to live in very cold water during winter time by spontaneously increasing the number of double bonds of the lipids in their cell membranes.

1.4.3 Effect of Ionic Condition

The ionic condition can also affect the phase transition behavior of biomembranes. Most biomembranes contain a certain amount of negatively charged phospholipids. These charged lipid bilayers are very sensitive to changes in pH value, ionic strength, and cation concentration. A higher pH value promotes ionization of these lipids in the membrane in favor of the fluid phase with lower T_t. It has been found that changes of pH value can induce shape transition of the cells (see later chapters), a fact related to the sensitivity of charged biomembranes with negatively charged lipids. Electric breaking of cell membrane is a cooperative effect of the external electric field and

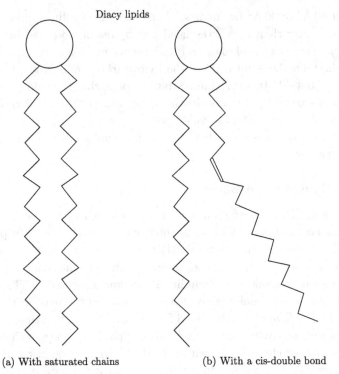

(a) With saturated chains   (b) With a cis-double bond

Fig. 1.13.   Schematic illustration of saturated *trans* chain (a) and chains with unsaturated *cis* double bond (b).

the charged membrane. This effect is an interesting subject in the theory of membrane shape.

1.4.4   Cholesterol Effect

The effect of cholesterol in biomembrane is very mysterious. In animal cell membranes, cholesterol is a very common constituent. Although why cholesterol has to be included in biomembrane is still an unsolved problem, the study of the incorporation of cholesterol in phospholipid bilayer has received much attention. It is said that the structure of biomembrane with phospholipids and cholesterol mixed in it can provide some clues to the biological function of the biomembrane. However, we are only concerned with their effect on the phase behavior in membrane.

The molecular structure of cholesterol is shown in Fig. 1.14. It has a polar hydroxyl group attached to a rigid planar body with four sterol rings.

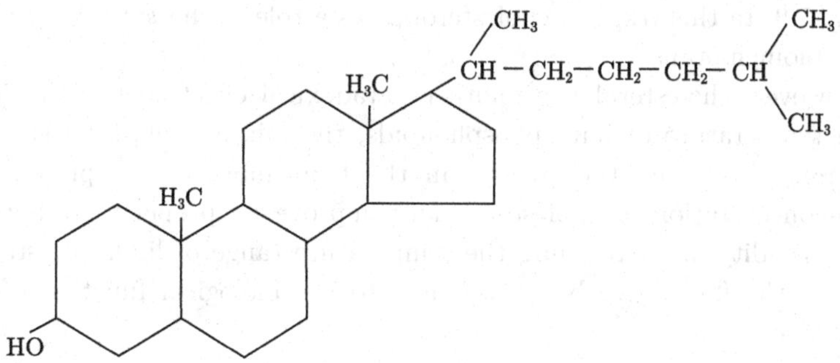

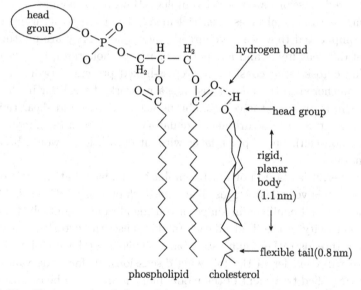

Fig. 1.14.   Molecular structure of cholesterol.

Fig. 1.15.   Hydrogen bond formed between a cholesterol and a phospholipid molecule.

The body of the rings is followed by a short flexible hydrocarbon chain. In phospholipid bilayer, the cholesterol molecule stays in parallel with the phospholipid molecule through hydrogen bond between its hydroxyl head and the carbonyl oxygen of the fatty acyl chain of the phospholipid (Fig. 1.15). Such an interaction thereby reduces the mobility of the molecules in the bilayer. This may well explain the disappearance of phase transition of the biomembrane with increasing cholesterol content as shown in Fig. 1.9. In this respect, cholesterol plays a role in the *solidifying* effect of the biomembrane.

However, cholesterol has another paradoxical effect: below the transition temperature of pure phospholipids, the mixed cholesterol shows a *fluidizing* effect. In other words, in the biomembrane, the presence of proto-concentration of cholesterol does improve its property by increasing its rigidity and extending the temperature range of liquid crystalline state. Both effects give big advantages to the biological function of cell membrane.

1.4.5   Complexity in Biomembrane

The above discussions are not yet complete. However, it should be pointed out that real biomembranes (i.e., cell membranes) are structurally much more complicated than we have already mentioned. They are mixtures of a lot of different lipids and components such as cholesterol, cytoskeleton, etc. The cytoskeleton consists of a cross-linked protein network, mainly spectin anchored to the bilayer just like a network of steel bar in the concrete shell. It provides the cell membrane with strength and shear rigidity. Under high stress or in a turbulent circulatory environment (e.g., membrane of red blood cell), pure lipid bilayer without cytoskeleton would break in deformation.

However, it is also important to emphasize the fact that the network of cytoskeleton is very flexible, just like a network of strings of beads. Therefore, the normal equilibrium shape of cell membrane is controlled mainly by the elasticity of the lipid bilayer. In this sense, the cytoskeleton would look like threads rather than "steel bars." This is the basic point of view that we will consider in the following discussions. In fact, the validity of such a simplified treatment of the model biomembrane can be confirmed in the famous shape problem of the red blood cells.

1.5   Artificial Bilayers and Vesicles

Many experimentalists work on artificial vesicles of pure phospholipid bilayer to test the validity of the elasticity theory of pure lipid bilayer biomembrane. Before discussing the theory of shape transition in lipid bilayer, it is helpful to know something about how artificial vesicles are prepared. The following outline of the preparation procedure is based on the review article by Sjoka and Papahadjopoulos.[22]

1.5.1 Lipids for Artificial Vesicles

Artificial vesicles can be prepared from a variety of lipids and lipid mixtures, among them, phospholipids are commonly used. In the above section, we have mentioned that for a given kind of phospholipids or a mixture of given phospholipids, there is a definite phase transition temperature T_t. Below T_t, the formed bilayers are in the gel state and above T_t they turn into liquid crystalline state. Here, the materials given in Table 1.1 are the so-called room temperature liquid crystal lipids. The artificial bilayers formed by them are in the liquid crystalline state at near room-temperature range.

1.5.2 Multilamellar Vesicles

The preparation of multilamellar vesicles needs no special technique. They form an equilibrium lamellar phase spontaneously, as described in Sec. 1.2. The practical procedure is as follows: First, place a droplet of lipid dissolved in organic solvent on the test-disc. Evaporation of the solvent leaves pure dry lipids (formed in some lamellar structure) deposited on the wall and the bottom surface of the disc as a thin film. An aqueous buffer is now added to the sample. At a temperature above T_t, with gentle shaking, the dry lipid hydrates. Apparently, the aqueous buffer is intercalated between the lipid layers. It separates the lamellae to form swelling lipid phase. The time to complete the hydration process runs from several minutes to several hours depending on the kind of lipid. After hydration, the solution contains onion skin-like multilamellar lipid vesicles.

In encapsulation, the size of the multilamellar vesicles depends on the time of hydration, the kind of lipid, and the way of shaking. For example, brief sonication can be used to obtain smaller and more uniform

Table 1.1. Some lipids used in preparation of vesicles chosen from Ref. [22].

Lipid	Abbreviation	$T_t/^\circ C$
Egg phosphatidylcholine	EPC	−1.5 to −7
Dilauryloylphosphatidylcholine (C12:0)	DLPC	−1.8
Dimyristoylphosphatidylcholine (C14:0)	DMPC	23
Dilauryloylphosphatidylglycerol	DLPG	4
Dimyristoylphosphatidylglycerol	DMPG	23
Dipalmitoylphosphatidylglycerol	DPPG	41
Dimyristoylphosphatidylserine	DMPS	38

multilamellar vesicles. Macromolecules such as DNA and drugs can be trapped in multilamellar vesicles. Therefore, the preparation of multilamellar vesicles is important in the study of drug delivery and bioengineering.

1.5.3 Single-Bilayered Vesicles

In the process of preparation, strong sonication may break the multilamellar vesicle to form small single-bilayered vesicle aggregates. The solution now appears optically clear with single-bilayered vesicle suspensions in it.

Besides sonication, several other methods have been used to obtain single-bilayered vesicles. Among them, the ethanol injection technique has been used very often. This method is very convenient. By dissolving lipids in ethanol and rapidly injecting it in an aqueous buffer, the lipids will form single-bilayered vesicles spontaneously.

Either by sonication or by ethanol injection, the obtained single-bilayered vesicles have nearly the same internal volume, about $0.5\,L$ per mol of lipid. The typical size of the vesicle is in the range of $10^2 - 10^3\,Å$. The distribution of the vesicle size and the average size depend upon the details of the formation procedure. It was found that different concentrations of lipids in ethanolic solution give vesicles with different diameters. At low lipid concentration ($\sim 3\,mM$), vesicles of about $300\,Å$ in diameter can be obtained, whereas at high concentration ($\sim 36\,mM$) vesicles of $1100\,Å$ in diameter can be formed.

The size and the shape transition of vesicles are the most important subjects in the study of elasticity theory of lipid bilayer. Such kind of vesicles (typically $1 - 10\,\mu m$ in diameter) can be examined and recorded by phase-contrast video microscope. The shape transitions of vesicles are induced by adjusting the temperature[23] and/or the osmotic pressure.[24] The shape transition of single-bilayered vesicles is the main subject in our discussion. For simplicity, the term "vesicle" will be used to mean single-bilayered vesicle in the following discussions.

References

[1] S. Freiberg, *Lyotropic Liquid Crystals and the Structure of Biomembrane* (Am. Chem. Soc, 1976); G. H. Brown and J. J. Wolken, *Liquid Crystals and Biological Structures* (Academic Press, 1979); D. B. Datta, *Introduction to Membrane Biochemistry*, Chap. 4 (Floral Publishing, Madison, 1987); G. Cevc and D. Marsh, *Phospholipid Bilayers: Physical Principles and Models* (Wiley, New York, 1987).

[2] P. G. de Gennes, *The Physics of Liquid crystals* (Clarendon, Oxford, 1975).

[3] G. W. Gray, *Molecular Structure and the Properties of Liquid Crystals* (Academic Press, London, 1962); G. W. Gray and J. W. Goodby, *Smectic Liquid Crystals: Textures and Structures* (Heiden and Sons, 1984); S. Chandrasekhar, *Liquid crystals* (Heiden and Sons, 1980).

[4] G. Friedel, *Ann. Phys.* (Paris) **2** (1922) 273.

[5] A. W. Adamson, *Physical Chemistry of Surfaces*, 3rd edn. (Wiley, New York, 1976).

[6] C. Tanford, *The Hydrophobic Effect, Formation of Micelles and Biological Membranes* (Wiley, New York, 1973).

[7] J. Meunier, D. Langivan, and N. Boccara, *Physics of Amphiphilic Layers* (Springer, Berlin, 1987).

[8] G. S. Hartley, *Aqueous Solutions of Paraffin-Chain Salts* (Hermann et Cie., Paris, 1936).

[9] J. W. Gibbs, *Trans. Conn. Acad.* **3** (1876) 108; *ibid.* **3** (1878) 343.

[10] I. Langmuir, *J. Chem. Phys.* **1** (1933) 756.

[11] S. Marcelza, *Biochem. Biophys. Acta* **367** (1974) 165.

[12] A. Sugimura, M. Iwamoto, and Ou-Yang Zhong-can, *Phys. Rev. E* **50** (1994) 614.

[13] N. Israelachivili, *Intehrmolecular and Surface Forces* (Academic Press, London, 1992).

[14] A. Skoulius, *Adv. Colloid and Interface Sci.* **1** (1967) 79.

[15] S. I. Singer and G. L. Nicholson, *Science* **175** (1972) 720.

[16] V. Luzzati, in *Biological Membranes*, ed. D. Chapman (Acad. New York, 1968), Ch. 3.

[17] L. Saunders, J. Perrin, and D. Gammack, *J. Pharmacol.* **14** (1962) 567.

[18] B. D. Ladbroke, R. M. Williams, and D. Chapman, *Biochim. Biophys. Acta* **150** (1968) 333.

[19] M. J. Janick, D. M. Small, and G. G. Shipley, *Biochemistry* **15** (1976) 4575.

[20] D. Marsh, A. Watts, and P. F. Knowles, *Biochemistry* **15** (1976) 3570.

[21] L. L. M. van Deenen, *Pure and Appl. Chem.* **25** (1971) 25.

[22] F. Szoka and D. Papahadjopoulos, *Ann. Rev. Biophys. Bioeng.* **9** (1980) 467.

[23] E. Evans and W. Rawicz, *Phys. Rev. Lett.* **64** (1990) 2094.

[24] H. Hotani, *J. Mod. Biol.* **178** (1984) 113.

2

Curvature Elasticity of Fluid Membranes

In this chapter, we trace back to a long-standing problem in physiology: Why is the normal shape of red blood cells in human bodies always in a biconcave discoidal shape? We show that both the isotropic fluid film model and the solid shell model of biomembranes cannot explain the biconcave discoidal shape. Through this way, the cell membrane which consists mainly of lipid-bilayer is naturally explored to the liquid crystalline origin. This chapter derives the bending energy expression of Helfrich spontaneous curvature model with two methods: One follows the original derivation of Helfrich in 1973; the other is a more rigorous calculation with 2D differential geometry. We will extensively focus on the second method in this chapter. Both methods are based on the analogy with curvature elasticity theory of liquid crystals. The important concept of the Helfrich model, the spontaneous curvature, is discussed in several sources of chemical physics of the membranes and their aqueous environment.

2.1   Shape Problem in Red Blood Cell

2.1.1   Membranes in Cell

Physics plays an important role in the development of biological science. The following historical facts vividly showed this process. With a simple microscope, in 1665, Hooke recognized the cell as the unit of life and suggested that the cell cannot exist without cell membrane. The development of the compound microscope in the 1830s made it possible to observe individual cells. In 1858, Virchow recognized that each cell originates from another cell.[1] This discovery marked the beginning of the science of cellular systems.

The introduction of the electronic microscope in the early 1940s made it possible to observe directly the fine structure of the cell membrane and brought us closer to a molecular description of the cellular structures. It is clear that the elucidation of the structure of the cell has gone hand in hand with the development in optics and electronics.

In molecular biology, studies show that the cell membrane acts not only as a partition between the cell and its environment in order to keep the interior of the cell in a steady state but also regulates the directional transport of materials, of energy, and of information with the environment of the cell. Furthermore, protein molecules and enzyme molecules will perform their biochemical functions only when they are embedded on the cell membrane. On account of these "biological requests", the structures of most cell organelles are in membrane structures.

With their biological functions, the membrane of the cell can be distinguished roughly in several types. As shown schematically in Fig. 2.1, each cell is confined by a thin film to separate two aqueous media, the cytoplasm within the cell and the extracellular medium. This film, the *plasma* membrane, is a bilayer of phospholipids associated with cholesterol, proteins and enzymes as described in Chap. 1. Its configuration, i.e., the shape of the cell vesicle, is the main topic discussed in this book. On the other hand, most cells consist of a nucleus and cytoplasm confined

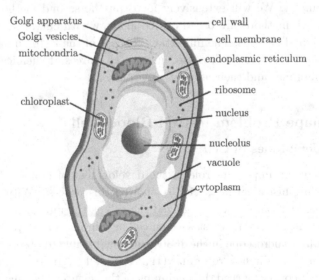

Fig. 2.1.   Membrane structure in cell.

in the aforementioned plasma membranes. It should be noted that the nucleus itself is also enveloped by a double membrane with the structure of bilayer of phospholipids similar to the plasma membrane. But the *nucleus* membrane has a more complex chemical composition and is a set of two parallel bilayers. It is worthy to point out that the same situation occurs in most other organelles[2] made up of membrane (also phospholipid bilayers). These can be seen from examples like mitochondria, golgi apparatus, endochondria, etc. In other words, in principle, the present study on the shape of plasma membranes can be applied to membrane systems of most cell organelles. It is believed that the beautiful and mystic shapes in cell and its organelles are natural experimental demonstrations of the "plateau problem" in biomembranes.

2.1.2 High Deformability of Cell Membranes

The plasma membrane has been considered as a thin film with elastic properties comparable to those of rubber, but this is not compatible with large and rapid deformations of shape observed in small cells. Among them, the typical one is the red blood cell. These cells are highly deformable in the environment of the circulation. At rest, normal human red cells are typically in a biconcave discoidal shape as shown in Fig. 2.2 and are called *discocytes*.

For passing through narrow blood vessels, the flat region of the discocytes can transform into extremely long cylindrical extension. Such large

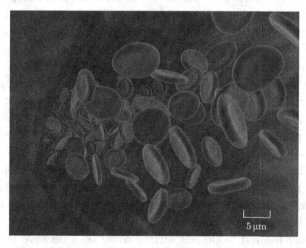

Fig. 2.2. The biconcave discoidal shape of human red blood cell at rest.

deformation would be impossible with thin rubber, since it would cause a large reduction in its thickness. On the contrary, electron microscopy of the cells shows a constant thickness of the membranes irrespective of whether they are in biconcave shape or in elongated finger shape. The high deformability of cell membranes can be seen in the micro-circulation of the blood cells where cells of about $8\,\mu m$ in diameter under normal conditions can be squeezed through inter-endothelia slits that seldom exceed 0.2–$0.5\,\mu m$ in width.[3]

On the other hand, in their circulation through large blood vessels, the red cells change their biconcave discoidal shape into an ellipsoid with long axis parallel to the direction of flow. However, at rest, e.g., in the quiescent environment of a drop of plasma on a glass slide, normal human red cells return to their biconcave discoidal shape. All these mean that the cell membranes possess not only high deformability but also elasticity. Both properties are the sources of the rich shape transitions in red blood cells.

2.1.3 Difficulty in the Explanation of Discocyte Shape

The biconcave discoidal shape of a human red blood cell at rest has attracted considerable investigative interests in biomechanics. The basic reason is that red blood cells have no nuclei and their shapes are regulated only by the balance of forces acting on the cell membranes. In other words, the study on the shape transition in red blood cells can simply enable us to expose the mechanical properties of the membrane. A correct model of membrane elasticity thereby should be able to answer, first of all, the question: Why are red blood cells at rest always in a biconcave discoidal shape?

In 1948, the physiologist Ponder[4] suggested that biconcave discoid is the optimal configuration for oxygen circulation. However, it cannot explain why red blood cells change their (discoid) shapes into slipper shapes in blood capillary where exchange of gases takes place. In literature, a series of approaches were offered by scientists working on biomechanics to explain the biconcave discoidal configuration. For example, Fung and Tong[5] assumed the red blood cell as a fluid-filled shell limited by membranes and suggested that the thickness of the membrane varies from region to region in order to regulate the biconcave shape. This model contradicts with the observations under electron microscope that the thickness of the membrane is uniform. Lopez et al.[6] suggested that the difference in electric charge distribution over the red blood cell surface could be a decisive factor in the shape formation. However, the measurement of Greer and Baker[7] showed a uniform distribution of charge over the surface of red blood cells. Murphy[8] attempted to relate the red blood cell shape to regional variations in the distribution of cholesterol in the cell membrane, but it is not supported by the experiment performed later by Seeman *et al.*[9] Besides the aforementioned models, so far, there have been a lot of theories in literature concerning the problem of the shape of red blood cell, but none of them have been successful. The failure of these models comes basically from the fact that none of them recognized nor considered the correct material state of the cell membranes. In fact, the cell membrane is in the liquid crystal state. In the following sections, we shall show clearly that without considering the liquid crystal feature there is no adequate explanation for the biconcave discoidal shape of red blood cell.

2.2 Classic Differential Geometry of Surface

Before getting into the actual calculation of the shape of red blood cells and vesicles of artificial membranes, we shall introduce some definitions and mathematical tools for the shape problem.

2.2.1 Lipid-Bilayer Vesicle Viewed as a Closed Surface

The shape of a vesicle is defined as the configuration of a closed lipid bilayer in equilibrium between two aqueous media with different osmotic pressures. The study of the shapes of vesicles is the focus of this book. As we have discussed in Chapter 1, the thickness of lipid bilayer (about 5 nm) is much smaller than the size of the vesicle (about $1–10\,\mu m$ in diameter). This feature enables us to deal with the bilayer as a mathematical surface (of zero thickness), and the configuration of the bilayer vesicle as a closed surface. Therefore, it is necessary for us to have a brief summary of the differential geometry of surfaces. We shall discuss how to calculate the area and the curvature of a surface and the volume in a closed surface. These problems are important in the study of the energy of the vesicle which depends on the area (through *surface tension*), the curvature (through *bending energy*), and the volume (through osmotic pressure) of the closed vesicle surface.

The following contents are mainly based on the rather old but very useful book on differential geometry by Weatherburn.[10] From our personal experience, this book makes it very easy to learn for students of theoretical physics even if they have not had previous training on this subject. Another suitable and also easily understandable textbook is by do Carmo.[11]

2.2.2   Space Curve

To understand the curvature of a surface (which is a 2D subject), one should first learn about the curvature of a curve (a 1D subject). In differential geometry, a curve in space is described by a position vector $\mathbf{r} = \mathbf{r}(u)$, where u is a scalar parameter to denote the points along the curve. If u is taken as the arc-length, s, of the curve then the parameter is called a *natural* parameter. In what follows, s will be defined as the arc-length parameter without further indication.

To describe the curved feature of a space curve at point s, one needs to know the change of the following triple unit vectors:

$$\mathbf{t}(s) = \frac{\mathrm{d}}{\mathrm{d}s}\mathbf{r}(s) \equiv \mathbf{r}_s, \tag{2.1}$$

$$\mathbf{m}(s) = \frac{\mathrm{d}^2}{\mathrm{d}s^2}\mathbf{r}(s) \bigg/ \left|\frac{\mathrm{d}^2}{\mathrm{d}s^2}\mathbf{r}(s)\right| \equiv \mathbf{r}_{ss}/|\mathbf{r}_{ss}|, \tag{2.2}$$

$$\mathbf{b}(s) = \mathbf{t}(s) \times \mathbf{m}(s), \tag{2.3}$$

where $\mathbf{t}(s)$ is the unit tangent vector and $\mathbf{m}(s)$ and $\mathbf{b}(s)$ are the main normal and *binormal* vectors, respectively. From Eqs. (2.1)–(2.3) one can prove the mutual orthogonality of the triple vectors as schematized in Fig. 2.3.

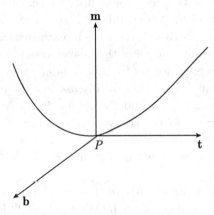

Fig. 2.3.   Triple unit vectors form a Frenet frame at point s of a space curve.

The *curvature* of the curve at point s is now defined by the rate of change of the tangent vector as

$$k(s) = \left|\frac{dt}{ds}\right| = |\mathbf{r}_{ss}|. \tag{2.4}$$

In analogy with dynamics of a vector, the geometric meaning of curvature is just the angular speed of rotation of $\mathbf{t}$ with respect to s, i.e.,

$$k(s) = \lim_{\Delta s \to 0} \left|\frac{\Delta\phi}{\Delta s}\right|, \tag{2.5}$$

where $\Delta\phi$ is the angle between $\mathbf{t}(s + \Delta s)$ and $\mathbf{t}(s)$. Obviously, $k(s)$ reflects just the bending strength of the curve at point s.

The *tangent plane* of the space curve at s, often denoted as $\sigma(s)$, is defined as the plane consisting of the vectors $\mathbf{t}(s)$ and $\mathbf{m}(s)$ (see Fig. 2.3). Thus, the normal vector of the plane is the binormal $\mathbf{b}(s)$. Thereby, for a planar curve, we always have $\mathbf{b}(s) = \mathbf{b}_0$, where $\mathbf{b}_0$ is a constant unit vector.

For a space curve, the curve is not only bending but also twisting along s. In geometry, the twisting strength of a curve means the rotational rate of the tangent plane along s. This is just the angular speed of the binormal $\mathbf{b}(s)$ and is called the *torsion* of the curve which is defined by

$$|\tau(s)| = \left|\frac{db(s)}{ds}\right| = \lim_{\Delta s \to 0} \left|\frac{\Delta\psi}{\Delta s}\right|, \tag{2.6}$$

where $\Delta\psi$ is the angle between $\mathbf{b}(s + \Delta s)$ and $\mathbf{b}(s)$. From Eqs. (2.1)–(2.6), one can derive the following relations:

$$\mathbf{t}_s = \frac{dt}{ds} = k(s)\mathbf{m}, \tag{2.7}$$

$$\mathbf{m}_s = \frac{dm}{ds} = -k(s)\mathbf{t} - \tau(s)\mathbf{b}, \tag{2.8}$$

$$\mathbf{b}_s = \frac{db}{ds} = \tau(s)\mathbf{m}. \tag{2.9}$$

These relations are called the *Frenet formulas* of a curve in space. This is the *basic theorem* of space curve. For given curvature $k(s)$ and torsion $\tau(s)$, one can integrate Eqs. (2.7)–(2.9) to get the equation of the curve, except for a difference of rigid rotation or translation.

One has to note that the torsion τ defined in Eq. (2.6) has two possibilities in sign. The sign of τ in the formulas shown in Eqs. (2.7)–(2.9) follows the notations given in Ref. [11], while in the book by Weatherburn, τ must change into $-\tau$ in the Frenet formulas.[10]

2.2.3   Surface and Parametric Curves

We have seen that a curve is the locus of points of one dimension. It is now easy to define a surface as a 2D locus of points represented by a positional vector $\mathbf{r} = \mathbf{r}(u,v)$ where u and v are two independent parameters, since the Cartesian coordinates x, y, z on the surface are known functions of u, v, i.e.,

$$\mathbf{r}(u,v) = (x(u,v), y(u,v), z(u,v)). \tag{2.10}$$

Elimination of the two parameters u, v leads to a single relation between x, y, and z:

$$F(x,y,z) = 0. \tag{2.11}$$

This is the oldest form of definition of the surface by Gauss. It is called the *equation of the surface.*

If one can solve z from Eq. (2.11), then one obtains the equation of the surface in Monge's form:

$$z = f(x,y). \tag{2.12}$$

2.2.4   First Fundamental Form

To describe the curved feature of a surface at point $\mathbf{r}(u,v)$, neither Gauss's form Eq. (2.11) nor Monge's form Eq. (2.12) is convenient. It is necessary to derive it directly from $\mathbf{r}(u,v)$ with tensor calculus. Let suffix 1 denote u and suffix 2 denote v. We define

$$(u,v) = (u^1, u^2),$$

$$\mathbf{r}_1 = \frac{\partial \mathbf{r}}{\partial u}, \quad \mathbf{r}_2 = \frac{\partial \mathbf{r}}{\partial v}, \tag{2.13}$$

$$\mathbf{r}_{11} = \frac{\partial^2 \mathbf{r}}{\partial u^2}, \quad \mathbf{r}_{12} = \frac{\partial^2 \mathbf{r}}{\partial u \partial v}, \quad \mathbf{r}_{22} = \frac{\partial^2 \mathbf{r}}{\partial v^2},$$

and so on.

Consider two neighboring points on the surface with position vectors $\mathbf{r}$ and $\mathbf{r} + d\mathbf{r}$, corresponding to the parameters (u,v) and $(u + du, v + dv)$, respectively. Then, with summation convention, i.e., repeated indices imply summation over them, we have

$$d\mathbf{r} = \mathbf{r}(u + du, v + dv) - \mathbf{r}(u,v) = \mathbf{r}_1 du + \mathbf{r}_2 dv = \mathbf{r}_i du^i, \tag{2.14}$$

and the length ds of $|\mathrm{d}\mathbf{r}|$ can be expressed by

$$I = \mathrm{d}s^2 = \mathrm{d}\mathbf{r} \cdot \mathrm{d}\mathbf{r} = E\mathrm{d}u^2 + 2F\mathrm{d}u\mathrm{d}v + G\mathrm{d}v^2, \qquad (2.15)$$

where

$$E = \mathbf{r}_1^2, \quad F = \mathbf{r}_1 \cdot \mathbf{r}_2, \quad G = \mathbf{r}_2^2. \qquad (2.16)$$

Formula (2.15) is called the *first fundamental form* and E, F, G are called the coefficients of the first fundamental form or the *fundamental magnitudes of the first order* in older textbooks.[10] For the convenience of derivation, we also use the following tensor notations:

$$g_{11} = E, \quad g_{12} = F, \quad g_{22} = G, \qquad (2.17)$$

or write $g_{ij} = \mathbf{r}_i \cdot \mathbf{r}_j$.

The geometric meanings of F, F, G are of greatest importance throughout the remainder of this book. For example, if we map out the surface by the doubly infinite set of *parametric curves* on the surface, among which either u or v remains constant, then parameters u, v represent a system of curvilinear coordinates for points on the surface (see Fig. 2.4). The unit tangential vectors of the system for point (u, v) then read as

$$\mathbf{a} = \mathbf{r}_1/\sqrt{E}, \quad \mathbf{b} = \mathbf{r}_2/\sqrt{G}. \qquad (2.18)$$

The angle between $\mathbf{a}$ and $\mathbf{b}$, ω, can be found as

$$\cos\omega = \mathbf{a} \cdot \mathbf{b} = F/\sqrt{EG}. \qquad (2.19)$$

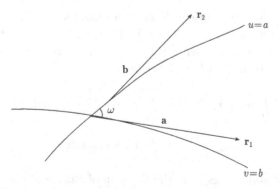

Fig. 2.4.   System of curvilinear coordinates in surface.

Therefore,

$$\sin\omega = \sqrt{1 - \cos^2\omega} = \sqrt{\frac{EG - F^2}{EG}} = \frac{\sqrt{g}}{\sqrt{EG}}, \tag{2.20}$$

where the determinant g is determined by

$$g \equiv \begin{vmatrix} g_{11} & g_{12} \\ g_{21} & g_{22} \end{vmatrix} = EG - F^2. \tag{2.21}$$

Since $\sin\omega = |\mathbf{a} \times \mathbf{b}| = |\mathbf{r}_1 \times \mathbf{r}_2|/\sqrt{EG}$, it follows that

$$\sqrt{g} = |\mathbf{r}_1 \times \mathbf{r}_2|. \tag{2.22}$$

Obviously, $F = 0$ is the condition of $\omega = \pi/2$, i.e., the parametric curves of $u = $ const. and $v = $ const. form an *orthogonal system*.

2.2.5 Area

Another metric problem that can be treated by the first fundamental form is the computation of the area of the surface. From Fig. 2.4, it is easy to see that the area element of the region bounded by the four curves of curvilinear coordinates $u = a$, $u = a + du$, $v = b$ and $v = b + dv$ is given by

$$dA = |\mathbf{r}_1 \times \mathbf{r}_2| du dv = \sqrt{g} du dv. \tag{2.23a}$$

Therefore, the total area of the bounded region D on the surface is given by

$$A = \int_D \sqrt{g} du dv = \int_D |\mathbf{r}_1 \times \mathbf{r}_2| du dv. \tag{2.23b}$$

Obviously, the measurement of the area should not depend on the choice of parameterization of u, v, and so for Eq. (2.23). In fact, if we use the Jacobian of the transformation between parameters of (u, v) and $(\bar{u}, \bar{v})$, $\partial(u, v)/\partial(\bar{u}, \bar{v})$, then from Eq. (2.23), we have

$$\begin{aligned} \int_{\bar{D}} |\mathbf{r}_{\bar{u}} \times \mathbf{r}_{\bar{v}}| d\bar{u} d\bar{v} &= \int_{\bar{D}} |\mathbf{r}_u \times \mathbf{r}_v| \left| \frac{\partial(u, v)}{\partial(\bar{u}, \bar{v})} \right| d\bar{u} d\bar{v} \\ &= \int_D |\mathbf{r}_u \times \mathbf{r}_v| du dv, \end{aligned} \tag{2.24}$$

where the last step comes from the theorem of change of variables in multiple integrals.

2.2.6   The Normal and the Tangent Plane

The normal to the surface at any point is a unit vector perpendicular to every tangent line through that point, i.e., perpendicular to both vectors $\mathbf{r}_1$ and $\mathbf{r}_2$. Therefore, it must be parallel to the vector $\mathbf{r}_1 \times \mathbf{r}_2$. According to Eq. (2.22), the unit normal $\mathbf{n}$ is given by

$$\mathbf{n} = \mathbf{r}_1 \times \mathbf{r}_2/\sqrt{g}. \qquad (2.25)$$

Obviously, the normal $\mathbf{n}$ given in Eq. (2.25) depends only on the first derivatives of $\mathbf{r}$ with respect to u and v.

At the point $\mathbf{r}(u,v)$, the plane which is perpendicular to the normal $\mathbf{n}(u,v)$ at that point is called the tangent plane, $\Pi(u,v)$, of the surface at that point. If $\mathbf{R}(X,Y,Z)$ denotes any point on the tangent plane, then we have

$$[\mathbf{R} - \mathbf{r}(u,v)] \cdot \mathbf{n}(u,v) = 0.$$

This is the equation of the tangent plane $\Pi(u,v)$.

2.2.7   Second Fundamental Form

So far, all the geometric quantities discussed above for a surface concerns only the first derivatives of $\mathbf{r}(u,v)$ with respect to u and v. However, as we have seen in Sec. 2.2.2, the curvature of a space curve $\mathbf{r}(s)$, Eq. (2.4), involves the second derivative of $\mathbf{r}(s)$ with respect to s. Therefore, before discussing the surface curvature, we introduce the second derivatives of $\mathbf{r}(u,v)$ with respect to u and v as

$$\mathbf{r}_{11} = \frac{\partial^2 \mathbf{r}}{\partial u^2}, \quad \mathbf{r}_{12} = \frac{\partial^2 \mathbf{r}}{\partial u \partial v}, \quad \mathbf{r}_{22} = \frac{\partial^2 \mathbf{r}}{\partial v^2}. \qquad (2.26)$$

To understand the geometric meaning of $\mathbf{r}_{ij}$, let us consider the relation between two neighboring points on the surface, P, $\mathbf{r}(u,v)$ and Q, $\mathbf{r}(u + du, v + dv)$. The distance $d\mathbf{r}$ between P and Q is equal to

$$d\mathbf{r} = \overrightarrow{PQ} = \mathbf{r}_1 du + \mathbf{r}_2 dv + (1/2)(\mathbf{r}_{11}du^2 + 2\mathbf{r}_{12}dudv + \mathbf{r}_{22}dv^2) + \cdots.$$
$$(2.27)$$

The distance from the point Q to the tangent plane at P, which demonstrates the curved strength of the surface at P, is then equal to

$$\mathbf{n} \cdot d\mathbf{r} = \frac{1}{2}(L_{11}du^2 + 2L_{12}dudv + L_{22}dv^2), \qquad (2.28a)$$

where

$$L_{11} = \mathbf{n} \cdot \mathbf{r}_{11}, \quad L_{12} = \mathbf{n} \cdot \mathbf{r}_{12}, \quad L_{22} = \mathbf{n} \cdot \mathbf{r}_{22}. \qquad (2.28b)$$

The quadratic form,

$$II = 2\mathbf{n} \cdot d\mathbf{r} = L_{11}du^2 + 2L_{12}dudv + L_{22}dv^2 \tag{2.29}$$

is called the *second fundamental form* of surface $\mathbf{r}(u, v)$ at P, and L_{11}, $L_{12} = L_{21}$, and L_{22} are called the coefficients of the second fundamental form. In tensor form, Eq. (2.28b) becomes

$$L_{ij} = \mathbf{n} \cdot \mathbf{r}_{ij}, \tag{2.30}$$

where $i, j = 1, 2$. Thus, the second fundamental form can be written as $II = L_{ij}du^i du^j$ with $du^1 = du$ and $du^2 = dv$. As stated in Eq. (2.14), here and in the following, the repeated indices imply summation over them.

2.2.8   Christoffel Symbols

According to Eq. (2.30) and the fact that $\mathbf{n}_i = \partial n/\partial u^i$ is normal to $\mathbf{n}$ on account of $|\mathbf{n}| = 1$, we may express the second-order derivatives of $\mathbf{r}$ and the first-order derivatives of $\mathbf{n}$ with respect to u, v (i.e., u^1 and u^2) in the following tensor product equations

$$\mathbf{r}_{ij} = \Gamma_{ij}^k \mathbf{r}_k + L_{ij}\mathbf{n},$$

$$\mathbf{n}_i = \mu_i^j \mathbf{r}_j. \tag{2.31}$$

In what follows, we are going to determine the two quantities Γ_{ij}^k, the so-called *Christoffel symbols* and μ_i^j.

From Eq. (2.17), the derivative of $g_{ij} = \mathbf{r}_i \cdot \mathbf{r}_j$ with respect to u^l is given by

$$\frac{\partial g_{ij}}{\partial u^l} = \mathbf{r}_{il} \cdot \mathbf{r}_j + \mathbf{r}_i \cdot \mathbf{r}_{jl},$$

similarly,

$$\frac{\partial g_{jl}}{\partial u^i} = \mathbf{r}_{ji} \cdot \mathbf{r}_l + \mathbf{r}_j \cdot \mathbf{r}_{li}.$$

Since $\mathbf{r}_{ij} = \mathbf{r}_{ji}$, we then have from Eq. (2.31)

$$\frac{1}{2}\left(\frac{\partial g_{il}}{\partial u^j} + \frac{\partial g_{lj}}{\partial u^i} - \frac{\partial g_{ij}}{\partial u^l}\right) = \mathbf{r}_{ij} \cdot \mathbf{r}_l = \Gamma_{ij}^k g_{kl}. \tag{2.32}$$

Let (g^{ij}) be the inverse matrix of (g_{ij}), i.e., $g^{ij}g_{jl} = \delta_l^i$. From Eq. (2.32), we obtain the expression of the Christoffel symbols as

$$\Gamma_{ij}^k = \frac{1}{2}g^{kl}\left(\frac{\partial g_{il}}{\partial u^j} + \frac{\partial g_{lj}}{\partial u^i} - \frac{\partial g_{ij}}{\partial u^l}\right). \tag{2.33}$$

Next, we study the term μ_i^j in Eq. (2.31). With Eq. (2.30) and $\mathbf{n} \cdot \mathbf{r}_i = 0$, we have

$$\frac{\partial}{\partial u^k}(\mathbf{n} \cdot \mathbf{r}_i) = \mathbf{n}_k \cdot \mathbf{r}_i + \mathbf{n} \cdot \mathbf{r}_{ik} = 0,$$

or $\mathbf{n}_k \cdot \mathbf{r}_i = -\mathbf{n} \cdot \mathbf{r}_{ik} = -L_{ik}$. Therefore, from Eq. (2.30), we obtain

$$-L_{ik} = \mathbf{n}_i \cdot \mathbf{r}_k = \mu_i^l g_{lk}.$$

Multiplication of this equation with g^{kj} gives

$$\mu_i^j = -L_{ik} g^{kj}. \tag{2.34}$$

The two equations in Eq. (2.31) now take the forms

$$\begin{aligned}
\mathbf{r}_{ij} &= \Gamma_{ij}^k \mathbf{r}_k + L_{ij}\mathbf{n}, \\
\mathbf{n}_i &= -L_{ik} g^{kj} \mathbf{r}_j.
\end{aligned} \tag{2.35}$$

They are the basic equations for the surface. The first equation in Eq. (2.35) is called the *Gauss equation* while the second one is called the *Weingarten equation*.

2.2.9 Curves and Directions on a Surface

The basic equations (2.35) give the differential relations between the derivatives of $\mathbf{r}_i$ and $\mathbf{n}$. From the theorem of existence of solution of any linear differential equation, one can prove from Eq. (2.35) that for a given set of (g_{ij}) and (L_{ij}) the corresponding surface is completely determined except for its rotation and translation in space. The equations can also serve for the study of the curved properties of the curves on the surface.

Any curve on the surface is determined by

$$\mathbf{r}(u,v) = \mathbf{r}(u(s), v(s)). \tag{2.36}$$

It is a vector function of the positions on the surface, Eq. (2.10), and $u(s)$, $v(s)$ are the parametric equations of the given curve with s being an arc length parameter.

Elimination of s gives another form of the curve equation as $f(u,v) = 0$. According to the general definition of the unit tangent vector for space curves, Eq. (2.1), we can obtain its tangent vector at point P, $\mathbf{r}(u,v)$, by

$$\mathbf{t}(s) = \mathbf{r}_1 \frac{du}{ds} + \mathbf{r}_2 \frac{dv}{ds}, \tag{2.37}$$

where $\mathbf{r}_1 = \partial \mathbf{r}/\partial u$ and $\mathbf{r}_2 = \partial \mathbf{r}/\partial v$. Obviously, $\mathbf{n} \cdot \mathbf{t} = 0$ means that any tangent direction of any curve on the surface has to be in the tangent plane $\Pi(u, v)$ of the surface at that point.

From the first fundamental form Eq. (2.15), we have

$$ds = (E du^2 + 2F du dv + G dv^2)^{1/2}.$$

Substitution of ds into Eq. (2.37) reveals the important conclusion that any direction of the surface at point P, $t(s)$ is completely determined by the ratio of $du : dv$, i.e., du/dv at that point. Therefore, for two directions with du/dv and $\delta u/\delta v$, or writing as two increments of vector

$$d\mathbf{r} = \mathbf{r}_1 du + \mathbf{r}_2 dv, \quad \delta \mathbf{r} = \mathbf{r}_1 \delta u + \mathbf{r}_2 \delta v,$$

the inclination angle ψ between the two directions is given by

$$\cos \psi = \frac{d\mathbf{r}}{ds} \cdot \frac{\delta \mathbf{r}}{\delta s} = E \frac{du}{ds}\frac{\delta u}{\delta s} + F \left(\frac{du}{ds}\frac{\delta v}{\delta s} + \frac{dv}{ds}\frac{\delta u}{\delta s} \right) + G \frac{dv}{ds}\frac{\delta v}{\delta s}, \qquad (2.38)$$

where ds, δs are the lengths of $d\mathbf{r}$ and $\delta \mathbf{r}$, respectively. Equation (2.38) yields the condition for the two directions being perpendicular to each other, $\psi = \pi/2$ as

$$E \frac{du}{dv}\frac{\delta u}{\delta v} + F \left(\frac{du}{dv} + \frac{\delta u}{\delta v} \right) + G = 0. \qquad (2.39)$$

For a given differential equation of the second order

$$P(u, v) du^2 + Q(u, v) du dv + R(u, v) dv^2 = 0, \qquad (2.40)$$

it determines two directions on the surface at P, i.e., the two roots of Eq. (2.40) du/dv and $\delta u/\delta v$. From the quadratic form of Eq. (2.40), we have $du/dv + \delta u/\delta v = -Q/P$ and $(du/dv)(\delta u/\delta v) = R/P$. Substituting these two relations into Eq. (2.39), we have the relation

$$ER - FQ + GP = 0, \qquad (2.41)$$

if the two directions given by (P, Q, R) are perpendicular to each other.

2.2.10   Normal Curvature of a Curve on a Surface

For the curve given in Eq. (2.36), the Frenet formula (2.7) gives

$$\frac{d^2}{ds^2}\mathbf{r}(s) = \kappa(s)\mathbf{m},$$

where $\kappa(s)$ is the curvature of the curve at the point $\mathbf{r}(u, v)$. Let θ be the angle between the surface normal $\mathbf{n}$ and the main normal of the curve, we have

$$\mathbf{n} \cdot \frac{\mathrm{d}^2}{\mathrm{d}s^2}\mathbf{r}(s) = \kappa(s)\cos\theta. \tag{2.42}$$

On the other hand, $\mathrm{d}^2\mathbf{r}/\mathrm{d}s^2 = \mathrm{d}(\mathbf{r}_i \mathrm{d}u^i/\mathrm{d}s)/\mathrm{d}s = \mathbf{r}_{ij}(\mathrm{d}u^i/\mathrm{d}s)(\mathrm{d}u^j/\mathrm{d}s) + \mathbf{r}_i\mathrm{d}^2u^i/\mathrm{d}s^2$. Therefore, from Eq. (2.31) and $\mathbf{n} \cdot \mathbf{r}_i = 0$, Eq. (2.42) gives

$$\kappa_n(s) \equiv \kappa(s)\cos\theta = \frac{L_{ij}\mathrm{d}u^i\mathrm{d}u^j}{g_{ij}\mathrm{d}u^i\mathrm{d}u^j} = \frac{II}{I}. \tag{2.43}$$

Both the first and the second fundamental forms Eq. (2.15) and Eq. (2.29) have been used in the derivation.

Equation (2.43) is determined by the ratio of $\mathrm{d}u/\mathrm{d}v$, i.e., the given tangent direction. It leads to the *theorem of Meusnier* that all curves lying on a surface which have the same tangent line at a given point on the surface have the same $\kappa(s)\cos\theta = \kappa_n(s)$. $\kappa_n(s)$ is defined as the normal curvature of the curve at that point. Figure 2.5 shows the feature of the theorem. In Fig. 2.5, the vector $\mathbf{v}$ is a given tangent direction of the surface at point P. The plane containing $\mathbf{v}$ and $\mathbf{n}$ is called the normal section of the surface at P along the direction $\mathbf{v}$. The intersection of the surface with the normal section is a plane curve C_n whose main normal $\mathbf{m}$ is parallel to $\mathbf{n}$. The curvature of C_n at P is just the normal curvature κ_n.

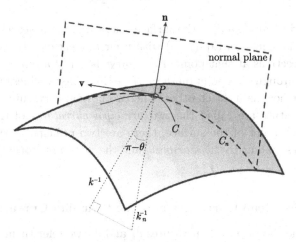

Fig. 2.5.   The normal section at a point on a surface.

2.2.11   Principal Directions, Line of Curvature and Principal Curvatures

Equation (2.43) implies that the normal curvature κ_n is a determined function of a given direction at P, that is, a function of the ratio du/dv:

$$\kappa_n = \frac{L_{11}du^2 + 2L_{12}dudv + L_{22}dv^2}{g_{11}du^2 + 2g_{12}dudv + g_{22}dv^2} \equiv \kappa_n\left(\frac{du}{dv}\right). \qquad (2.44)$$

A straightforward calculation of $\partial\kappa_n/\partial(du/dv) = 0$ gives the following equation:

$$(g_{11}L_{12} - g_{12}L_{11})du^2$$

$$+ (g_{11}L_{22} - g_{22}L_{11})dudv + (g_{12}L_{22} - g_{22}L_{12})dv^2 = 0. \qquad (2.45)$$

As stated in Sec. 2.2.9, Eq. (2.40) determines two mutually perpendicular directions satisfying Eq. (2.41) at the point P. We call the two directions described by Eq. (2.45) *the principal directions* on the surface at P. Since $\partial\kappa_n/\partial(du/dv) = 0$, Eq. (2.45) is also the equation that determines the directions of maximum and minimum normal curvature at the point P.

It is interesting to note that if

$$g_{11} : g_{12} : g_{22} = L_{11} : L_{12} : L_{22}, \qquad (2.46)$$

then Eq. (2.45) becomes an identity. In other words, it means that all directions are principal directions and κ_n has the same value along all the directions at the point [see also Eq. (2.44)]. Such a point is called an *umbilical point.*

As stated, at any point on the surface, there are two curves which satisfy the differential equation Eq. (2.45) with their tangent directions along the principal directions at that point. Each curve is called a *line of curvature.* Obviously, through each point on the surface there are two lines of curvature which are orthogonal to each other. The two normal curvatures of the two lines of curvature are called the two *principal curvatures* of the surface. Thus, at each point of the surface, there are two principal curvatures, c_1 and c_2, corresponding to the *maximum* and the *minimum* normal curvature at that point.

2.2.12   The Mean Curvature and the Gaussian Curvature

To determine the principal curvatures c_1 and c_2, we refer them to the two parametric curves (see Sec. 2.2.3) $u = $ const. and $v = $ const. on the surface as lines of curvature. This means that the differential equation Eq. (2.45) reduces to

$$dudv = 0. \tag{2.47}$$

Now, we must have

$$g_{11}L_{12} - g_{12}L_{11} = g_{12}L_{22} - g_{12}L_{12} = 0, \tag{2.48}$$

and

$$g_{11}L_{22} - g_{22}L_{11} \neq 0. \tag{2.49}$$

The multiplication of Eq. (2.49) with L_{12} gives

$$(g_{11}L_{22} - g_{22}L_{11})L_{12} = g_{11}L_{12}L_{22} - g_{22}L_{12}L_{11}. \tag{2.50}$$

From Eq. (2.48), we have $g_{11}L_{12} = g_{12}L_{11}$ and $g_{22}L_{12} = g_{12}L_{22}$. Substituting them into Eq. (2.50) we find

$$(g_{11}L_{22} - g_{22}L_{11})L_{12} = g_{12}L_{11}L_{22} - g_{12}L_{22}L_{11} \equiv 0. \tag{2.51}$$

Similarly we can find

$$(g_{11}L_{22} - g_{22}L_{11})g_{12} \equiv 0. \tag{2.52}$$

Equations (2.49), (2.51), and (2.52) give that

$$L_{12} = g_{12} = 0. \tag{2.53}$$

These are the necessary and sufficient conditions for the parametric curves to be the lines of curvature. Under such a parametric system, Eq. (2.44) becomes simply

$$\kappa_n = \frac{L_{11}du^2 + L_{22}dv^2}{g_{11}du^2 + g_{22}dv^2}. \tag{2.54}$$

For u-curves ($dv = 0$), κ_n gives the principal curvature c_1 and

$$c_1 = \frac{L_{11}}{g_{11}}. \tag{2.55}$$

Similarly, for v-curves ($du = 0$), κ_n gives the principal curvature c_2 and

$$c_2 = \frac{L_{22}}{g_{22}}. \tag{2.56}$$

To determine the value of κ_n along any direction of du/dv that makes an angle ψ with the principal direction $\delta v = 0$, we must turn back to Sec. 2.2.9 by letting the direction $\delta \mathbf{r}$ be the principal direction of $\delta v = 0$ and $d\mathbf{r}$ the discussed direction. Substitution of $F = g_{12} = 0$ (See Eq. (2.53)) and $\delta v = 0$ in Eq. (2.38) gives

$$\cos \psi = \frac{E du}{ds} \frac{\delta u}{\delta s} = \frac{g_{11} du \delta u}{\sqrt{g_{11} du^2 + g_{22} dv^2} \sqrt{g_{11} \delta u^2}}. \tag{2.57}$$

The last step comes from the use of the first fundamental form. Therefore, we have

$$\cos^2 \psi = \frac{g_{11} du^2}{g_{11} du^2 + g_{22} dv^2} \tag{2.58}$$

and

$$\sin^2 \psi = 1 - \cos^2 \psi = \frac{g_{22} dv^2}{g_{11} du^2 + g_{22} dv^2}. \tag{2.59}$$

On the other hand, Eq. (2.54) can be written in the form:

$$\kappa_n = \frac{L_{11}}{g_{11}} \frac{g_{11} du^2}{g_{11} du^2 + g_{22} dv^2} + \frac{L_{22}}{g_{22}} \frac{g_{22} dv^2}{g_{11} du^2 + g_{22} dv^2}.$$

From Eqs. (2.55)–(2.59), we have

$$\kappa_n = c_1 \cos^2 \psi + c_2 \sin^2 \psi. \tag{2.60}$$

This is the *Euler's theorem* on the normal curvature.

The formulas of c_1 and c_2 are valid only under the special choice of the lines of curvature given by Eqs. (2.55) and (2.56). Sometimes, it is convenient to calculate both c_1 and c_2 on any parametric curve other than $u = $ const and $v = $ const. To get the expression of c_1 and c_2 in general parametric systems, we write Eq. (2.44) as

$$\kappa_n(x)(g_{11}x^2 + 2g_{12}x + g_{22}) = L_{11}x^2 + 2L_{12}x + L_{22},$$

where $x \equiv du/dv$. Along the principal directions, we have $\partial \kappa_n / \partial x = 0$. Taking the derivative of the above equation with respect to x, we get the relation

$$\kappa_n(2g_{11}x + 2g_{12}) = 2L_{11}x + 2L_{12},$$

or

$$x = \frac{du}{dv} = \frac{g_{12}\kappa_n - L_{22}}{L_{11} - g_{11}\kappa_n}. \tag{2.61}$$

Substitution of Eq. (2.61) into Eq. (2.45) leads to the following quadratic equation for κ_n:

$$(g_{11}g_{22} - g_{12}^2)\kappa_n^2 - (L_{22}g_{11} - 2L_{12}g_{12} + L_{11}g_{22})\kappa_n + (L_{11}L_{22} - L_{12}^2) = 0. \tag{2.62}$$

Obviously, the two roots of this equation are the principal curvatures c_1 and c_2. From the properties of quadratic equation, we have

$$K \equiv c_1 c_2 = \frac{L_{11}L_{22} - L_{12}^2}{g_{11}g_{22} - g_{12}^2} \tag{2.63}$$

and

$$H = \frac{1}{2}(c_1 + c_2) = \frac{L_{22}g_{11} - 2L_{12}g_{12} + L_{11}g_{22}}{2(g_{11}g_{22} - g_{12}^2)}. \tag{2.64}$$

Here, K is called the *Gaussian curvature* and H the *mean curvature* at any point on the surface. Let $L = \det(L_{ij})$, and according to the definition of g, Eq. (2.21), we may write Eqs. (2.63) and (2.64) as

$$\begin{aligned} K &= L/g, \\ H &= (1/2)g^{ij}L_{ij}, \end{aligned} \tag{2.65}$$

where (g^{ij}) is the inverse matrix of (g_{ij}) as defined in Sec. 2.2.8.

2.3   Differential Invariants on a Surface

In Sec. 2.2, we have given a brief account of the quantities which are used to describe a surface and the curves on it. In this section, we want to discuss both the scalar and the vector fields on a surface. The fields on a surface are defined as functions of position on the surface. The values of the functions are determined by the coordinates of the surface, u and v, or denoted as u^1, u^2. The differential invariants for the functions, the scalar and the vector fields, on the surface play an important role in the discussion of physical problems connected with the curved surface, especially for the derivation of the membrane elasticity that we are concerned with.

2.3.1 Gradient of a Scalar Field

The gradient or the slope of a function in Euclidean three space, $f(x, y, z)$, is defined as a vector function

$$\nabla f(x, y, z) = \left(\frac{\partial f}{\partial x}, \frac{\partial f}{\partial y}, \frac{\partial f}{\partial z} \right). \tag{2.66}$$

Its direction represents the direction of maximum rate of increase of f in space, and its magnitude gives the maximum rate of increase of f. Similarly, the gradient or slope of a scalar function $\phi(u, v)$ at any point P (denoted as $\mathbf{r}(u, v)$) on a surface is defined as a vector function, its direction is the direction on the surface at P which gives the maximum arc-rate of increase of ϕ, and its magnitude is this maximum rate of increase of ϕ. Without proof, we put forward the definition

$$\nabla' \phi = g^{ij} \mathbf{r}_i \partial_j \phi, \tag{2.67}$$

where $(g^{ij}) = (g_{ij})^{-1}$ has been defined before and $\partial_j = \partial/\partial u^j$. Here, the summation convention for repeated indices is used as before. The detailed form of the 2D gradient in curved coordinates u and v is then given by

$$\nabla' = \frac{1}{g} \mathbf{r}_1 \left(g_{22} \frac{\partial}{\partial u} - g_{12} \frac{\partial}{\partial v} \right) + \frac{1}{g} \mathbf{r}_2 \left(g_{11} \frac{\partial}{\partial v} - g_{12} \frac{\partial}{\partial u} \right) = g^{ij} \mathbf{r}_i \partial_j. \tag{2.68}$$

The 2D gradient has all the properties of the usual 3D gradient in Eq. (2.66). For example, the tensor summation form in Eq. (2.68) shows its invariance with respect to the choice of coordinate systems, i.e., it is independent of the parameters. Especially, one can prove that if C is a curve joining two points A and B on the surface, the definite integral from A to B of that part of $\nabla' \phi$ that is tangential to the curve is equal to

$$\int_A^B (\nabla' \phi) \cdot \mathbf{t} ds = \phi_B - \phi_A, \tag{2.69}$$

where $\mathbf{t}$ is the unit tangent vector along the curve C and ϕ_A and ϕ_B are the values of ϕ at points A and B, respectively. It is easy to prove Eq. (2.69) as follows: With Eqs. (2.37) and (2.67), the left-handed side of Eq. (2.69) becomes

$$\int_A^B (t \cdot \nabla' \phi) \mathrm{d}s = \int_A^B \left(\frac{\mathbf{r}_1 \mathrm{d}u}{\mathrm{d}s} + \frac{\mathbf{r}_2 \mathrm{d}v}{\mathrm{d}s} \right) \cdot g^{ij} \mathbf{r}_i \partial_j \phi \mathrm{d}s$$

$$= \int_A^B g^{ij} \left(\frac{g_{i1} \mathrm{d}u}{\mathrm{d}s} + \frac{g_{i2} \mathrm{d}v}{\mathrm{d}s} \right) \partial_j \phi \mathrm{d}s$$

$$= \int_A^B g^{ij} g_{il} \frac{\mathrm{d}u^l}{\mathrm{d}s} \partial_j \phi \mathrm{d}s$$

$$= \int_A^B \delta_l^j \frac{\mathrm{d}u^l}{\mathrm{d}s} \partial_j \phi \mathrm{d}s$$

$$= \int_A^B \frac{\mathrm{d}u^j}{\mathrm{d}s} \partial_j \phi \mathrm{d}s = \int_A^B \partial_j \phi \mathrm{d}u^j$$

$$= \int_A^B \mathrm{d}\phi = \phi_B - \phi_A.$$

When the curve C is a closed curve (i.e., $\phi_B = \phi_A$), we then have

$$\oint \nabla' \phi \cdot \mathrm{d}\mathbf{r} = 0, \tag{2.70}$$

where $\mathrm{d}\mathbf{r} = t \mathrm{d}s$. Equation (2.70) gives the theorem on the 2D gradient where $\oint \nabla' \phi \cdot \mathrm{d}\mathbf{r}$ vanishes for every closed curve drawn on the surface.

2.3.2 Divergence of a Vector Field

A vector field $\mathbf{F}$ on a surface $\mathbf{r}(u,v)$ is defined as a vector function on the surface in the form

$$\mathbf{F} = P(u,v)\mathbf{r}_1 + Q(u,v)\mathbf{r}_2 + R(u,v)\mathbf{n}. \tag{2.71}$$

The 2D-differential operator ∇' given in Eq. (2.68) can be applied to the vector field $\mathbf{F}$ in different ways. The first is a scalar differential invariant called the *divergence* of $\mathbf{F}$

$$\mathrm{div}\,\mathbf{F} = \nabla' \cdot \mathbf{F}. \tag{2.72}$$

With the help of the Weingarten equation given in Eq. (2.35), $\mathbf{n}_i = -L_{ik} g^{kj} \mathbf{r}_j$, we can calculate the divergence of the unit normal $\mathbf{n}$ on the surface as

$$\mathrm{div}\,\mathbf{n} = g^{ij} \mathbf{r}_i \cdot \partial_j \mathbf{n} = -g^{ij} \mathbf{r}_i \cdot L_{jk} g^{kl} \mathbf{r}_l = -g^{ij} g_{il} L_{jk} g^{kl}$$

$$= -\delta_l^j L_{jk} g^{kl} = -L_{lk} g^{lk} = -2H, \tag{2.73}$$

where H is the mean curvature. The last equation follows from the result of Eq. (2.65). It is easy to show that

$$\text{div}(R\mathbf{n}) = \nabla'R \cdot \mathbf{n} + R\text{div}\,\mathbf{n} = R\text{div}\,\mathbf{n} = -2HR, \qquad (2.74)$$

since the gradient $\nabla'R$ is normal to $\mathbf{n}$.

It is convenient to derive the expressions for both the divergences of $P\mathbf{r}_1$ and $Q\mathbf{r}_2$. From the definition of ∇', Eq. (2.68), and the Gauss equation, Eq. (2.35), we have

$$\text{div}(P\mathbf{r}_1) = \nabla' \cdot (P\mathbf{r}_1) = g^{ij}\mathbf{r}_i \cdot \partial_j(P\mathbf{r}_1) = g^{ij}\mathbf{r}_i \cdot (\mathbf{r}_1 P_j + P\mathbf{r}_{1,j})$$

$$= g^{ij}(g_{i1}P_j + P\mathbf{r}_i \cdot \mathbf{r}_{1,j}) = \delta_1^j P_j + Pg^{ij}\Gamma_{1j}^k g_{ik}$$

$$= P_1 + P\Gamma_{1j}^j.$$

With the definition of g, Eq. (2.21), and that of the Christoffel symbols, Eq. (2.33), it is now easy to show that

$$\text{div}(P\mathbf{r}_1) = P_1 + [P/(2g)](g_{11}\partial_1 g_{22} + g_{22}\partial_1 g_{11} - 2g_{12}\partial_1 g_{12})$$

$$= P_1 + [P/(2g)]\partial_1 g = (1/\sqrt{g})\partial_1(\sqrt{g}P). \qquad (2.75)$$

Similarly, we have

$$\text{div}(Q\mathbf{r}_2) = (1/\sqrt{g})\partial_2(\sqrt{g}Q). \qquad (2.76)$$

Finally, from Eqs. (2.74)–(2.76), we obtain the formula for the divergence of $\mathbf{F}$ as

$$\text{div}\,\mathbf{F} = \frac{1}{\sqrt{g}}\partial_i(\sqrt{g}F^i) - 2HR, \qquad (2.77)$$

with $F^1 = P$ and $F^2 = Q$.

2.3.3   Laplace-Beltrami Operator on a Scalar Function

As we have indicated, there are many ways to construct the differential invariants for a surface by using the 2D gradient operator ∇' given in Eq. (2.68). Among them, the most important one, which concerns the shape equation of equilibrium membranes, is the *Laplace–Beltrami* operator $\nabla'^2 = \nabla' \cdot \nabla'$ operated on a scalar function $\phi(u,v)$. It is straightforward just by inserting in Eq. (2.77) the vector $\mathbf{F} = \nabla'\phi$ given by Eq. (2.67), with

$F^1 = P = g^{1j}\partial_j\phi$, $F^2 = Q = g^{2j}\partial_j\phi$, and $R = 0$. Then we have

$$\nabla'^2\phi = \frac{1}{\sqrt{g}}\partial_i(g^{ij}\sqrt{g}\partial_j\phi). \tag{2.78}$$

When the parametric curves are orthogonal (i.e., $g_{12} = 0$), Eq. (2.78) becomes

$$\nabla'^2\phi = \frac{1}{\sqrt{g}}\left[\frac{\partial}{\partial u}\left(\sqrt{\frac{g_{22}}{g_{11}}}\phi_1\right) + \frac{\partial}{\partial v}\left(\sqrt{\frac{g_{11}}{g_{22}}}\phi_2\right)\right]. \tag{2.79}$$

2.3.4 Two-Dimensional Curl of a Vector

Application of the 2D-gradient ∇' to the vector function $\mathbf{F}$ given in Eq. (2.71) to give the vector product $\nabla' \times \mathbf{F}$ yields another differential invariant vector, which we call the curl of $\mathbf{F}$

$$\text{curl}\,\mathbf{F} \equiv \nabla' \times \mathbf{F} = g^{ij}\mathbf{r}_i\partial_j \times \mathbf{F}. \tag{2.80}$$

With the help of the Weingarten equation given in Eq. (2.35), from Eq. (2.80) we find the curl of the component $R\mathbf{n}$ of $\mathbf{F}$ as

$$\text{curl}(R\mathbf{n}) = (\nabla'R) \times \mathbf{n} + R\nabla' \times \mathbf{n}.$$

With the operator ∇' represented by Eqs. (2.68) and (2.35), one has that

$$\nabla' \times \mathbf{n} = g^{ij}\mathbf{r}_i \times \mathbf{n}_j = -g^{ij}L_{jk}g^{kl}\mathbf{r}_i \times \mathbf{r}_l.$$

Simultaneous interchange of the pair of indices i and l and the pair of indices j and k leads to

$$\nabla' \times \mathbf{n} = -g^{lk}L_{kj}g^{ji}\mathbf{r}_l \times \mathbf{r}_i.$$

These two relations have opposite signs. Thus, we have

$$\nabla' \times \mathbf{n} = 0, \tag{2.81}$$

and consequently

$$\text{curl}(R\mathbf{n}) = (\nabla'R) \times \mathbf{n}. \tag{2.82}$$

It is also easy to prove that

$$\nabla' \times \mathbf{r} = 0, \tag{2.83}$$

where $\mathbf{r}$ is the surface position vector, simply by using Eq. (2.68) and followed by an interchange of the indices i and j. With the help of the two relations given in Eq. (2.35), we can show that

$$\nabla' \times (P\mathbf{r}_1) = \frac{1}{\sqrt{g}}[\partial_1(g_{22}P) - \partial_2(g_{11}P)]\mathbf{n} - \frac{P}{\sqrt{g}}(L_{12}\mathbf{r}_1 - L_{11}\mathbf{r}_2), \quad (2.84)$$

$$\nabla' \times (Q\mathbf{r}_2) = \frac{1}{\sqrt{g}}[\partial_1(g_{22}Q) - \partial_2(g_{12}Q)]\mathbf{n} + \frac{Q}{\sqrt{g}}(L_{22}\mathbf{r}_1 - L_{12}\mathbf{r}_2). \quad (2.85)$$

Finally, the summation of Eqs. (2.82), (2.84) and (2.85) gives that

$$\nabla' \times \mathbf{F} = \frac{1}{\sqrt{g}}[\partial_1(g_{12}P + g_{22}Q) - \partial_2(g_{11}P + g_{12}Q)]\mathbf{n} + (\nabla'R) \times \mathbf{n}$$

$$+ \frac{1}{\sqrt{g}}(L_{12}P + L_{22}Q)\mathbf{r}_1 - \frac{1}{\sqrt{g}}(L_{11}P + L_{12}Q)\mathbf{r}_2. \quad (2.86)$$

2.3.5   Other Differential Invariants

Based on the definition of the 2D-gradient operator ∇' given by Eq. (2.68), the basic surface equations, the Gauss equation and the Weingarten equation, we can find many differential invariants concerning a surface. In the following, we give a list of the invariants that will be needed in our derivation of the elasticity and the variation of the fluid membranes in the following sections and chapters:

$$\nabla'^2\mathbf{r} = 2H\mathbf{n}, \quad (2.87)$$

$$\mathbf{n} \cdot \nabla'^2\mathbf{n} = 2K - 4H^2, \quad (2.88)$$

$$2K = \mathbf{n} \cdot \nabla'^2\mathbf{n} - \nabla' \cdot \nabla'^2\mathbf{r}. \quad (2.89)$$

If ϕ is a scalar function and $\mathbf{F}$, $\mathbf{U}$ are vector functions on the surface, similar to the usual 3D-gradient operator ∇, we have

$$\nabla' \cdot (\phi\mathbf{F}) = (\nabla'\phi) \cdot \mathbf{F} + \phi\nabla' \cdot \mathbf{F}, \quad (2.90)$$

$$\nabla' \times (\phi\mathbf{F}) = (\nabla'\phi) \times \mathbf{F} + \phi\nabla' \times \mathbf{F}, \quad (2.91)$$

$$\nabla' \cdot (\mathbf{U} \times \mathbf{F}) = \mathbf{F} \cdot \nabla' \times \mathbf{U} - \mathbf{U} \cdot \nabla' \times \mathbf{F}. \quad (2.92)$$

Some special results concerning the differentials of the surface vector $\mathbf{r}(u, v)$ are

$$\nabla' \cdot \mathbf{r} = 2, \quad (2.93)$$

$$\nabla' \cdot (H\mathbf{r}) = \mathbf{r} \cdot \nabla'H + 2H, \quad (2.94)$$

$$\nabla' \times (H\mathbf{r}) = (\nabla'H) \times \mathbf{r}. \quad (2.95)$$

With these mathematical tools in hand, we are going to use them to study the curvature elasticity of fluid membranes.

2.4 Curvature Elasticity of Fluid Membranes in Liquid Crystal Phase

2.4.1 Fluid Membranes Viewed as Liquid Crystals

The correlation between biomembranes and liquid crystals has long been noticed. In 1854, while studying the structure of living cells, Virchow had already given descriptions of the myeline figures.[12] Such myelin figures have been characterized as liquid crystal structures. It was during the 1933 Faraday Society Meeting that the viewpoint of liquid crystalline nature of biological structures emerged. By the time of the First International Liquid Crystal Conference at Kent University in 1965, liquid crystals and biological structures became a hot subject in discussion. During the discussions at the *5th International Liquid Crystal Conference* at Stockholm in 1974, it was established that long range order of orientation of the molecules is the common factor in both the biomembrane system and the lyotropic liquid crystal system. Based upon the *Proceedings of the 5th International Liquid Crystal Conference*, Freiberg edited a book entitled *Lyotropic Liquid Crystals and the Structure of Biomembranes.*[13] Later on, the publication of the book entitled *Liquid Crystals and Biological Structures*[14] highly promoted the study of the role of liquid crystals in living systems. However, even up to 1979, the mentioned discussions on biomembranes and liquid crystals still remained on a level that is simply analogical and qualitative. It is Professor Helfrich of Freie Universität in Berlin who, based on the curvature elasticity theory of liquid crystals, first proposed the quantitative theory of fluid membranes.[15]

2.4.2 Helfrich's Approach

As we have discussed in Sec. 2.1, the interest that attracted people to investigate the bending elasticity of fluid membranes comes from the difficulty in explaining the discocyte shape of the red blood cells.

In the literature, there have been two major approaches to the problem of the red blood cell shape. The first was proposed by a mechanical engineer who treated the problem of the shape in terms of mechanical tension, stress, strain, etc.[16] However this approach frequently requires changes in the area and the thickness of the membrane.[5] To overcome this defect, in 1970, Canham,[17] a specialist in biomechanics, proposed a hypothesis that the shape of red blood cell was determined solely by the bending elasticity, i.e., bending energy per unit area of the membrane in the form

$$g_b = \frac{1}{2}kH^2. \tag{2.96}$$

However, subsequently it was found that this model leads to dumbbell-like shapes not observed in red blood cells.[18] On the other hand, workers in the field of mechanics tried to explain the deformation of red blood cells with a thin-shell theory, but were unable to define the state of zero stress.[19] Furthermore, the assumption of solid shell is inconsistent with the kind of caterpillar motion of the red blood cells along the wall of the blood vessel. For quite some time, the problem of the shape of red blood cells had remained unsolved.

The second approach was proposed by physicists who related biomembrane to certain condensed matter known in physics. Helfrich noticed that polar lipid bilayer is the basic structural unit of the cell membranes. The thickness of the bilayer is of the order of 5 nm with each layer of the bilayer about 2.5 nm in thickness. Phospholipids are major molecules in lipid bilayer. They possess strong polar molecules, with their non-polar portions oriented away from the polar environment. On the interface between polar liquid, like water, and non-polar liquid, like benzene, the charged hydrophilic heads of the lipids stay in the water while the hydrophobic fatty acid chains of the lipids stay in the non-polar liquid side. In this way, a monolayer of lipids is formed. In water alone, since the hydrophobic chains refuse to get in contact with water, two lipid monolayers will group together to form a lipid bilayer. Under certain concentration, the lipid bilayer will bend automatically to form a closed vesicle. Such is the basic model of the cell membrane. In the bilayer membrane, the alignment of the hydrocarbon chains has a certain degree of orderliness. In the normal physiological temperature range, the preferred orientation of the hydrocarbon chains is normal to the surfaces of the bilayer membrane. Helfrich recognized that if we treat the *hydrocarbon chains* as the *director of uniaxial liquid crystal*, the fluid membrane is just like a *homeotropic nematic liquid crystal* cell of thickness twice the length of the lipid molecule. Based on the Frank free energy density expression for uniaxial liquid crystals,[20] with the normal to the membrane as the director of liquid crystal, Helfrich[15] deduced the elastic energy of curvature per unit area of the membrane as

$$g = (1/2)k(c_1 + c_2 - c_0)^2 + \bar{k}c_1 c_2, \tag{2.97}$$

where c_1 and c_2 are the two principal curvatures of the surface of the membrane as discussed in Secs. 2.2.11 and 2.2.12, and the constant c_0 is called the spontaneous curvature of the membrane surface.[15] The constant c_0 takes care of the asymmetry of the layers or the environment and is closely related to the permanent splay (sometimes called spontaneous splay) s_0 of the liquid crystal.[20] The constant $k(> 0)$ is the bending rigidity and $\bar{k}(> 0 \text{ or } < 0)$ is the elastic modulus of the Gaussian curvature $c_1 c_2$. By comparison with the curvature elasticity theory of liquid crystals, both k and $\bar{k}$ are found to be of the order of the product of the elastic constants of liquid crystals and the thickness of the membrane; i.e., of the order of 10^{-19} J. Equation (2.97) is called the Helfrich free energy of fluid membranes and is generally recognized as the basic quantity in dealing with the mechanical behavior of biomembranes in the L_α phase.

The original approach in the derivation of Eq. (2.97) in 1973 by Helfrich is based on the analogy of the derivation of the curvature elastic energy of liquid crystals introduced by Frank.[20] He considered a local Cartesian coordinate system, with the z-axis of the system parallel to the director $\mathbf{n}$ of the liquid crystal. With $\mathbf{n}$ as a function of x and y, $\mathbf{n} = (n_x, n_y, \sqrt{1 - n_x^2 - n_y^2})$ in the local system, and with uniaxial symmetry of rotation, the splay and the saddle splay deformations as introduced by Frank[20] are given by

$$\frac{\partial n_x}{\partial x} + \frac{\partial n_y}{\partial y}, \tag{2.98}$$

and

$$\frac{\partial n_x}{\partial x} \frac{\partial n_y}{\partial y} - \frac{\partial n_x}{\partial y} \frac{\partial n_y}{\partial x}, \tag{2.99}$$

respectively. Obviously, up to quadratic terms, the curvature-elastic energy density can be written as

$$g = \frac{1}{2} k \left(\frac{\partial n_x}{\partial x} + \frac{\partial n_y}{\partial y} - c_0 \right)^2 + \bar{k} \left(\frac{\partial n_x}{\partial x} \frac{\partial n_y}{\partial y} - \frac{\partial n_x}{\partial y} \frac{\partial n_y}{\partial x} \right). \tag{2.100}$$

Let the director $\mathbf{n}$ be the normal of the fluid membrane, then we have the Helfrich curvature-elastic energy formula Eq. (2.97) with

$$c_1 + c_2 = \frac{\partial n_x}{\partial x} + \frac{\partial n_y}{\partial y}, \tag{2.101}$$

and

$$c_1 c_2 = \frac{\partial n_x}{\partial x} \frac{\partial n_y}{\partial y} - \frac{\partial n_x}{\partial y} \frac{\partial n_y}{\partial x}. \tag{2.102}$$

In comparison with the 2D-differential invariant given in Eq. (2.73), $\nabla' \cdot \mathbf{n} = -2H$, here we find that $c_1 + c_2$ is equal to $-2H$, a result that has been used in a paper by Ou-Yang and Helfrich.[21] In Eq. (4) of that paper and in the following chapters of this book, the definition $H = -(1/2)(c_1 + c_2) = (1/2)g^{ij}L_{ij}$ comes directly from the original definition Eq. (2.101) given by Helfrich.

In Eq. (2.100), the bend and the twist terms, $\mathbf{n} \times \nabla \times \mathbf{n}$ and $\mathbf{n} \cdot \nabla \times \mathbf{n}$, in Frank free energy of liquid crystals are not involved. It was stated by Helfrich[15] that the rotation, i.e., the curl of the vector field $\mathbf{n}(x, y)$ must vanish since it is normal to a uniquely defined surface. Therefore, there is the relation

$$\frac{\partial n_x}{\partial y} - \frac{\partial n_y}{\partial x} = 0, \qquad (2.103)$$

which is just Eq. (2.81), the result of the 2D-curl of $\mathbf{n}$.

Later on, in Ref. [22], Helfrich gave an easy way to prove Eq. (2.97). Bending deformations can be described in terms of the curvature of the membrane surface. As we have seen in Secs. 2.2 and 2.3, the curved properties of a surface at any point are described by the two principal curvatures c_1 and c_2. Therefore, up to quadratic terms of c_1 and c_2, the bending energy expression is a complex combination of all the linear and the quadratic invariants, i.e., those immutable forms under an exchange of c_1 and c_2. One finds that in linear form only $c_1 + c_2$ and in quadratic form $(c_1 + c_2)^2$, $c_1 c_2$, $c_1^2 + c_2^2$, and $(c_1 - c_2)^2$ satisfy this requirement. Among the latter four terms, only two are independent. Taking the sum of the linear term $c_1 + c_2$ and the first two quadratic terms $(c_1 + c_2)^2$ and $c_1 c_2$, we can construct an energy density expression just like Eq. (2.97).

The most important contribution of the original approach by Helfrich,[15] the analogy between fluid membranes and liquid crystals, is that it gives a correct estimate of the elastic moduli k and $\bar{k}$ for fluid membranes by multiplying the known curvature moduli of liquid crystals ($\approx 10^{-11}$N) by the thickness of the bilayer (≈ 5 nm), i.e., $k, |\bar{k}| \approx 5 \times 10^{-20}$ J. A more detailed comparison of Eq. (2.97) with the curvature elastic energy density expression of liquid crystal by Frank,[20]

$$q_{\mathrm{LC}} = (1/2)[k_{11}(\nabla \cdot \mathbf{n} - s_0)^2 + k_{22}(\mathbf{n} \cdot \nabla \times \mathbf{n} + t_0)^2 + k_{33}(\mathbf{n} \cdot \nabla \mathbf{n})^2]$$

$$- k_{12}(\nabla \cdot \mathbf{n})(\mathbf{n} \cdot \nabla \times \mathbf{n})$$

$$- (1/2)(k_{22} + k_{24})[(\nabla \cdot \mathbf{n})^2 + (\nabla \times \mathbf{n})^2 - \nabla \mathbf{n} : \nabla \mathbf{n}], \qquad (2.104)$$

shows that the value of k and $\bar{k}$ may be estimated by[23]

$$k = k_{11}d,$$
$$\bar{k} = -(k_{22} + k_{24})d,$$

(2.105)

where d is the thickness of the membrane. The splay elastic constant k_{11} of liquid crystals is of the order of 10^{-11}N. With these values, for the first time, Helfrich estimated that $k = 5 \times 10^{-20}$ J.[15] This is close to the experimental values of 0.9×10^{-19} J to 2.3×10^{-19} J measured on lecithin bilayers.[24] Such good agreement suggests that fluid membranes and liquid crystals are highly analogous and can be treated with similar mathematical formalism. However, the second equation of Eq. (2.105) may cause some puzzle. The Maier–Saupe model of liquid crystals[25] gives $k_{24} = -3k_{22}/2$ which leads to $\bar{k} = k_{22}d/2$. For liquid crystals, k_{22} is always positive. Thus, it suggests that $\bar{k}$ is always positive. However, it is found[26] that for membranes in large amount of water there exist three different equilibrium configurations: a lattice of passages,[27] a single big vesicle, and many small vesicles. The last two configurations correspond to negative values of $\bar{k}$. Therefore, it is necessary to take a closer examination on the problem. In the next section, instead of comparing with Frank's theory of liquid crystals, we shall use the more comprehensive theory developed by Nehring and Saupe[28] for the analogical calculation.

In Frank's theory as well as in Helfrich's theory, only the first-order derivatives of the director $\mathbf{n}$ or the surface normal $\mathbf{n}$ are taken into consideration. However, Nehring and Saupe demonstrated that the second derivatives of $\mathbf{n}$ must also be considered.[29] They showed that, besides the terms given in Eq. (2.104), the elastic free energy density expression has two more terms

$$\delta g_{\mathrm{LC}} = k_{13}\mathbf{n} \cdot \nabla\nabla \cdot \mathbf{n} - k_{23}\mathbf{n} \cdot \nabla(\mathbf{n} \cdot \nabla \times \mathbf{n}).$$

(2.106)

In fact, from $\mathbf{n} \cdot \mathbf{n} = 1$, one may easily prove that the contribution of the $n_{i,jm}$ to g_{LC} is comparable with that of the terms $n_{i,j}n_{j,m}$. Therefore, δg_{LC} cannot be neglected. Here, let us examine the terms analogous to k_{13} and k_{23} in the curvature elasticity theory of membrane. In 1978, Mitov argued that, in contrast to the case of nematics, there is no term analogous to k_{13} in the second-order elastic free energy of bilayers.[29] In Sec. 2.4.3, we shall show that the consideration of terms associated with k_{13} and k_{23} will not change the form of the Helfrich free energy Eq. (2.97). However, the bending rigidity k and the Gaussian curvature modulus $\bar{k}$ will now take the forms

$$k = (k_{11} - 2k_{13})d,$$
$$\bar{k} = (2k_{13} - k_{22} - k_{24})d. \tag{2.107}$$

Till date, there are many discussions on the terms associated with k_{13} and k_{24} in the theory of liquid crystals.[30] These terms are difficult to understand. In Sec. 2.4.4, we are going to present a Landau–de Gennes theory[31] for the curvature elasticity of membranes. We find that k_{13}, k_{24} and c_0 are closely related to the orientational order of the long chains of the molecules in the bulk as well as the two interfaces of the fluid membrane. In particular, the spontaneous curvature c_0 is closely related to the difference of the order parameters on the two side-surfaces of the membrane. It describes just the effects of the asymmetry of the membrane or its environment. Section 2.4.5 will give a comparison with some known experimental results and will suggest some further experiments.

2.4.3 A Derivation by Way of 2D Differential Invariants

In this section, we will give a complete derivation of Helfrich's curvature elastic energy, Eq. (2.97), by way of rigorous 2D differential invariants discussed in Sec. 2.3. Before getting into the actual calculation, let us briefly summarize the physical model of fluid membranes discussed in the first chapter and the present one.

Generally, fluid membranes (amphiphilic bilayers) may be idealized as 2D surfaces in an aqueous solution with each membrane being made up of a double layer of long molecules.[32] The long axes (long chains) of the amphiphilic molecules in the membrane are oriented more or less along the direction of the normal of the surfaces. The integrity of the membrane is kept by the interaction between the aqueous solution and the hydrophilic heads as well as the hydrophobic chains of the molecules. At the same time, the interaction also maintains the normal orientation of the molecules in the membrane. The curved fluid membrane may thus be treated as a bending homeotropic liquid crystal cell with uniaxial molecular order. The director $\mathbf{n}$ of the liquid crystal cell corresponds simply to the unit normal vector of the membrane surface. For simplicity, we may take the thickness d of the membrane as uniform.

In order to find the expression of the curvature elastic free energy per unit area of the membrane, let us define the two side-surfaces of the imagined "liquid crystal cell" by the vectors $\mathbf{Y}(u,v)$ and $\mathbf{Y}(u,v) + d\mathbf{n}$

respectively, where u and v are two real parameters, i.e., material coordinates. To derive calculations free of coordinates, we use differential geometry formalism as discussed in Sec. 2.2 and introduce the following quantities:

$$\mathbf{Y}_i = \partial_i \mathbf{Y}, \qquad \mathbf{Y}_{ij} = \partial_i \partial_j \mathbf{Y}, \qquad g_{ij} = \mathbf{Y}_i \cdot \mathbf{Y}_j,$$

$$g = \det(g_{ij}), \qquad g^{ij} = (g_{ij})^{-1}, \qquad L_{ij} = \mathbf{Y}_{ij} \cdot \mathbf{n}, \qquad (2.108)$$

$$L = \det(L_{ij}), \qquad L^{ij} = (L_{ij})^{-1}, \qquad (i,j = 1,2),$$

where $\partial_1 = \partial_u$, $\partial_2 = \partial_v$, and g_{ij} and L_{ij} are associated with the first and the second fundamental forms of the surface $\mathbf{Y}$, respectively. Obviously, we have used $\mathbf{Y}$ to replace the vector $\mathbf{r}$ in Sec. 2.2. For convenience, we call $\mathbf{Y}$ the inside surface and $\mathbf{Y} + d\mathbf{n}$ the outside surface. The outward normal unit vector $\mathbf{n}$ is then defined by

$$\mathbf{n} = g^{-1/2}(\mathbf{Y}_1 \times \mathbf{Y}_2). \qquad (2.109)$$

The 2D-gradient operator ∇' on the surface $\mathbf{Y}$ is defined in Eq. (2.68) as:

$$\nabla' = g^{ij}\mathbf{Y}_i\partial_j, \qquad (2.110)$$

and $\mathbf{n}\partial_n$ is the gradient operator along the direction of $\mathbf{n}$. The 3D-gradient operator ∇ in the free energy density formulas Eqs. (2.104) and (2.106) for liquid crystals is now given by[10]

$$\nabla = \nabla' + \mathbf{n}\partial_n. \qquad (2.111)$$

The invariant form of ∇', Eq. (2.110), plays an important role in our calculation. If we denote the mean curvature H and the Gaussian curvature K as given in Sec. 2.3.2,

$$H = -(1/2)(c_1 + c_2) = (1/2)g^{ij}L_{ij},$$
$$K = c_1 c_2 = L/g, \qquad (2.112)$$

respectively, with definitions Eqs. (2.108) and (2.109) we have shown in Eqs. (2.74) and (2.86) that

$$\nabla' \cdot (\phi\mathbf{n}) = -2H\phi,$$

$$\nabla' \times \mathbf{F} = g^{-1/2}\{[\partial_1(g_{12}f_1 + g_{22}f_2) - \partial_2(g_{11}f_1 + g_{12}f_2)]\mathbf{n}$$

$$+ (f_1 L_{12} + f_2 L_{22})\mathbf{Y}_1 - (f_1 L_{11} + f_2 L_{21})\mathbf{Y}_2\} + (\nabla'f_3) \times \mathbf{n},$$

$$(2.113)$$

where $\phi(u,v)$, $f_1(u,v)$, $f_2(u,v)$, $f_3(u,v)$ are scalar functions and $\mathbf{F} = f_1\mathbf{Y}_1 + f_2\mathbf{Y}_2 + f_3\mathbf{n}$. Here, we use the notation (f_1, f_2, f_3) to replace (F^1, F^2, R) in Sec. 2.3.2. It follows directly from Eq. (2.113) that

$$\nabla' \cdot \mathbf{n} = -2H,$$
$$\nabla' \times \mathbf{n} = 0. \tag{2.114}$$

Although all the above formulas have been given in Sec. 2.3, we collect them here for the following derivations. Since the outward normal unit vector $\mathbf{n}'$ on the surface $\mathbf{Y} + d\mathbf{n}$ is related to $\mathbf{n}$ by $\mathbf{n}' = \mathbf{n} + O(d^2)$, we may easily show from Eqs. (2.108), (2.109) and (2.111) that

$$\mathbf{n} \cdot \nabla\mathbf{n} = \partial_n\mathbf{n} = 0. \tag{2.115}$$

It follows that the 3D divergence and curl of $\mathbf{n}$ become

$$\nabla \cdot \mathbf{n} = \nabla' \cdot \mathbf{n} = -2H,$$
$$\nabla \times \mathbf{n} = \nabla' \times \mathbf{n} = 0, \tag{2.116}$$

respectively. The second equation of Eq. (2.116) is well known for smectics.[31] It follows that

$$\mathbf{n} \cdot \nabla \times \mathbf{n} = 0,$$
$$(\nabla\mathbf{n})(\mathbf{n} \cdot \nabla \times \mathbf{n}) = 0, \tag{2.117}$$
$$\mathbf{n} \cdot \nabla(\mathbf{n} \cdot \nabla \times \mathbf{n}) = 0.$$

Together with Eqs. (2.104) and (2.106), we see that the elastic constants k_{12} and k_{23} do not enter in the free energy expression of the membranes. Furthermore, Eq. (2.115) shows that the bend elastic constant k_{33} is also ruled out.

However, the term associated with k_{13} in Eq. (2.106) needs further investigation. Since d is very small in comparison with the size of the membrane, we may replace $\partial_n H$ in the expression (see Eq. (2.116))

$$\mathbf{n} \cdot \nabla\nabla \cdot \mathbf{n} = \mathbf{n} \cdot \nabla(-2H) = -2\partial_n H \tag{2.118}$$

by

$$\partial_n H = \lim_{d\to 0}(\delta H/d), \tag{2.119}$$

where d is the thickness of the membrane and $\delta H = H' - H$ is the variation of the mean curvature between the two surfaces $\mathbf{Y} + d\mathbf{n}$ and $\mathbf{Y}$. Using the result $\partial_n H = 2H^2 - K$ obtained by Ou-Yang and Helfrich (cf. Ref. [21, Eq. (16)], which will be discussed thoroughly in the next chapter), we find that

$$\mathbf{n} \cdot \nabla\nabla \cdot \mathbf{n} = 2K - 4H^2. \tag{2.120}$$

As a direct check, we may consider the case of a sphere of radius r, for which $H = -r^{-1}$ and $K = r^{-2}$. Indeed, here we have that $\partial_n H = \partial_r H = r^{-2} = 2H^2 - K$.

Next, let us consider the $\nabla\mathbf{n} : \nabla\mathbf{n}$ in Eq. (2.104). With Eqs. (2.110), (2.111) and (2.115) we have that

$$\nabla\mathbf{n} : \nabla\mathbf{n} = \nabla'\mathbf{n} : \nabla'\mathbf{n} = g^{ij}\mathbf{Y}_i\mathbf{n}_j : g^{kl}\mathbf{Y}_l\mathbf{n}_k. \tag{2.121}$$

With the Weingarten Equation given in Eq. (2.35) in the form

$$\mathbf{n}_i = -L_{im}g^{mn}\mathbf{Y}_n, \tag{2.122}$$

Eq. (2.121) transforms into

$$\nabla\mathbf{n} : \nabla\mathbf{n} = g^{ij}\mathbf{Y}_i L_{jm}g^{mn}\mathbf{Y}_n : g^{kl}\mathbf{Y}_l L_{kp}g^{pq}\mathbf{Y}_q. \tag{2.123}$$

Using Eqs. (2.108) and (2.112) and the relationships[33]

$$\begin{aligned}
g^{ij}L_{ik}g^{kl} &= 2Hg^{il} - KL^{ji}, \\
L_{ij}g^{ik}L_{kl} &= 2HL_{jl} - Kg_{jl},
\end{aligned} \tag{2.124}$$

we find that Eq. (2.123) reduces simply to

$$\nabla\mathbf{n} : \nabla\mathbf{n} = 4H^2 - 2K. \tag{2.125}$$

In order to find the final expression of the curvature free energy per unit area of the membrane, we use the following approximation:

$$F = \int (g_{\mathrm{LC}} + \delta g_{\mathrm{LC}})\mathrm{d}V = d \oint (g_{\mathrm{LC}} + \delta g_{\mathrm{LC}})\mathrm{d}A, \tag{2.126}$$

where $\mathrm{d}V$ is the volume element of the bulk between the two sides of the membrane $\mathbf{Y}$ and $\mathbf{Y}+d\mathbf{n}$, and $\mathrm{d}A$ is the area element of $\mathbf{Y}$ surface. Since the thickness d of the membrane is about twice the length of the amphiphilic molecule and is negligible in comparison with the linear size of the membrane, the replacement of $\mathrm{d}V$ in Eq. (2.126) by the product of $\mathrm{d}A$ and the thickness d of the membrane is a good approximation. With Eqs. (2.104), (2.106), (2.116), (2.117), (2.118) and (2.120), we have

$$F = d \oint \left[\frac{1}{2}(k_{11} - 2k_{13})(2H)^2 \right.$$

$$\left. + (2k_{13} - k_{22} - k_{24})K + 2k_{11}s_0H + \frac{1}{2}k_{11}s_0^2 \right] dA \qquad (2.127)$$

On the other hand, Eq. (2.97) of Helfrich's theory gives

$$F = \oint g dA = \oint \left[\frac{1}{2}k(2H + c_0)^2 + \bar{k}K \right] dA. \qquad (2.128)$$

The two forms of F differ by a trivial constant. Comparison of Eqs. (2.127) and (2.128) gives not only Eq. (2.107) but also an expression for the spontaneous curvature:

$$c_0 = k_{11}s_0/(k_{11} - 2k_{13}). \qquad (2.129)$$

Frank interpreted s_0 as the spontaneous splay of the liquid crystal.[20] One feature of our calculation is that k_{13} enters both in the expression of k and in that of $\bar{k}$. This result differs from the argument given by Mitov[29] who concluded that, in contrast to the case of nematics, there is no term related to k_{13} in the second-order elastic free energy of bilayer. Mitov gave no details in his paper and we have no way to check his conclusion. But here we prove clearly that k_{13} is involved in the second-order elastic energy of both liquid crystals and bilayer. And it is the key clue to clear up the puzzles of only one-sign of $\bar{k}$ as discussed in Sec. 2.4.2 on Eq. (2.105).

2.4.4   The Spontaneous Curvature Viewed from Landau–de Gennes Theory

In the last section, we have seen from Eq. (2.129) that without taking account of k_{13}, c_0 simply reduces to s_0, the effect of the spontaneous splay of the liquid crystals. This gives nothing on why k_{13} exists. Therefore, the present process of involving k_{13} in c_0 may present its role in the understanding of the Helfrich spontaneous curvature c_0. However, this is quite a task even in the theory of liquid crystals.

In the free energy density of liquid crystals, the term

$$k_{13}\mathbf{n} \cdot \nabla\nabla \cdot \mathbf{n} \qquad (2.130)$$

was first introduced phenomenologically by Oseen around 1930.[34] Then, it was disregarded for almost 40 years. Not until 1971, Nehring and Saupe rediscovered this term.[28] In fact, this term is difficult to handle and poses somewhat as a long lasting problem. So far, the magnitude and even the sign of k_{13} are still unknown. A basic difficulty in studying k_{13} comes from the fact that, in the free energy expression, this term may be replaced by

$$k_{13}\nabla \cdot (\mathbf{n}\nabla \cdot \mathbf{n}), \qquad (2.131)$$

with the part $k_{13}(\nabla \cdot \mathbf{n})^2$ being absorbed in the splay free energy density term. In the form of Eq. (2.131), apparently, k_{13} contributes only to a part of the surface energy of the sample and cannot be measured experimentally from the equilibrium structure of the bulk liquid crystal. On the other hand, the expressions of k_{13} derived from the molecular approach by Nehring and Saupe[28] and recently by Barbero and Oldano,[25] respectively, involve too many terms to handle. Thus, it is practically almost impossible to solve the problem from the molecular approach either. Here, we would like to propose a compromise, i.e., to find an expression for k_{13} by using the Landau–de Gennes way of approach.[31]

In order to obtain a quantitative theory of the phase transitions of the second kind, Landau introduced the order parameter S of a system and expanded the free energy density of the system in a power series of the order parameter. It was de Gennes who generalized Landau theory to uniaxial liquid crystal systems. For a uniaxial liquid crystal system, the order parameter $S(\mathbf{r})$ at the point $\mathbf{r}$ may be defined as

$$S(\mathbf{r}) = \langle P_2(\cos\theta)\rangle, \qquad (2.132)$$

where P_2 is simply the Legendre polynomial of the second order and θ is the angle between the long axis of the molecule and the local director $\mathbf{n}(\mathbf{r})$ at the point $\mathbf{r}$. However, de Gennes' generalized order parameter $S_{\alpha\beta}$ is a symmetric traceless tensor of second rank, i.e.,

$$S_{\alpha\beta} = S(\mathbf{r})[n_\alpha(\mathbf{r})n_\alpha(\mathbf{r}) - \delta_{\alpha\beta}/3], \qquad (2.133)$$

where $\alpha, \beta = 1, 2, 3$ denote the components along the three orthogonal axes of the Cartesian coordinate system and $\delta_{\alpha\beta} = 1$ for $\alpha = \beta$ and zero otherwise. Up to terms of the order $O(S^2)$, the most general form of the inhomogeneous part of the free energy density g_{LC} is

$$g_{\mathrm{LC}} = \frac{1}{2}L_1(\nabla_\alpha S_{\beta\gamma})(\nabla_\alpha S_{\beta\gamma}) + \frac{1}{2}L_2(\nabla_\alpha S_{\alpha\gamma})(\nabla_\beta S_{\beta\gamma}), \qquad (2.134)$$

where L_1 and L_2 may be referred to as the elastic constants and, as a good first approximation they may be considered as constants.

Substituting Eqs. (2.132) and (2.133) into Eq. (2.134) we find that

$$g_{LC} = (1/3)[L_1 + (1/6)L_2](\nabla S)^2 + (1/6)L_2(\mathbf{n} \cdot \nabla S)^2$$
$$+ (1/3)L_2 S[2(\nabla \cdot \mathbf{n})(\mathbf{n} \cdot \nabla S) + (\mathbf{n} \times \nabla \times \mathbf{n}) \cdot \nabla S]$$
$$+ S^2[L_1 \nabla \mathbf{n} : \nabla \mathbf{n} + (1/2)L_2(\nabla \cdot \mathbf{n})^2 + (1/2)L_2(\mathbf{n} \cdot \nabla \mathbf{n})^2]. \quad (2.135)$$

In case of S being a constant, Eq. (2.135) reduces to the free energy density expression given by Frank–Nehring–Saupe with the surface energy contribution term being neglected and

$$k_{11} = k_{33} = 2S^2(L_1 + L_2/2),$$
$$k_{22} = 2S^2 L_1. \quad (2.136)$$

The elastic constants k_{12}, k_{13}, k_{23}, the spontaneous splay constant s_0, and the spontaneous twist constant t_0 now become zero. Therefore, in the case of constant S, the Landau–de Gennes free energy density expression is only qualitatively correct for the nematic phase.

Fluid membranes are curved surfaces with both the surface normal $\mathbf{n}$ and the order parameter S being functions of the space coordinate $\mathbf{r}$. Making use of Eqs. (2.115), (2.116) and (2.125), we see that the last term on the right-handed side of Eq. (2.135) may be written as

$$g_n = S^2[4(L_1 + L_2/2)H^2 - 2L_1 K]. \quad (2.137)$$

The other terms in Eq. (2.135) are directly related to ∇S. For asymmetric bilayers, we may make a simple assumption that

$$S(r) = S(u, v, z) = \begin{cases} S_0 & \text{for } d/2 \le z \le d, \\ S_i & \text{for } 0 \le z \le d/2, \end{cases} \quad (2.138)$$

where the constants S_0 and S_i are simply the orientational order parameters of the molecules in the two monolayers of the bilayer, respectively. Here, in Eq. (2.138), the z-axis of the local Cartesian coordinate system coincides with the local normal $\mathbf{n}$. With Eq. (2.111), we have

$$\nabla S(\mathbf{r}) = (\nabla' + \mathbf{n}\partial_z)S = (S_0 - S_i)\mathbf{n}\delta(z - d/2), \quad (2.139)$$

where $\delta(z - d/2)$ is the Dirac delta function. The curvature elastic free energy per unit area of the bilayer, G, is simply

$$G = \int_0^d g_{LC} dz. \quad (2.140)$$

The first two terms of Eq. (2.135), which are proportional to the square of ∇S, are equal to

$$G_1 = \frac{L_2}{3} \int_0^d 2(-2H)(S_0 - S_i)S\delta\left(z - \frac{d}{2}\right)dz$$

$$= (2L_2/3)(S_i^2 - S_0^2)H(d/2), \qquad (2.141)$$

where we have used Eqs. (2.116) and (2.117). Since $\partial_n H = 2H^2 - K$, we have

$$H(d/2) - H = d(2H^2 - K)/2 + O(d^2). \qquad (2.142)$$

It follows that

$$G_1 = (2L_2/3)(S_i^2 - S_0^2)[H + d(2H^2 - K)/2] + O(d^2). \qquad (2.143)$$

Similarly, the contribution of G_2 from the last term of Eq. (2.135) is

$$G_2 = \int_0^d g_n dz = \frac{d}{2}(S_i^2 + S_0^2)\left[4\left(L_1 + \frac{L_2}{2}\right)H^2 - 2L_1 K\right] + O(d^2).$$
$$\qquad (2.144)$$

Thus, we have

$$G = G_1 + G_2 = 2kH^2 + \bar{k}K + 2kc_0 H, \qquad (2.145)$$

where

$$k = d[S_i^2(L_1 + 5L_2/6) + S_0^2(L_1 + L_2/6)],$$
$$\bar{k} = -d[S_i^2(L_1 + L_2/3) + S_0^2(L_1 - L_2/3)], \qquad (2.146)$$
$$c_0 = (S_i^2 - S_0^2)L_2/(3k).$$

Apart from a trivial constant, Eq. (2.145) corresponds exactly to the Helfrich expression Eq. (2.128).

Comparison of Eqs. (2.146) with Eqs. (2.107) and (2.136), while replacing $2S^2$ by $S_0^2 + S_i^2$, we find that the Oseen–Frank–Nehring–Saupe elastic constants are given by

$$k_{11} = k_{33} = (S_i^2 + S_0^2)(L_1 + L_2/2),$$
$$k_{22} = (S_i^2 + S_0^2)L_1,$$
$$k_{13} = -(S_i^2 - S_0^2)L_2/6, \qquad (2.147)$$
$$k_{24} = 0.$$

In case of symmetric bilayers where $S_i = S_0$, G_1, k_{13} and c_0 all vanish. This confirms that the spontaneous curvature c_0 and the elastic constant k_{13} are directly related to the asymmetry property of the membrane.

2.4.5 Discussion of Helfrich Bending Energy on Liquid Crystal Point of View

Since k_{13} varies with the distortions of directors along the boundary near the surfaces of the fluid membrane,[25] consequently, as shown in Eqs. (2.107) and (2.129), k, $\bar{k}$ and c_0 are no longer constants. Examination of Eq. (2.146) shows that the order parameters S_i and S_0 depend not only on the polar properties of the membrane molecules but also on the forces acting on both sides of the boundary surfaces of the membrane. Helfrich's theory has predicted that the spontaneous curvature constant c_0 is not a constant. Here, we have shown that k and $\bar{k}$ are not constants too, and they may change with the chemical structure on the two sides of the membrane.

The bending rigidity k of lecithin bilayers has been measured in a number of laboratories. The experimental value of k is determined from the measured mean square amplitude of the shape fluctuations. Tubular vesicles were used by both the Helfrich group and the Webb group[24] for the measurement. The value of k obtained by both groups ranges from 0.9×10^{-19} J to 2.3×10^{-19} J. All these experimental values are higher than 0.5×10^{-19} J as predicted by Helfrich's theory.[15] The present calculated value of k from Eq. (2.107) would be higher than 0.5×10^{-19} J and is in better agreement with the available experimental data if k_{13} is negative.

Ou-Yang and Helfrich[21] have shown that the necessary condition for a fluctuating circular cylindrical vesicle of radius ρ_0 to be stable is

$$c_0 \rho_0 \geq 1. \tag{2.148}$$

The stability of circular cylindrical vesicle will be discussed in more detail in the next chapter. In connection with Eqs. (2.146) and (2.147), for cylindrical vesicles, we certainly have $k_{13} < 0$. This fact suggests that the present calculation is in better agreement with the experimental data, at least qualitatively. To check our calculation, it would be helpful to have experiments on the determination of the dependency of k and $\bar{k}$ on the chemical agents around the two sides of the membrane.

Obviously, the two-monolayer order parameters S_i and S_0 used in Sec. 2.4.4 can be termed as a simple assumption. In general, it may need more terms in the expansion of the Landau–de Gennes free energy. For example, we may take

$$S(\mathbf{r}) = \begin{cases} S_0 & \text{for } z \geq d/2, \\ S_m & \text{for } -d/2 < z < d/2, \\ S_i & \text{for } z \leq -d/2, \end{cases} \tag{2.149}$$

where S_0 and S_i specify the orientational orders of the outside and the inside monolayers of the vesicle. In this case, we will find that

$$k = 2S_m^2 d[(L_1 + L_2) + AL],$$

$$\bar{k} = -2S_m^2 d[L_1 + AL], \qquad (2.150)$$

$$c_0 = 2S_m(S_i - S_0)L_2/(3k),$$

where

$$A = (2/3)[1 - (S_0 + S_i)/(2S_m)]. \qquad (2.151)$$

Thus, S_m, S_i and S_0 describe the orientational order of the bulk and the orientational order of the two sides of the membrane, respectively. And the spontaneous curvature c_0 remains to describe the asymmetry property of the membrane. Further study on this model is still in progress.

2.4.6 Spontaneous Curvature and Flexoelectric Effect

As we have seen in Secs. 2.4.4 and 2.4.5, the essential concept of Helfrich model is that the spontaneous curvature c_0 is strongly related to the orientational order parameters on both sides of the membrane. In the mechanism, c_0 is strongly influenced by the changes in chemical structures of the molecules, the salinity and the temperature in the vesicle.[35] In simplest quantitative description, c_0 may be expressed by and linked[15] with the spontaneous splay s_0 of liquid crystal or the orientational order parameters S_0 and S_i of the long chains of the molecules in the outside layer and the inside layer of the bilayer-membrane.[23] However, the drawback is that the values of s_0, S_0 and S_i have not yet been measured. Therefore, we propose a new approach to investigate the meaning of c_0 macroscopically. From the above discussions, one can see that c_0 is a controlling parameter of the geometry as well as the size of the vesicle of membrane. On the other hand, Kaler et al.[36] have reported a general method for producing vesicles of well-defined average size by using the charge as a controlling parameter. Hence, there should be a close relation between c_0 and the elastic fields acting on the membrane. Phenomenologically, we may assume that the complex agents of the asymmetry of the membrane produce an electric field $\mathbf{E} = (\psi/d)\mathbf{n}$, where $\psi = \psi^{in} - \psi^{ext}$ is the electric potential across the membrane of thickness d. We then consider its effect on the bending of the membrane. Analogous to piezoelectricity in solids, Meyer[37] introduces the concept of curvature electricity in the field of liquid crystal. This effect gives rise to an additional term, ΔF, in the membrane energy,

$$\Delta F = - \oint dA \int_0^d \mathbf{P} \cdot \mathbf{E} dz, \qquad (2.152)$$

where $\mathbf{P}$ is the polarization induced by bending in the form

$$\mathbf{P} = e_{11} \mathbf{n} \nabla \cdot \mathbf{n} + e_{33} (\nabla \times \mathbf{n}) \times \mathbf{n}. \qquad (2.153)$$

The phenomenological $\mathbf{P}$ takes account of the possible combinations of the components of the tensor $\nabla \mathbf{n}$ up to linear order, and the values and signs of the flexoelectric constants e_{11} and e_{33} are related to the geometry of the molecules and the electric dipole moment carried by them. With $\nabla \times \mathbf{n} = 0$ and $\nabla \cdot \mathbf{n} = -2H$ given by Eq. (2.116), Eq. (2.152) reduces to

$$\Delta F = 2 e_{11} \psi \oint H dA. \qquad (2.154)$$

Incorporation of Eq. (2.154) and Eq. (2.97) yields a relation between c_0 and ψ:

$$c_0 = e_{11} \psi / k. \qquad (2.155)$$

This relation must be tested by an actual measurement of the membrane potential.

A strict test for the present theory is based on an examination of Eq. (2.155). The bending modulus k is well known to be of the order of 10^{-19} J from both theory[15] and experiment,[38] but up to now the flexoelectric constant e_{11} for membranes is still lacking. Fortunately, as in the estimate of k in Ref. [15], we may again make use of e_{11} in liquid crystals and set $e_{11} \simeq 10^{-4} \mathrm{dyne}^{1/2}$.[39] Putting these rough estimates into Eq. (2.155) and taking $c_0 R_0 \simeq -1.62$ with $R_0 = 3.25 \,\mu\mathrm{m}$ (the radius of a sphere with the same area of the surface of the red blood cell), the result of a calculation on c_0 of red blood cells,[40] we obtain for ψ, the red blood cell potential,

$$\psi \simeq -15.0 \,\mathrm{mV}. \qquad (2.156)$$

Measurement of transmembrane potential in a cell as small as the red blood cell by introducing a microelectrode seems technically impossible. Nevertheless, two groups[41] have impaled human red blood cells and reported to find cell potential of $-14 \,\mathrm{mV}$ (Lassen $et\ al.$) and of $-8 \,\mathrm{mV}$ (Jay and Burton) independently. The former measured value is in surprisingly good agreement with our prediction of Eq. (2.156). If we replace e_{11} by the Helfrich estimation[42] of $4 \times 10^{-5} \mathrm{dyne}^{1/2}$, the latter value, $-8 \,\mathrm{mV}$, is about the same as what is obtained from Eq. (2.156). Obviously, the experimental results obtained by these biologists support the present theory. The observed changes in the shape of red blood cell with changes of pH value[3] may provide another evidence because the cell potential ψ does vary with the pH value of the solution. Especially, the so-called magical glass effect of red blood cells appears now clearly to have the same origin, as chemical electrode with different potentials glasses will change the cell potential ψ accordingly, and so with c_0. The changes in c_0 cause the deformations in the shape of red blood cells associated with high-order spherical harmonics.[43]

In summary, our discussion shows that the membrane potential incorporated in the flexoelectric effect may be partially responsible for the physical mechanism of the spontaneous curvature of membranes. The rather good agreement with previous measurements on red blood cell appears to confirm our theoretical prediction.

References

[1] R. Virchow, *Die Cellularpathologie in Ihrer Begrüdung auf Physiologische und Pathologische Gewebelehre* (Hirschwald, 1858).

[2] D. W. Fawcett, *The Cell* (W. B. Saunders Co. 1981).

[3] M. Bessis, *Living Blood Cells and Their Ulstructure* (Springer-Verlag, 1973).

[4] E. Ponder, *Helmolysis and Related Phenomenona* (Grune and Stratton, New York, 1948).

[5] Y. C. Fung and P. Tong, *Biophys. J.* **8** (1968) 175.

[6] L. Lopez, I. M. Duck and W. A. Hunt, *Biophys. J.* **8** (1968) 1228.

[7] M. A. Greer and R. F. Baker, in *Congr. Int. Microsc. Electron.* 7th. edn. P. Favard, (Paris) **3** (1970) 31.

[8] J. R. Murphy, *J. Lab. Clin. Med.* **65** (1965) 756.

[9] P. Seeman, D. Cheng and G. H. lies, *J. Cell Biol.* **56** (1973) 519.

[10] C. E. Weatherburn, Differential geometry, in *Pure and Applied Mathematics*, Vol. XX, eds. R. Courant, L. Bers and J. J. Stoker (Wiley, New York, 1969).

[11] M. P. do Carmo, *Differential Geometry of Curves and Surfaces* (Prentice-Hall 1976).

[12] R. Virchow, *Virchows Arch. Parrhol. Anat. Physiol.* **6** (1854) 562.

[13] S. Freiberg, *Lyotropic Liquid Crystals and the Structure of Biomembrane* (American Chemical Society, 1976).

[14] G. H. Brown and J. J. Wolken, *Liquid Crystals and Biological Structures* (Academic Press, 1979).

[15] W. Helfrich, *Z. Naturforsch. C* **28** (1973) 693.

[16] For a review, see, E. A. Evans and R. Sklak, *Mechanics and Thermodynamic of Biomembranes* (CRC, Boca Raton, Florida, 1979).

[17] P. B. Canham, *J. Theor. Biol* **26** (1970) 61.

[18] W. Helfrich and H. J. Deuling, *J. Phys.* (Paris) **36** (1975) 327.

[19] H. J. Deuling and W. Helfrich, *J. Phys.* (Paris) **37** (1976) 1335.

[20] F. C. Frank, *Discuss. Faraday Soc.* **25** (1958) 19.

[21] Ou-Yang Zhong-can and W. Helfrich, *Phys. Rev. A* **39** (1989) 5280.

[22] W. Helfrich, an unpublished lecture.

[23] Ou-Yang Zhong-can, S. Liu and Xie Yu-zhang, *Mol. Cryst. Liq. Cryst.* **204** (1991) 143.

[24] R. M. Servuss, W. Harbich and W. Helfrich, *Biochim. Biophys. Ada* **436** (1976) 900; G. Bebik, R. M. Servuss and W. Helfrich, *J. Physique* **46** (1985) 1773; M. B. Schneider, J. T. Jenkins and W. W. Webb, *Biophys. J.* **45** (1984) 891; *J. Physique* **45** (1984) 1457; H. P. Duwe, H. Eugelhardt, A. Zilker and E. Sackmann, *Mol. Cryst. Liq. Cryst.* **152** (1987) 1.

[25] See, for example, G. Barbero and C. Oldano, *Mol. Cryst. Liq. Cryst.* **170** (1989) 99.

[26] W. Harbich and W. Helfrich, *Chem Phys. Lipids* **36** (1984) 39.

[27] W. Harbich, R. M. Servuss and W. Helfrich, *Z. Naturforsch. A* **33** (1978) 1013

[28] J. Nehring and A. Saupe, *J. Chem. Phys.* **54** (1971) 337.

[29] M. D. Mitov, *C. R. Acad. Bulgare Sci.* **31** (1978) 513.

[30] H. Hinov, *Mol. Cryst. Liq. Cryst.* **148** (1987) 197; **168** (1989) 7; C. Oldano and G. Barbero, *J. Phys. Lett.* (Paris) **46** (1985) L451; G. Barbero and C. Oldano, *Nuovo Cim. D* **6** (1986) 479.

[31] P. G. de Gennes, *Mol. Cryst. Liq. Cryst.* **12** (1971) 193.

[32] For a more detailed review, see *Physics of Amphiphilic Layers,* eds. J. Meunier, D. Langivan and N. Boccara (Springer, Berlin, 1987).

[33] J. J. Stoker, Differential Geometry, in *Pure and Applied Mathematics,* Vol. XX, eds. R. Courant, L. Bers and J. J. Stoker (Wiley, New York, 1969).

[34] C. W. Oseen, *Arkiv. Math. Astron. Fysik. A* **19** (1925) 1; *Trans. Faraday Soc.* **29** (1933) 883.

[35] A. G. Petrov and I. Bivas, *Prog. Surf. Sci.* **16** (1984) 389; S. A. Safran, L. A. Turkevich and P. A. Pincus, in *Surfactants in Solution,* eds. K. Mitta and B. Lindman (Plenum, 1984) p. 1177.

[36] E. W. Kaler *et al., Science* **245** (1989) 1371.

[37] R. B. Meyer, *Phys. Rev. Lett.* **22** (1969) 918.

[38] E. A. Evans, *Biophys. J.* **43** (1983) 27.

[39] D. Schmidt, M. Schadt and W. Helfrich, *Z. Naturforsch. A* **27** (1972) 277; L. Dozov, Ph. Martinot-Lagarde and G. Durand, *J. Phys. Lett.* (Paris) **43** (1982) L365.

[40] Ou-Yang Zhong-can, J.-G. Hu and J.-X. Liu, *Mod. Phys. Lett.* **B6** (1992) 1577.

[41] U. V. Lassen and O. Sten-Knudsen, *J. Physiol* (London) **195** (1968) 681; A W. L. Jay and A. C. Burton, *Biophys. J.* **9** (1969) 115.

[42] W. Helfrich, *Z. Naturforsch. A* **26** (1971) 833.

[43] Ou-Yang Zhong-can and W. Helfrich, *Phys. Rev. Lett.* **59** (1987) 2486.

3

Shape Equation of Lipid Vesicles and Its Solutions

In this chapter, the general shape equation of vesicles based on the Helfrich spontaneous curvature theory of membrane is derived. This equation is a nonlinear differential equation of high order. Its two simple solutions, the sphere and the circular cylinder, are discussed. With the help of the second and third variations of the Helfrich free energy, a systematic analysis of stability for two exact solutions to the equation, sphere and circular cylinder vesicles, is given. The exact solutions of Clifford torus and circular biconcave discoid are examples which show that the Helfrich spontaneous curvature theory can not only explain the existing shape of red blood cell, but can also predict new shapes of membrane. In addition to exact axisymmetrical solutions, an analytical solution of Dupin cyclide is also provided.

3.1 Mathematical Preliminary

A brief discussion on differential geometry of curves and surfaces has been introduced in the previous chapter. They are essential in all aspects of the following discussions. In this chapter, some of the results will be repeated in somewhat different forms and notations and some related variation terms will also be given.[1-3] Both vector forms, tensor forms, semi-tensor forms and component forms will be used at will. For readers not familiar with tensor notation, a brief introduction on this subject will be given in Appendix A.

In differential geometry, any point $\mathbf{Y}$ on a given surface may be expressed in terms of two independent parameters u and v, $\mathbf{Y} = \mathbf{Y}(u, v)$. In the Cartesian coordinate system, the line element ds between two neighboring points $\mathbf{Y}$ and $\mathbf{Y} + d\mathbf{Y}$ is given by

$$ds^2 = d\mathbf{Y} \cdot d\mathbf{Y} = dY^l dY_l = g_{ij} didj \quad (l = 1,2,3; \ i,j = u,v),$$

$$g_{ij}(u,v) = Y_{,i}^l Y_{l,j} = g_{ji}(u,v), \tag{3.1}$$

$$Y_{,i}^l = \partial Y^l / \partial i, \ Y_{,ij}^l = \partial^2 Y^l / \partial i \partial j \quad (i,j = u,v).$$

In the following discussions, any subscript after the comma sign of a quantity means differentiation of the quantity with respect to the subscript. The summation convention will be used in the whole discussions, i.e., repeated indices $i,j,k,\ldots$ in a single term imply summation over (1,2,3) or (u,v) as required. The contravariant tensor $g^{ij}(u,v)$ is defined by

$$g^{ij} g_{ik} = \delta_k^j, \tag{3.2}$$

where

$$\delta_k^j = \begin{cases} 1 & \text{if } j = k, \\ 0 & \text{if } j \neq k. \end{cases}$$

The determinate $|g_{ij}|$ is denoted by g, where

$$g = g_{uu} g_{vv} - g_{uv} g_{vu}. \tag{3.3}$$

It follows that

$$g^{uu} = g_{vv}/g, \quad g^{uv} = -g_{uv}/g, \quad g^{vv} = g_{uu}/g,$$

$$g^{uu} g_{uu} + g^{uv} g_{uv} = g^{uv} g_{uv} + g^{vv} g_{vv} = 1, \quad g^{ij} g_{ij} = 2. \tag{3.4}$$

The unit normal $\mathbf{n}$ of the surface is defined by

$$\mathbf{n} = \frac{\mathbf{Y}_{,u} \times \mathbf{Y}_{,v}}{\sqrt{g}}, \quad n^l = \frac{e^{lmn} Y_{m,u} Y_{n,v}}{\sqrt{g}}, \quad (l,m,n = 1,2,3), \tag{3.5}$$

where the permutation symbols (not tensors) e_{lmn} and e^{lmn} are defined by

$$e_{lmn} = e^{lmn} = \begin{cases} 1 & \text{if } (l,m,n) \text{ is an even permutation of (1,2,3)}, \\ -1 & \text{if } (l,m,n) \text{ is an odd permutation of (1,2,3)}, \\ 0 & \text{otherwise}, \end{cases} \tag{3.6}$$

and satisfy the relation[4]

$$e_{lmn} e^{lpq} = \delta_m^p \delta_n^q - \delta_m^q \delta_n^p. \tag{3.7}$$

In differential geometry, the positive direction of the normal is taken as toward the concave side of an open surface or the inside of a closed surface.

However, in the present discussion, the outward direction (i.e., toward the convex side of the open surface or the outside of the closed surface) is taken as the positive direction of the normal.

Tensors $L_{ij}(u,v), L^{ij}(u,v)$ and the determinant L are defined by

$$L_{ij} = \mathbf{n} \cdot \mathbf{Y}_{,ij} = n_l Y^l_{,ij} = n^l Y_{l,ij} = L_{ji}, \quad (l = 1, 2, 3; i, j = u, v),$$

$$L = L_{uu}L_{vv} - L_{uv}L_{vu}, \tag{3.8}$$

$$L^{ij}L_{ik} = \delta^i_k.$$

It follows that

$$L^{uu} = L_{vv}/L, \quad L^{uv} = -L_{uv}/L = L^{vu}, \quad L^{vv} = L_{uu}/L,$$

$$L^{uu}L_{uu} + L^{uv}L_{uv} = L^{uv}L_{uv} + L^{vv}L_{vv} = 1, \quad L^{ij}L_{ij} = 2. \tag{3.9}$$

With Eq. (3.7), it gives that:

$$g^{ij} = e_{3ik}e^{3jl}g_{kl}/g, \quad g_{ij} = g e_{3ik}e^{3jl}g^{kl},$$

$$L^{ij} = e_{3ik}e^{3jl}L_{kl}/L, \quad L_{ij} = L e_{3ik}e^{3jl}L^{kl} \qquad (i, j, k, l = u, v). \tag{3.10}$$

Since $\mathbf{n}$ is perpendicular to $\mathbf{Y}_{,i}$ $(i = u, v)$, it follows that

$$L_{ij} = \mathbf{n} \cdot \mathbf{Y}_{,ij} = -\mathbf{n}_{,i} \cdot \mathbf{Y}_{,j} = -\mathbf{n}_{,j} \cdot \mathbf{Y}_{,i} = -n_{l,i}Y^l_{,j},$$

$$(l = 1, 2, 3; i, j = u, v). \tag{3.11}$$

As a unit vector, it follows that $\mathbf{n}$ is perpendicular to $\mathbf{n}_{,i}$. In other words, $\mathbf{n}_{,i}$ is in the plane determined by $\mathbf{Y}_{,u}$ and $\mathbf{Y}_{,v}$ and can be expressed in the form

$$\mathbf{n}_{,i} = A_i\mathbf{Y}_{,u} + B_i\mathbf{Y}_{,v} \quad (i = u, v). \tag{3.12}$$

The four coefficients A_u, A_v, B_u and B_v can be determined by substituting Eq. (3.12) into Eq. (3.11). With Eq. (3.1), it gives the Weingarten formula[5]

$$\mathbf{n}_{,i} = -L_{ij}g^{jk}\mathbf{Y}_{,k} \quad (i, j, k = u, v). \tag{3.13}$$

The definition of L_{ij}, Eq. (3.8), implies that L_{ij} is the component of $Y_{,ij}$ along the $\mathbf{n}$ direction. In terms of its three components along the three directions $\mathbf{n}$, $\mathbf{Y}_{,u}$ and $\mathbf{Y}_{,v}$, $\mathbf{Y}_{,ij}$ can be written in the form

$$\mathbf{Y}_{,ij} = L_{ij}\mathbf{n} + l_{ij}\mathbf{Y}_{,u} + m_{ij}\mathbf{Y}_{,v}, \tag{3.14}$$

where l_{ij} and m_{ij} are yet undetermined coefficients. The different coefficients l_{ij} and m_{ij} can be determined by taking the dot product of Eq. (3.14) with $\mathbf{Y}_{,u}$ and $\mathbf{Y}_{,v}$, respectively. The result is

$$\mathbf{Y}_{,ij} = \Gamma_{ij}^k \mathbf{Y}_{,k} + L_{ij}\mathbf{n}, \tag{3.15}$$

where the Christoffel symbol Γ_{ij}^k is defined by[4,5]

$$\Gamma_{ij}^k = \frac{1}{2}g^{kl}(g_{il,j} + g_{jl,i} - g_{ij,l}). \tag{3.16}$$

At any point P on a surface, there are two directions on the surface, orthogonal to each other, such that the normal at a consecutive point in either of three directions meets the normal at P. These two directions are called the principal directions at P. A curve drawn on the surface and possessing the property where the normals to the surface at consecutive points intersect is called a line of curvature. The point of intersection of consecutive normals along a line of curvature at P is called a center of curvature of the surface and its distance from P measured along the normal $\mathbf{n}$ is called a principal radius of curvature of the surface. The sign of the principal radius of curvature and consequently the reciprocal of the principal radius of curvature, the principal curvature, depends upon the choice of the positive direction of $\mathbf{n}$. Let $\mathbf{n}$ be the outward normal of the surface that is directed toward the convex side of the surface, and c the corresponding principal curvature, then the corresponding center of curvature $\mathbf{s}$ is given by

$$\mathbf{s} = -\mathbf{n}/c + \mathbf{Y}.$$

Since at a consecutive point on the line of curvature, $\mathbf{s}$, as well as c, has the same value, hence

$$d\mathbf{s} = -d\mathbf{n}/c + d\mathbf{Y} = 0.$$

This is the vector form of the Rodrigues' formula.[2] In terms of u and v, it reads

$$(-c\mathbf{Y}_{,u} + \mathbf{n}_{,u})du + (-c\mathbf{Y}_{,v} + \mathbf{n}_{,v})dv = 0.$$

By taking the dot product of this equation with $\mathbf{Y}_{,u}$ and $\mathbf{Y}_{,v}$, respectively, and eliminating du and dv, it gives the relation

$$c^2 + g^{ij}L_{ij}c + L/g = 0.$$

This is a quadratic equation of c with two solutions c_1 and c_2, the negative value of two principal curvatures gives

$$2H = -(c_1 + c_2) = g^{ij}L_{ij},$$
$$K = c_1c_2 = L/g.$$
(3.17)

H is called the mean curvature and K the Gaussian curvature of the surface at P.

Now consider another surface $\mathbf{Y}'(u, v)$ defined by

$$\mathbf{Y}'(u, v) = \mathbf{Y}(u, v) + \psi(u, v)\mathbf{n},$$
(3.18)

where $\psi(u, v)$ is a sufficiently small smooth function of u and v, $\mathbf{Y}'(u, v)$ is simply a slightly distorted surface from $\mathbf{Y}(u, v)$. Now consider the variation of different quantities related to $\mathbf{Y}(u, v)$ arising from the small distortion $\psi(u, v)\mathbf{n}$. It is clear that

$$\mathbf{Y}'_{,i} = \mathbf{Y}_{,i} + \psi_{,i}\mathbf{n} + \psi\mathbf{n}_{,i}, \quad (i = u, v).$$
(3.19)

The first interesting quantity is the variation δg_{ij} of g_{ij}. It is given by

$$\delta g_{ij} = \mathbf{Y}'_{,i} \cdot \mathbf{Y}'_{,j} - \mathbf{Y}_{,i} \cdot \mathbf{Y}_{,j}.$$

With Eqs. (3.1), (3.13) and the orthogonality between $\mathbf{n}$ and $\mathbf{Y}_{,i}$, it is found that

$$\delta g_{ij} = -2L_{ij}\psi + \psi_{,i}\psi_{,j} + \psi^2\mathbf{n}_{,i} \cdot \mathbf{n}_{,j}.$$

With the help of Eqs. (3.2), (3.4), (3.9), (3.17) and the Weingarten formula (3.13), the explicit form of $\mathbf{n}_{,i} \cdot \mathbf{n}_{,j}$ for different combinations of i and j gives that

$$\mathbf{n}_{,i} \cdot \mathbf{n}_{,j} = -\mathbf{n} \cdot \mathbf{n}_{,ij} = L_{ik}L_{jl}g^{kl} = 2HL_{ij} - Kg_{ij}.$$
(3.20)

Thus, δg_{ij} is given by

$$\delta g_{ij} = -2L_{ij}\psi + \psi_{,i}\psi_{,j} + (2HL_{ij} - Kg_{ij})\psi^2.$$
(3.21)

The variation of g is given by

$$\delta g = [(g_{uu} + \delta g_{uu})(g_{vv} + \delta g_{vv}) - (g_{uv} + \delta g_{uv})^2] - (g_{uu}g_{vv} - g_{uv}^2).$$

With Eq. (3.21), it gives that

$$\delta g = g[-4H\psi + g^{ij}\psi_{,i}\psi_{,j} + (4H^2 + 2K)\psi^2$$
$$- 2KL^{ij}\psi\psi_{,i}\psi_{,j} - 4HK\psi^3] + O(\psi^4), \tag{3.22}$$

where $O(\psi^4)$ refers to terms of order higher than the third order of ψ. The variation of $\sqrt{g}$ is given by

$$\delta g^{1/2} = (g + \delta g)^{1/2} - g^{1/2} = g^{1/2}\left[\frac{1}{2}\frac{\delta g}{g} - \frac{1}{8}\left(\frac{\delta g}{g}\right)^2 + \frac{1}{16}\left(\frac{\delta g}{g}\right)^3 + \cdots\right].$$

With Eq. (3.22), it can be shown easily that

$$\delta g^{1/2} = g^{1/2}\left[-2H\psi + \frac{1}{2}g^{ij}\psi_{,i}\psi_{,j} + K\psi^2 + (Hg^{ij} - KL^{ij})\psi\psi_{,i}\psi_{,j}\right]$$
$$+ O(\psi^4). \tag{3.23}$$

From Eq. (3.10), it follows that

$$\delta g^{ij} = (1/g)e_{3ik}e^{3jl}\{\delta g_{kl} - (g_{kl} + \delta g_{kl})(\delta g/g)[1 - (\delta g/g) + (\delta g/g)^2 - \cdots]\}.$$

Equations (3.21) and (3.22) now lead to

$$\delta g^{ij} = 2(2Hg^{ij} - KL^{ij})\psi + [(1/g)e_{3ik}e^{3jl} - g^{ij}g^{kl}]\psi_{,k}\psi_{,l} + 3(4H^2g^{ij}$$
$$- 2HKL^{ij} - Kg^{ij})\psi^2 - 2[4Hg^{ij}g^{kl} - Kg^{ij}L^{kl} - Kg^{kl}L^{ij}$$
$$- (2H/g)e_{3ik}e^{3jl}]\psi\psi_{,k}\psi_{,l} + 4(8H^3g^{ij} - 4H^2KL^{ij}$$
$$- 4HKg^{ij} + K^2L^{ij})\psi^3 + O(\psi^4). \tag{3.24}$$

The variation of $\mathbf{n}$, $\delta\mathbf{n}$, is given by

$$\delta\mathbf{n} = [-(\delta g/(2g)) + (3/8)(\delta g/g)^2 - (5/16)(\delta g/g)^3 + \cdots]\mathbf{n}$$
$$+ g^{-1/2}[1 - (\delta g/(2g)) + (3/8)(\delta g/g)^2 - (5/16)(\delta g/g)^3 + \cdots]$$
$$\times \{(\mathbf{Y}_{,u} \times \mathbf{n})\psi_{,v} + (\mathbf{n} \times \mathbf{Y}_{,v})\psi_{,u} + [\mathbf{Y}_{,u} \times \mathbf{n}_{,v} + \mathbf{n}_{,u} \times \mathbf{Y}_{,v}$$
$$+ (\mathbf{n} \times \mathbf{n}_{,v})\psi_{,u} + (\mathbf{n}_{,u} \times \mathbf{n})\psi_{,v}]\psi + (\mathbf{n}_{,u} \times \mathbf{n}_{,v})\psi^2\}.$$

It is not difficult to show that

$$(\mathbf{Y}_{,u} \times \mathbf{n})\psi_{,v} + (\mathbf{n} \times \mathbf{Y}_{,v})\psi_{,u} = e_{3kl}(\mathbf{n} \times \mathbf{Y}_{,l})\psi_{,k} = -g^{1/2}g^{kl}\mathbf{Y}_{,k}\psi_{,l},$$
$$\mathbf{Y}_{,u} \times \mathbf{n}_{,v} + \mathbf{n}_{,u} \times \mathbf{Y}_{,v} = -2g^{1/2}H\mathbf{n},$$
$$(\mathbf{n} \times \mathbf{n}_{,v})\psi_{,u} + (\mathbf{n}_{,u} \times \mathbf{n})\psi_{,v} = e_{3kl}(\mathbf{n} \times \mathbf{n}_{,l})\psi_{,k} = g^{1/2}KL^{kl}\mathbf{Y}_{,k}\psi_{,l}, \tag{3.25}$$
$$\mathbf{n}_{,u} \times \mathbf{n}_{,v} = g^{1/2}K\mathbf{n}.$$

With Eq. (3.22), it is found that

$$\delta \mathbf{n} = -g^{kl}\mathbf{Y}_{,k}\psi_{,l} - (1/2)g^{kl}\mathbf{n}\psi_{,k}\psi_{,l} - (2Hg^{kl} - KL^{kl})\mathbf{Y}_{,k}\psi\psi_{,l}$$
$$+ (1/2)g^{kl}g^{mn}\mathbf{Y}_{,k}\psi_{,l}\psi_{,m}\psi_{,n} - (2Hg^{kl} - KL^{kl})\mathbf{n}\psi\psi_{,k}\psi_{,l}$$
$$- (4H^2 g^{kl} - 2HKg^{kl} - Kg^{kl})\mathbf{Y}_{,k}\psi^2\psi_{,l} + O(\psi^4). \qquad (3.26)$$

The variation of $L_{ij}, \delta L_{ij}$, is given by

$$\delta L_{ij} = \mathbf{Y}_{,ij} \cdot \delta \mathbf{n} + (\psi \mathbf{n})_{,ij} \cdot \mathbf{n} + (\psi \mathbf{n})_{,ij} \cdot \delta \mathbf{n}.$$

With the help of Eqs. (3.4), (3.9) and (3.17) and with proper grouping of terms, it is not difficult to show the following relations:

$$\Gamma^l_{ij}Kg_{kl}L^{km}\psi_{,m} = 2\Gamma^l_{ij}H\psi_{,l} - \Gamma^m_{ij}g^{kl}L_{km}\psi_{,l},$$
$$Kg_{kp}L^{kl}(g^{pq}L_{ip})_{,j}\psi_{,l} = 2H(g^{pl}L_{ip})_{,j}\psi_{,l} - g^{kl}L_{kq}(g^{pq}L_{ip})_{,j}\psi, l. \qquad (3.27)$$

With Eqs. (3.15), (3.26), and (3.27), it is found that

$$\mathbf{Y}_{,ij} \cdot \delta \mathbf{n} = -\Gamma^l_{ij}\psi_{,l} - (1/2)g^{kl}L_{ij}\psi_{,k}\psi_{,l} - \Gamma^m_{ij}g^{kl}L_{km}\psi\psi_{,l}$$
$$+ (1/2)\Gamma^l_{ij}g^{mn}\psi_{,l}\psi_{,m}\psi_{,n} - (2Hg^{kl} - KL^{kl})L_{ij}\psi\psi_{,k}\psi_{,l}$$
$$+ (\Gamma^l_{ij}K - 2\Gamma^m_{ij}Hg^{kl}L_{km})\psi^2\psi_{,l} + O(\psi^4),$$

$$(\psi \mathbf{n})_{,ij} \cdot \mathbf{n} = \psi_{,ij} - (2HL_{ij} - Kg_{ij})\psi,$$

$$(\psi \mathbf{n})_{,ij} \cdot \delta \mathbf{n} = g^{lm}(L_{jm}\delta^k_i + L_{im}\delta^k_j)\psi_{,k}\psi_{,l} + [\Gamma^l_{jq}g^{pq}L_{ip}$$
$$+ (g^{pl}L_{ip})_{,j}]\psi\psi_{,l} - (1/2)g^{kl}\psi_{,k}\psi_{,l}\psi_{,ij} + [(2Hg^{lm}$$
$$- KL^{lm})(L_{jm}\delta^k_i + L_{im}\delta^k_j) + (1/2)(2HL_{ij}$$
$$- Kg_{ij})g^{kl}]\psi\psi_{,k}\psi_{,l} + [(g^{pq}L_{ip})_{,j}g^{kl}L_{kq}$$
$$+ \Gamma^k_{jq}g^{pq}L_{ip}g^{lm}L_{km}]\psi^2\psi_{,l} + O(\psi^4).$$

With Eq. (3.27), it finally gives

$$\delta L_{ij} = \delta^{(1)}L_{ij} + \delta^{(2)}L_{ij} + \delta^{(3)}L_{ij} + O(\psi^4),$$

where

$$\delta^{(1)}L_{ij} = \nabla_i\psi_{,j} - (2HL_{ij} - Kg_{ij})\psi,$$

$$\delta^{(2)}L_{ij} = [g^{lm}(L_{jm}\delta^k_i + L_{im}\delta^k_j) - (1/2)g^{kl}L_{ij}]\psi_{,k}\psi_{,l} + [(g^{pl}L_{ip})_{,j}$$
$$+ \Gamma^l_{jq}g^{pq}L_{ip} - \Gamma^m_{ij}g^{kl}L_{km}]\psi\psi_{,l},$$

$$\delta^{(3)} L_{ij} = -(1/2) g^{kl} \psi_{,k} \psi_{,l} \nabla_i \psi_{,j} + [(2H g^{lm} - KL^{lm})(L_{jm} \delta_i^k + L_{im} \delta_j^i)$$

$$- H g^{kl} L_{ij} + K(L^{kl} L_{ij} - g_{ij} g^{kl}/2)] \psi \psi_{,k} \psi_{,l}$$

$$+ [(g^{pq} L_{ip})_{,j} g^{kl} L_{kq} + \Gamma_{jq}^k g^{pq} L_{ip} g^{lm} L_{km} - 2\Gamma_{ij}^m H g^{kl} L_{km}$$

$$+ \Gamma_{ij}^l K] \psi^2 \psi_{,l}, \qquad (3.28)$$

and the covariant derivative of $\psi_{,j}$, $\nabla_i \psi_{,j}$ is defined by

$$\nabla_i \psi_{,j} = \psi_{,ij} - \Gamma_{ij}^k \psi_{,k}. \qquad (3.29)$$

Here, the notation $\delta^{(n)} f$ ($n = 1, 2, 3$) represents the part of the variation of δf which is of the nth order of ψ.

With δg^{ij} and δL_{ij}, it is now possible to calculate the variation of the mean curvature H, δH, a quantity which is important in later calculations. Since δg^{ij} and δL_{ij} have no zeroth-order terms, hence,

$$\delta^{(1)} H = g^{ij} \delta^{(1)} L_{ij} + [\delta^{(1)} g^{ij}] L_{ij},$$

$$\delta^{(2)} H = g^{ij} \delta^{(2)} L_{ij} + [\delta^{(2)} g^{ij}] L_{ij} + [\delta^{(1)} g^{ij}][\delta^{(1)} L_{ij}],$$

$$\delta^{(3)} H = g^{ij} \delta^{(3)} L_{ij} + [\delta^{(3)} g^{ij}] L_{ij} + [\delta^{(2)} g^{ij}][\delta^{(1)} L_{ij}] + [\delta^{(1)} g^{ij}][\delta^{(2)} L_{ij}].$$

It can be shown easily that

$$g^{ij} g^{lm} (L_{jm} \delta_i^k + L_{im} \delta_j^k) \psi_{,k} \psi_{,l}$$

$$= 2 g^{ij} g^{lm} L_{jm} \psi_{,i} \psi_{,l} = 2(2H g^{kl} - KL^{kl}) \psi_{,k} \psi_{,l},$$

$$L_{ij} e_{3ik} e^{3jl} \psi_{,k} \psi_{,l} = LL^{ij} \psi_{,i} \psi_{,j}, \qquad (3.30)$$

$$g_{ij} e_{3ik} e^{3jl} \psi_{,k} \psi_{,l} = gg^{ij} \psi_{,i} \psi_{,j}.$$

Besides, the following relations and other similar relations can be obtained by properly regrouping their component terms:

$$\Gamma_{ij}^m g^{kl} L_{km} \psi_{,l} = 2\Gamma_{ij}^l H \psi_{,l} - \Gamma_{ij}^k K g_{km} L^{lm} \psi_{,l},$$

$$g^{lm} L^{ij} L_{jm} \psi_{,i} \psi_{,l} = g^{kl} \psi_{,k} \psi_{,l}, \qquad (3.31)$$

$$\Gamma_{jq}^l g^{ij} g^{pq} L_{ip} = \Gamma_{ij}^l (2H g^{ij} - KL^{ij}),$$

$$\Gamma_{jq}^l g^{pq} L^{ij} L_{ip} = \Gamma_{ij}^l g^{ij}.$$

With the help of Eqs. (3.30) and (3.31), it is shown that

$$\delta^{(1)} H = (1/2) g^{ij} \nabla_i \psi_{,j} + (2H^2 - K) \psi,$$

$$\delta^{(2)} H = (2H g^{ij} - KL^{ij}) \psi \nabla_i \psi_{,j} + (1/2)(H g^{kl} - KL^{kl}) \psi_{,k} \psi_{,l}$$

$$+ (1/2)[g^{ij}(g^{pl} L_{ip})_{,j} - \Gamma_{ij}^m g^{ij} g^{kl} L_{km}] \psi \psi_{,l} + (4H^3 - 3HK) \psi^2,$$

$$\delta^{(3)}H = (1/2)[(1/g)e_{3ik}e^{3jl} - (3/2)g^{ij}g^{kl}]\psi_{,k}\psi_{,l}\nabla_i\psi_{,j} + (3/2)(4H^2g^{ij}$$

$$- 2HKL^{ij} - Kg^{ij})\psi^2\nabla_i\psi_{,j} + (3H^2g^{kl} - 2HKL^{kl} - Kg^{kl})\psi\psi_{,k}\psi_{,l}$$

$$+ [(2Hg^{ij} - KL^{ij})(g^{pl}L_{ip})_{,j} + (1/2)(g^{pq}L_{ip})_{,j}g^{ij}g^{kl}L_{kq}$$

$$+ (1/2)\Gamma^l_{ij}(8H^2g^{ij} - 4HKL^{ij} - Kg^{ij})$$

$$- (1/2)\Gamma^m_{ij}(4Hg^{ij} - KL^{ij})g^{kl}L_{km}]\psi^2\psi_{,l}$$

$$+ (8H^4 - 8H^2K + K^2)\psi^3. \tag{3.32}$$

According to Eq. (3.1), the line elements ds_u and ds_v along the $u =$ const. and $v =$ const., curves are given by $ds_u = g_{vv}^{1/2}dv$ and $ds_v = g_{uu}^{1/2}du$, respectively. The direction cosines of the tangents to the u-curves and the v-curves are simply $g_{vv}^{-1/2}Y^l_{,u} = g_{vv}^{-1/2}Y_{l,u}$ and $g_{uu}^{-1/2}Y^l_{,v} = g_{uu}^{-1/2}Y_{l,v}$ ($l = 1, 2, 3$), respectively. If ω is the angle between the u-curve and the v-curve at a point on the surface, then

$$\cos\omega = (g_{uu}g_{vv})^{-1/2}Y^l_{,u}Y_{l,v} = g_{uv}(g_{uu}g_{vv})^{-1/2},$$

$$\sin\omega = [g/(g_{uu}g_{vv})]^{1/2}.$$

Figure 3.1 shows an area element dA of the surface formed by the quadrilateral with vertices (u, v), $(u + du, v)$, $(u, v + dv)$, $(u + du, v + dv)$.

Clearly, there is

$$dA = ds_u ds_v \sin\omega = g^{1/2}dudv,$$

$$A = \oint g^{1/2}dudv. \tag{3.33}$$

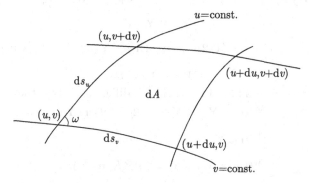

Fig. 3.1. Area element (not in the real scale).

From Eq. (3.23), the variation of the area element $\delta(dA)$ and the variation of the total area δA, respectively, are

$$\delta(dA) = g^{1/2}[-2H\psi + (1/2)g^{ij}\psi_{,i}\psi_{,j} + K\psi^2$$
$$+ (Hg^{ij} - KL^{ij})\psi\psi_{,i}\psi_{,j}]dudv$$
$$= [-2H\psi + (1/2)g^{ij}\psi_{,i}\psi_{,j} + K\psi^2 + (Hg^{ij} - KL^{ij})\psi\psi_{,i}\psi_{,j}]dA$$
$$+ O(\psi^4), \tag{3.34}$$

$$\delta A = \oint [-2H\psi + (1/2)g^{ij}\psi_{,i}\psi_{,j} + K\psi^2 + (Hg^{ij} - KL^{ij})\psi\psi_{,i}\psi_{,j}]dA$$
$$+ O(\psi^4).$$

The volume element dV is given by

$$dV = \frac{1}{3}\mathbf{Y} \cdot \mathbf{n}dA = \frac{1}{3}g^{\frac{1}{2}}(\mathbf{Y} \cdot \mathbf{n})dudv,$$

and the variation of the volume element $\delta(dV)$ is given by

$$\delta(dV) = \frac{1}{3}(g + \delta g)^{1/2}(\mathbf{Y} + \mathbf{n}\psi) \cdot (\mathbf{n} + \delta\mathbf{n})dudv - \frac{1}{3}g^{1/2}(\mathbf{Y} \cdot \mathbf{n})dudv.$$

With Eq. (3.25), $\mathbf{n} + \delta\mathbf{n}$ may be written as

$$\mathbf{n} + \delta\mathbf{n} = (g + \delta g)^{-1/2}(\mathbf{Y} + \mathbf{n}\psi)_{,u} \times (\mathbf{Y} + \mathbf{n}\psi)_{,v}$$
$$= (g + \delta g)^{-1/2}[g^{1/2}(1 - 2H\psi + K\psi^2)\mathbf{n}$$
$$+ e^{3ij}(\mathbf{n} \times \mathbf{Y}_{,j})\psi_{,i} + e^{3ij}(\mathbf{n} \times \mathbf{n}_{,j})\psi\psi_{,i}].$$

It follows that:

$$\delta(dV) = (1/3)[g^{1/2}(\psi - 2H\psi^2 + K\psi^3) - g^{1/2}(2H\psi - K\psi^2)(\mathbf{n} \cdot \mathbf{Y})$$
$$+ e^{3ij}\mathbf{n} \cdot (\mathbf{Y}_{,j} \times \mathbf{Y}\psi_{,i}) + e^{3ij}\mathbf{n} \cdot (\mathbf{n}_{,j} \times \mathbf{Y}\psi\psi_{,i})]dudv,$$

where

$$e^{3ij}\mathbf{n} \cdot (\mathbf{Y}_{,j} \times \mathbf{Y}\psi_{,i}) = e^{3ij}\{[\mathbf{n} \cdot (\mathbf{Y}_{,j} \times \mathbf{Y}\psi)]_{,i} - \mathbf{n}_{,i} \cdot (\mathbf{Y}_{,j} \times \mathbf{Y}\psi)$$
$$- \mathbf{n} \cdot (\mathbf{Y}_{,j} \times \mathbf{Y}_{,i})\psi\},$$

$$e^{3ij}\mathbf{n} \cdot (\mathbf{n}_{,j} \times \mathbf{Y}\psi\psi_{,i}) = (1/2)e^{3ij}\{[\mathbf{n} \cdot (\mathbf{n}_{,j} \times \mathbf{Y}\psi^2)]_{,i} - (\mathbf{n}_{,i} \times \mathbf{n}_{,j}) \cdot \mathbf{Y}\psi^2$$
$$- \mathbf{n} \cdot (\mathbf{n}_{,j} \times \mathbf{Y}_{,i}\psi^2)\}. \tag{3.35}$$

With Weingarten formula (3.13), it is not difficult to show that

$$e^{3ij}\mathbf{n}_{,i} \cdot (\mathbf{Y}_{,j} \times \mathbf{Y}) = -2g^{1/2}H(\mathbf{n} \cdot \mathbf{Y}),$$

$$e^{3ij}\mathbf{n} \cdot (\mathbf{Y}_{,j} \times \mathbf{Y}_{,i}) = -2g^{1/2},$$

$$e^{3ij}(\mathbf{n}_{,i} \times \mathbf{n}_{,j}) \cdot \mathbf{Y} = 2g^{1/2}K(\mathbf{n} \cdot \mathbf{Y}),$$

$$e^{3ij}\mathbf{n} \cdot (\mathbf{n}_{,j} \times \mathbf{Y}_{,i}) = 2g^{1/2}H. \tag{3.36}$$

The substitution of Eq. (3.36) into Eq. (3.35) gives $\delta(\mathrm{d}V)$ and δV, respectively, as

$$\delta(\mathrm{d}V) = (1/3)\{g^{1/2}(3\psi - 3H\psi^2 + K\psi^3) + e^{3ij}[\mathbf{n} \cdot (\mathbf{Y}_{,j} \times \mathbf{Y}\psi)]_{,i}$$

$$+ (1/2)e^{3ij}[\mathbf{n} \cdot (\mathbf{n}_{,j} \times \mathbf{Y}\psi^2)]_{,i} + O(\psi^4)\}\mathrm{d}u\mathrm{d}v, \qquad (3.37)$$

$$\delta V = \oint \left(\psi - H\psi^2 - \frac{1}{3}K\psi^3\right)\mathrm{d}A + O(\psi^4).$$

3.2   General Shape Equation[6,7]

In the equilibrium state, the energy of any physical system must be at its minimum, i.e., the equilibrium energy of the system must be less than that in a slightly varied state. In analogy with smectic-A liquid crystal bilayer, Helfrich proposed that the shape energy of a vesicle be given by

$$F = F_c + \Delta p \int \mathrm{d}V + \lambda \oint \mathrm{d}A$$

$$= \frac{1}{2}k_c \oint (2H + c_0)^2 \mathrm{d}A + \Delta p \int \mathrm{d}V + \lambda \oint \mathrm{d}A, \qquad (3.38)$$

where F_c is the curvature-elastic energy and k_c the bending rigidity of the vesicle membrane. The variation of the surface element $\mathrm{d}A$ and the volume element $\mathrm{d}V$ are given by Eqs. (3.34) and (3.37), respectively. H is the mean curvature of the membrane surface and the spontaneous curvature c_0 takes account of the asymmetry effect of the membrane or its environment. $\Delta p = p_{\mathrm{out}} - p_{\mathrm{in}}$ may be considered as the pressure difference between the outside and the inside of the membrane and λ may be considered as the tensile stress acting on the membrane. Mathematically, Δp and λ may be considered as Lagrange multipliers. The variations of different quantities under the variation $\psi\mathbf{n}$ of the vesicle surface have been calculated and listed in the last section. In order to find the shape equation of the vesicle, it is necessary to calculate the first variation $\delta^{(1)}F$ of F, i.e., terms proportional to ψ in δF.

The first variation $\delta^{(1)}F_c$ of F_c is given by

$$\delta^{(1)} F_c = \frac{1}{2}k_c \oint (2H + c_0)^2 \delta^{(1)}\mathrm{d}A + \frac{1}{2}k_c \oint 4(2H + c_0)(\delta^{(1)}H)\mathrm{d}A.$$

$$(3.39)$$

To the first order of ψ, Eqs. (3.29), (3.32), (3.34) and (3.37) give that

$$\delta^{(1)}\mathrm{d}A = -2H\psi g^{1/2}\mathrm{d}u\mathrm{d}v, \quad \delta^{(1)}\mathrm{d}V = \psi g^{1/2}\mathrm{d}u\mathrm{d}v,$$

$$\delta^{(1)} H = (2H^2 - K)\psi + (1/2)g^{ij}(\psi_{,ij} - \Gamma_{ij}^k \psi_{,k}).$$

$$(3.40)$$

It follows that

$$\delta^{(1)}F_c = k_c \oint [(2H + c_0)(2H^2 - c_0 H - 2K)\psi$$

$$+ g^{ij}(2H + c_0)\psi_{,ij} - g^{ij}\Gamma_{ij}^k(2H + c_0)\psi_{,k}]g^{1/2}\mathrm{d}u\mathrm{d}v.$$

It should be noted that for any two functions $f(u,v)$ and $\phi(u,v)$, the following two relations exist:

$$\oint f\phi_{,i}\mathrm{d}u\mathrm{d}v = \oint (f\phi)_{,i}\mathrm{d}u\mathrm{d}v - \oint f_{,i}\phi\mathrm{d}u\mathrm{d}v = -\oint f_{,i}\phi\mathrm{d}u\mathrm{d}v,$$

$$\oint f\phi_{,ij}\mathrm{d}u\mathrm{d}v = \oint (f\phi)_{,ij}\mathrm{d}u\mathrm{d}v - \oint f_{,j}\phi_{,i}\mathrm{d}u\mathrm{d}v - \oint f_{,i}\phi_{,j}\mathrm{d}u\mathrm{d}v$$

$$- \oint f_{,ij}\phi\mathrm{d}u\mathrm{d}v$$

$$= \oint (f\phi)_{,ij}\mathrm{d}u\mathrm{d}v - \oint (f_{,j}\phi)_{,i}\mathrm{d}u\mathrm{d}v + \oint f_{,ij}\phi\mathrm{d}u\mathrm{d}v \qquad (3.41)$$

$$- \oint (f_{,i}\phi)_{,j}\mathrm{d}u\mathrm{d}v + \oint f_{,ij}\phi\mathrm{d}u\mathrm{d}v - \oint f_{,ij}\phi\mathrm{d}u\mathrm{d}v$$

$$= \oint f_{,ij}\phi\mathrm{d}u\mathrm{d}v,$$

where both i and j stand for u and v. With Eq. (3.41), it gives that

$$\delta^{(1)}F_c = k_c \oint \{(2H + c_0)(2H^2 - c_0 H - 2K)g^{1/2} + [g^{1/2}g^{ij}(2H + c_0)]_{,ij}$$

$$+ [g^{1/2}g^{ij}(2H + c_0)\Gamma_{ij}^k]_{,k}\}\psi\mathrm{d}u\mathrm{d}v.$$

The term $[g^{1/2}g^{ij}(2H + c_0)]_{,ij}$ can be written as $[(g^{1/2}g^{ij})_{,j}(2H + c_0)]_{,i} + [g^{1/2}g^{ij}(2H + c_0)_{,j}]_{,i}$. With Eqs. (3.1), (3.4) and (3.16), for any function $f(u,v)$, the direct expansion gives that

$$[(g^{1/2}g^{ij})_{,j}f]_{,i} = -(\Gamma_{ij}^k g^{1/2}g^{ij}f)_{,k}, \qquad (3.42)$$

It is well known that[5]

$$\nabla^2 = g^{-1/2}\frac{\partial}{\partial i}\left(g^{1/2}g^{ij}\frac{\partial}{\partial j}\right), \qquad (3.43)$$

where ∇^2 is the 2D Laplacian operator. Therefore, $[g^{1/2}g^{ij}(2H + c_0)_{,j}]_{,i}$ is simply equal to $g^{1/2}\nabla^2(2H + c_0)$. Thus, $\delta^{(1)}F_c$ is given by

$$\delta^{(1)}F_c = k_c \oint [(2H + c_0)(2H^2 - c_0 H - 2K) + \nabla^2(2H + c_0)]\psi g^{1/2}\mathrm{d}u\mathrm{d}v,$$

$$(3.44)$$

and $\delta^{(1)}F$ becomes

$$\delta^{(1)}F = \oint [\Delta p - 2\lambda H + k_c(2H + c_0)(2H^2 - c_0 H - 2K)$$

$$+ k_c \nabla^2 (2H + c_0)]\psi g^{1/2} du dv. \qquad (3.45)$$

Since ψ is an arbitrary, sufficiently small and very smooth function of u and v, the vanishing of the first variation of F requires that

$$\Delta p - 2\lambda H + k_c(2H + c_0)(2H^2 - c_0 H - 2K) + k_c \nabla^2 (2H + c_0) = 0.$$

This is just the general shape equation of the vesicle membrane.[8,9] Unless the symmetry effect of the membrane and its environment varies from point to point (i.e., $c_0 = c_0(u, v)$), generally, c_0 is a constant and the general shape equation reduces to

$$\Delta p - 2\lambda H + k_c(2H + c_0)(2H^2 - c_0 H - 2K) + 2k_c \nabla^2 H = 0. \qquad (3.46)$$

The vanishing of the first variation $\delta^{(1)}F$ gives the condition that the energy of the vesicle has an extreme value. However, it does not mean that the energy of the vesicle is necessarily at its minimum. In other words, it does not mean that the solution of the shape equation gives necessarily a stable vesicle solution. But only stable vesicles are observable in experiments. A stable solution requires that the second variation $\delta^{(2)}F$, i.e., terms of $O(\psi^2)$ in δF, must be positively definite. The second variation $\delta^{(2)}F$ is given by

$$\delta^{(2)}F = \Delta p \delta^{(2)} \int dV + \lambda \delta^{(2)} \oint dA + \frac{1}{2} k_c \delta^{(2)} \oint (2H + c_0)^2 dA, \qquad (3.47)$$

where

$$\delta^{(2)} \int dV = -\oint H\psi^2 g^{1/2} du dv,$$

$$\delta^{(2)} \oint dA = \oint \left(\frac{1}{2} g^{ij} \psi_{,i} \psi_{,j} + K\psi^2 \right) g^{1/2} du dv. \qquad (3.48)$$

The last term in Eq. (3.47) is somewhat involved and is given by

$$\delta^{(2)} F_c = \frac{k_c}{2} \oint 4[(2H + c_0)\delta^{(2)}H + (\delta^{(1)}H)^2]g^{1/2} du dv$$

$$+ \frac{k_c}{2} \oint 4(2H + c_0)(\delta^{(1)}H)\delta^{(1)} dA + \frac{k_c}{2} \oint (2H + c_0)^2 \delta^{(2)} dA. \qquad (3.49)$$

The substitution of Eqs. (3.32) and (3.48) into Eq. (3.49) gives that

$$\delta^{(2)}F_c = k_c \oint \{(1/2)(g^{ij}\nabla_i\psi_{,j})^2 - 2(K + c_0H)g^{ij}\psi\nabla_i\psi_{,j}$$

$$+ 2(2H + c_0)(2Hg^{ij} - KL^{ij})\psi\psi_{,ij} + (H + c_0/2)[3Hg^{ij}$$

$$+ (1/2)c_0g^{ij} - 2KL^{ij}]\psi_{,i}\psi_{,j} + (2H + c_0)[g^{ij}(g^{kl}L_{ik})_{,j}$$

$$- \Gamma_{ij}^m g^{ij}g^{kl}L_{km} - \Gamma_{ij}^l(2Hg^{ij} - KL^{ij})]\psi\psi_{,l} + [2(2H^2 - K)^2$$

$$- 2K(H^2 - c_0^2/4)]\psi^2\}g^{1/2}dudv. \tag{3.50}$$

With

$$\psi\psi_{,l} = \frac{1}{2}(\psi^2)_{,l} \quad \text{and} \quad \psi\psi_{,ij} = \frac{1}{2}(\psi^2)_{,ij} - \psi_{,i}\psi_{,j}, \tag{3.51}$$

Eq. (3.50) can be written as:

$$\delta^{(2)}F_c = k_c \oint \{(1/2)(g^{ij}\nabla_i\psi_{,j})^2 - 2(K + c_0H)g^{ij}\psi\nabla_i\psi_{,j}$$

$$+ (H + c_0/2)[(H + c_0/2)g^{ij}$$

$$+ 2(KL^{ij} - 3Hg^{ij})]\psi_{,i}\psi_{,j} + (2H + c_0)(2Hg^{ij} - KL^{ij})(\psi^2)_{,ij}$$

$$+ (H + c_0/2)[g^{ij}(g^{kl}L_{ik})_{,j} - \Gamma_{ij}^m g^{ij}g^{kl}L_{km}$$

$$- \Gamma_{ij}^l(2Hg^{ij} - KL^{ij})](\psi^2)_{,l}$$

$$+ [2(2H^2 - K)^2 - 2K(H^2 - c_0^2/4)]\psi^2\}g^{1/2}dudv. \tag{3.52}$$

With Eq. (3.41), it finally becomes

$$\delta^{(2)}F_c = k_c \oint \{(1/2)(g^{ij}\nabla_i\psi_{,j})^2 - 2(K + c_0H)g^{ij}\psi\nabla_i\psi_{,j}$$

$$+ (H + c_0/2)[(H + c_0)g^{ij} + 2(KL^{ij} - 3Hg^{ij})]\psi_{,i}\psi_{,j}$$

$$+ g^{-1/2}[g^{1/2}(2H + c_0)(2Hg^{ij} - KL^{ij})]_{,ij}\psi^2$$

$$- g^{-1/2}[g^{1/2}(H + c_0/2)g^{ij}(g^{kl}L_{ik})_{,j}$$

$$- g^{1/2}(H + c_0/2)\Gamma_{ij}^m g^{ij}g^{kl}L_{km}$$

$$- g^{1/2}(H + c_0/2)\Gamma_{ij}^l(2Hg^{ij} - KL^{ij})]_{,l}\psi^2$$

$$+ 2[(2H^2 - K)^2 - K(H^2 - c_0^2/4)]\psi^2\}g^{1/2}dudv. \tag{3.53}$$

Then the second variation of F can be expressed as

$$\delta^{(2)}F = \delta^{(2)}F_c + \oint \left[\frac{1}{2}\lambda g^{ij}\psi_{,i}\psi_{,j} + (\lambda K - \Delta pH)\psi^2\right]g^{1/2}dudv. \tag{3.54}$$

For a stable vesicle, it is necessary that $\delta^{(2)}F$ is positively definite for any $\psi \neq 0$. In general, this requires that the eigenvalues of an operator acting on ψ and associated with Eq. (3.54) satisfy certain condition. However, although Eq. (3.54) provides a tool for the numerical stability analysis of any equilibrium shape of vesicles, nevertheless, it is not easy to carry out such an analysis in general. Only certain simple cases may be evaluated analytically. Besides, the success of the application of Eq. (3.54) depends upon the proper choice of the function ψ.

3.3 Spherical Vesicles

In dealing with the stability and the deformation energies of vesicles, the area of the membrane is usually taken to be constant. In such a case, spheres can only be deformed if the enclosed volume is a variable. They can be stabilized by a pressure difference or by a spontaneous curvature. The former case may be produced osmotically or by means of a micro-pipette that opens into the inside of the vesicle. The sphere may be described by

$$\mathbf{Y} = R(\sin\theta\cos\phi, \sin\theta\sin\phi, \cos\theta). \tag{3.55}$$

It follows that the non-vanishing basic terms are

$$g_{\theta\theta} = R^2, \quad g_{\phi\phi} = R^2\sin^2\theta, \quad g = R^4\sin^2\theta, \quad g^{\theta\theta} = 1/R^2,$$

$$g^{\phi\phi} = 1/(R^2\sin^2\theta), \quad L_{\theta\theta} = -R, \quad L_{\phi\phi} = -R\sin^2\theta, \quad L = R^2\sin^2\theta,$$

$$L^{\theta\theta} = -1/R, \quad L^{\phi\phi} = -1/(R\sin^2\theta), \quad L^{ij} = -Rg^{ij}, \quad \Gamma^\theta_{\phi\phi} = -\sin\theta\cos\theta,$$

$$\Gamma^\phi_{\theta\phi} = \cot\theta, \quad H = -1/R, \quad K = 1/R^2. \tag{3.56}$$

The Laplacian operator ∇^2 now becomes

$$\nabla^2 = \frac{1}{R^2\sin\theta}\frac{\partial}{\partial\theta}\left(\sin\theta\frac{\partial}{\partial\theta}\right) + \frac{1}{R^2\sin^2\theta}\frac{\partial^2}{\partial\phi^2} \tag{3.57}$$

and

$$g^{ij}\nabla_i\phi_{,j} = -(1/R)L^{ij}\nabla_i\psi_{,j} = \nabla^2\psi. \tag{3.58}$$

The shape equation (3.46) shows that the sphere is a solution if the parameters satisfy the relation

$$\Delta pR^3 + 2\lambda R^2 - k_cc_0R(2 - c_0R) = 0, \tag{3.59}$$

a relation that can be used to express λ in terms of Δp and R. The variations $\delta^{(1)}F_c$, $\delta^{(2)}F_c$, and $\delta^{(2)}F$ given by Eqs. (3.44), (3.50) and (3.54), respectively, now become

$$\delta^{(1)}F_c = k_c \oint [(c_0 R - 2)(c_0 + R\nabla^2)\psi] \sin\theta d\theta d\phi, \tag{3.60}$$

$$\delta^{(2)}F_c = \frac{1}{2}k_c \oint \left[R^2(\nabla^2\psi)^2 + \left(2 + 2c_0 R - \frac{1}{2}c_0^2 R^2\right)\psi\nabla^2\psi \right.$$

$$\left. + c_0^2\psi^2 \right] \sin\theta d\theta d\phi, \tag{3.61}$$

$$\delta^{(2)}F = \frac{k_c}{2} \oint \left[R^2(\nabla^2\psi)^2 + \left(\frac{\Delta p R^3}{2k_c} + c_0 R + 2\right)\psi\nabla^2\psi \right.$$

$$\left. + \left(\frac{\Delta p R}{k_c} + \frac{2c_0}{R}\right)\psi^2 \right] \sin\theta d\theta d\phi, \tag{3.62}$$

where Eqs. (3.41), (3.51), (3.59) and the relation

$$\oint [g^{ij}g^{1/2}(\psi^2)_{,ij} - (g^{ij}g^{1/2}\Gamma_{ij}^l)(\psi^2)_{,l}]d\theta d\phi$$

$$= \oint [(g^{ij}g^{1/2})_{,ij} + (g^{ij}g^{1/2}\Gamma_{ij}^l)_{,l}]\psi^2 d\theta d\phi = 0$$

for spherical surfaces have been used.

It is well known that the spherical harmonics $Y_{lm}(\theta,\phi)$ satisfy the differential equation[10]

$$\left[\frac{1}{\sin\theta}\frac{\partial}{\partial\theta}\left(\sin\theta\frac{\partial}{\partial\theta}\right) + \frac{1}{\sin^2\theta}\frac{\partial^2}{\partial\phi^2}\right]Y_{lm}(\theta,\phi) = -l(l+1)Y_{lm}(\theta,\phi) \tag{3.63}$$

with

$$\sqrt{4\pi}Y_{00}(\theta,\phi) = 1,$$

and the orthogonal relation

$$\int_0^\pi \int_0^{2\pi} Y_{lm}^*(\theta,\phi)Y_{l'm'}(\theta,\phi)\sin\theta d\theta d\phi = \delta_l^{l'}\delta_m^{m'}, \tag{3.64}$$

where $Y_{lm}^*(\theta,\phi)$ is the complex conjugate function of $Y_{lm}(\theta,\phi)$. In other words, for spherical vesicles, $Y_{lm}(\theta,\phi)$ satisfies the relation

$$\nabla^2 Y_{lm} = -l(l+1)Y_{lm}/R^2. \tag{3.65}$$

The set $Y_{lm}(\theta, \phi)$ now provides a convenient basis for the computation of $\delta^{(1)}F_c$, $\delta^{(2)}F_c$, and $\delta^{(2)}F$. Let ψ be written as

$$\psi = \sum_{l,m} a_{lm} Y_{lm}(\theta, \phi), \tag{3.66}$$

with $l = 0, 1, 2, \ldots$ and $|m| \leq l$. The requirement that $a_{l,m}^* = a_{l,-m}$ will ensure ψ is a real function. Equation (3.60) is of first order of ψ. Taking Eq. (3.63) into consideration, one may multiply the right-hand side of Eq. (3.60) with $\sqrt{4\pi}Y_{00}$ to take advantage of the orthogonality property Eq. (3.64). The results are

$$\delta^{(1)}F_c = k_c(4\pi)^{1/2} a_{00} c_0 (c_0 R - 2), \tag{3.67}$$

$$\delta^{(2)}F_c = [k_c/(2R^2)] \sum_{l,m} |a_{lm}|^2 [l^2(l+1)^2$$

$$- l(l+1)(2 + 2c_0 R - c_0^2 R^2/2) + c_0^2 R^2], \tag{3.68}$$

$$\delta^{(2)}F = \sum_{l,m} |a_{lm}|^2 \{(1/2)\Delta p R + k_c c_0/R - l(l+1)[\Delta p R/4$$

$$+ k_c(c_0 R + 2)/(2R^2)] + k_c l^2(l+1)^2/(2R^2)\}$$

$$= (k_c/2) \sum_{l,m} (|a_{lm}|/R)^2 [l(l+1) - 2][l(l+1) - c_0 R - \Delta p R^3/(2k_c)]. \tag{3.69}$$

Regardless of whether area, volume or radius of the sphere is kept constant, Eq. (3.69) holds generally. This can be seen in the following way: up to $O(\psi^2)$, the variations of area and volume calculated from Eqs. (3.24) and (3.27), respectively, are

$$\delta A = 2(4\pi)^{1/2} a_{00} R + \sum_{l,m} |a_{lm}|^2 \left[1 + \frac{1}{2}l(l+1)\right],$$

$$\delta V = (4\pi)^{1/2} a_{00} R^2 + \sum_{l,m} |a_{lm}|^2 R. \tag{3.70}$$

In case of constant area, i.e., $\delta A = 0$, one obtains

$$(4\pi)^{1/2} a_{00} = -\frac{1}{2} \sum_{l,m} |a_{lm}|^2 \left[1 + \frac{1}{2}l(l+1)\right] R^{-1},$$

which leads to $\delta F = \delta^{(1)}F_c + \delta^{(2)}F_c + \Delta p \delta V$, the same expression given by Eq. (3.69). On the other hand, for constant volume, i.e., $\delta V = 0$, one obtains

$$(4\pi)^{1/2}a_{00} = -\sum_{l,m}|a_{lm}|^2 R^{-1},$$

which leads to $\delta F = \delta^{(1)}F_c + \delta^{(2)}F_c + \lambda\delta A$, the same expression of Eq. (3.69) again. It should be noted that a volume reservoir should exist in case of $\delta A = 0$ and an area reservoir in case of $\delta V = 0$. In both cases, the Y_{00} mode cannot be excited separately. The trivial case $l = 1$, characterized by $\delta^{(2)}F = 0$, means a translation of the sphere. In case of constant radius, both volume and area reservoir are needed and it requires that $a_{00} = 0$. In this case, $\sum_{l,m}|a_{lm}|^2 = \sum_{l,m}|a_{lm}|^2 l(l+1) = 0$ and $\delta^{(2)}F = \delta^{(1)}F_c + \delta^{(2)}F_c = 0$. Again, it is contained in Eq. (3.69).

Equation (3.69) shows that the coefficient of $|a_{jm}|^2$ for $1 < j \le l$ will be negative for Δp greater than a threshold pressure difference Δp_l given by

$$\Delta p_l = (2k_c/R^3)[l(l+1) - c_0 R], \tag{3.71}$$

which depends only on l but not on m. Obviously, in this case, nonzero a_{jm} will reduce the curvature energy and a deformation corresponding to higher spherical harmonics than Y_{lm} can occur.

The spherical harmonics Y_{l0} are rotationally symmetric about the polar-axis, i.e., the cross-section of the shape along the axis will show some lth polygon symmetry. Deuling and Helfrich[11] has calculated a large variety of rotationally symmetric shapes of vesicles. Among them, contours having triangular, pentagonal, and heptagonal cross-sections are examples of the deformation of nearly spherical vesicles. Ou-Yang and Helfrich[8] made a check and found that the deformations correspond exactly to the modes of $l = 3$, 5 and 7, respectively. The agreement with the theory and the numerical calculation on Δp_l are very good. The observed changes in the shape of red blood cells may provide another example. The so-called triconcave and quadriconcave red blood cells produced by hypotonic media[12] may be the deformation associated with $l = 3$ and 4, respectively. In a sense, the normal red blood cell, the so-called discocyte, represents a branch of $l = 2$ because of its biconcave shape. A calculated contour of this type is also given in Ref. [11]. Equation (3.71) also predicts that under given Δp, the deformation may be induced by a change of c_0. The so-called glass effect of red blood cells, which become echinocytic when they are near a glass rod[12] seems to have this origin: Glass may increase the value of c_0 by its chemical effect and, therefore, cause l to be increased at a constant Δp. As a result, the deformations represent spherical harmonics of very high order.

However, the shape may also be influenced by the shear elasticity of the red blood cell membrane.

3.4 Nearly Spherical Vesicles and Third-Order Energy Variation

In the previous section, it has been pointed out that at Δp_l a spherical vesicle becomes unstable and begins to deform. The deformed shape can be described by a linear combination of the functions Y_{2m}. In the rotationally symmetric case, the deformation of positive and negative amplitude is not physically equivalent, being prolate and oblate ellipsoids, respectively. However, both of them have the same second-order energy as shown by Eq. (3.69). To break this symmetry, it is necessary to consider the third-order energy variation, $\delta^{(3)}F$. The third-order variation of F_c is given by

$$\delta^{(3)}F_c = \frac{1}{2}k_c c_0^2 \delta^{(3)} \oint dA + 2k_c c_0 \delta^{(3)} \oint H dA + 2k_c \delta^{(3)} \oint H^2 dA.$$

where the first term, according to Eq. (3.34), is given by

$$\frac{1}{2}k_c c_0^2 \delta^{(3)} \oint dA = \frac{1}{2}k_c c_0^2 \oint (Hg^{ij} - KL^{ij})\psi\psi_{,i}\psi_{,j}dA.$$

It vanishes on account of Eq. (3.56). The middle term is given by

$$2k_c c_0 \delta^{(3)} \oint H dA = 2k_c c_0 \left(\oint \delta^{(3)} H dA + \oint \delta^{(2)} H \delta^{(1)} dA \right.$$

$$\left. + \oint \delta^{(1)} H \delta^{(2)} dA + \oint H \delta^{(3)} dA \right).$$

It is not difficult to prove the following relation

$$(g^{pq}L_{ip})_{,j}g^{kl}L_{kq}\psi_{,l} = 2H(g^{pl}L_{ip})_{,j}\psi_{,l} - K(g^{pk}L_{ip})_{,j}g_{km}L^{lm}\psi_{,l}. \tag{3.72}$$

With Eqs. (3.31), (3.32), (3.34) and (3.72), it is found that

$$\delta^{(3)} \oint H dA = \{(1/2)[(1/g)e_{3ik}e^{3jl} - g^{ij}g^{kl}]\psi_{,k}\psi_{,l}\nabla_i\psi_{,j} + (2H^2 g^{ij}$$

$$- HKL^{ij} - Kg^{ij})\psi^2\nabla_i\psi_{,j}$$

$$+ [4H^2 g^{ij} - 2HKL^{ij} - (3/2)Kg^{ij}]\psi\psi_{,i}\psi_{,j}$$

$$+ [-(1/2)\Gamma_{ij}^l Kg^{ij} + (2Hg^{ij} - KL^{ij})(g^{pl}L_{ip})_{,j}$$

$$- (1/2)K(g^{pk}L_{ip})_{,j}g^{ij}g_{km}L^{lm}$$

$$+ \Gamma_{ij}^k(Hg^{ij} - KL^{ij}/2)Kg_{km}L^{lm}]\psi^2\psi_{,l}\}dA. \tag{3.73}$$

Similarly, one finds that

$$
\begin{aligned}
\delta^{(3)} \oint H^2 \mathrm{d}A = \oint &\{[(3/2)Hg^{ij} - KL^{ij}]\psi\nabla_i\psi_{,j}\nabla^2\psi + (1/2)[(2H/g)e_{3ik}e^{3jl} \\
&- Hg^{ij}g^{kl} - Kg^{ij}L^{kl}]\psi_{,k}\psi_{,l}\nabla_i\psi_{,j} + (1/2)[g^{ij}(g^{pl}L_{ip})_{,j} \\
&+ \Gamma^m_{ij}Kg^{ij}g_{km}L^{kl} - \Gamma^l_{ij}KL^{ij}]\psi\psi_{,l}\nabla^2\psi \\
&+ (12H^3g^{ij} - 6H^2KL^{ij} - 7HKg^{ij} + 2K^2L^{ij})\psi^2\nabla_i\psi_{,j} \\
&+ (9H^3g^{kl} - 5H^2KL^{kl} - 4HKg^{kl} + K^2L^{kl})\psi\psi_{,k}\psi_{,l} \\
&+ [(6H^2g^{ij} - 2HKL^{ij} - Kg^{ij})(g^{pl}L_{ip})_{,j} \\
&- HK(g^{pk}L_{ip})_{,j}g^{ij}g_{km}L^{lm} \\
&+ \Gamma^m_{ij}(4H^2Kg^{ij} - HK^2L^{ij} - K^2g^{ij})g_{km}L^{kl} \\
&- \Gamma^l_{ij}(2H^2KL^{ij} + HKg^{ij} - K^2L^{ij})]\psi^2\psi_{,l} \\
&+ (8H^5 - 12H^3K + 4HK^2)\psi^3\}\mathrm{d}A. \qquad (3.74)
\end{aligned}
$$

For deformed spherical vesicles, if it is limited to ellipsoids of revolution only, ψ may be taken as

$$
\psi = R[a_0 + a_2 P_2(\cos\theta)] \quad \text{with} \quad P_2(\cos\theta) = \frac{1}{2}(3\cos^2\theta - 1). \qquad (3.75)
$$

Up to $O(\psi^3)$, in other words, up to $O(a_2^3)$, Eqs. (3.34), (3.37), (3.44) and (3.52) now become

$$
\begin{aligned}
\delta A &= 4\pi R^2(2a_0 + a_0^2 + 4a_2^2/5), \\
\delta V &= 4\pi R^3(a_0 + a_0^2 + a_2^2/5 + a_0^3/3 + a_0a_2^2/5 + 2a_2^3/105), \\
\delta^{(1)}F_c &= 4\pi k_c(c_0^2R^2 - 2c_0R)a_0, \\
\delta^{(2)}F_c &= 4\pi k_c[c_0^2R^2a_0^2/2 + (12 - 6c_0R + 2c_0^2R^2)a_2^2/5],
\end{aligned} \qquad (3.76)
$$

respectively. Similarly, Eqs. (3.73) and (3.74) now lead to

$$
\begin{aligned}
\delta^{(3)} \oint H\mathrm{d}A &= 4\pi R\left(\frac{3}{5}a_0a_2^2 - \frac{6}{35}a_2^3\right), \\
\delta^{(3)} \oint H^2\mathrm{d}A &= -4\pi\left(\frac{12}{5}a_0a_2^2 + \frac{12}{35}a_2^3\right).
\end{aligned} \qquad (3.77)
$$

Therefore, up to $O(\psi^3)$,

$$\delta^{(3)}F_c = 4\pi k_c[(6c_0R/5 - 24/5)a_0a_2^2 - (12c_0R/35 + 24/35)a_2^3],$$

$$\delta F_c = \delta^{(1)}F_c + \delta^{(2)}F_c + \delta^{(3)}F_c$$

$$= 4\pi k_c[(c_0^2R^2 - 2c_0R)a_0 + c_0^2R^2a_0^2/2$$

$$+ (12/5 - 6c_0R/5 + 2c_0^2R^2/5)a_2^2$$

$$+ (6c_0R/5 - 24/5)a_0a_2^2 - (12c_0R/35 + 24/35)a_2^3]. \tag{3.78}$$

The conservation of area, i.e., $\delta A = 0$, up to $O(a_2^3)$ gives $a_0 = -(2/5)a_2^2$ and leads to the condition

$$a_2^3 - \frac{21}{2}a_2^2 - \frac{35}{2}\eta = 0, \tag{3.79}$$

in which $\eta = \delta V/(4\pi R^3/3)$, the relative volume variation, is very small and is negative in value. In such a case, up to $O(a_2^3)$, Eq. (3.78) reduces to

$$\delta F_c = -4\pi k_c[(24 + 20c_0R)a_2^2/5 + 6(c_0R + 2)\eta]. \tag{3.80}$$

Let $a_2 = 7(b + 1/2)$, Eq. (3.79) reduces to the standard cubic equation

$$b^3 - \frac{3}{4}b - \frac{1}{4}\left(1 + \frac{10}{49}\eta\right) = 0.$$

In order to find the solution of the equation, it is convenient to let

$$1 + \frac{10}{49}\eta = \cos 3\theta_0, \tag{3.81}$$

where θ_0 is a very small angle (nearly equal to zero). Cardan's formula gives to three roots of b as $A^{1/3} + B^{1/3}$, $(1/2)[(-1 + i\sqrt{3})A^{1/3} + (-1 - i\sqrt{3})B^{1/3}]$, and $(1/2)[(-1 - i\sqrt{3})^3A^{1/3} + (-1 + i\sqrt{3})^3B^{1/3}]$, where

$$A = \frac{1}{8}\cos 3\theta_0 + \left[\frac{1}{64}\cos^2 3\theta_0 - \frac{1}{64}\right]^{1/2} = \frac{1}{8}e^{i3\theta_0},$$

and

$$B = \frac{1}{8}\cos 3\theta_0 - \left[\frac{1}{64}\cos^2 3\theta_0 - \frac{1}{64}\right]^{1/2} = \frac{1}{8}e^{-i3\theta_0},$$

Therefore, the three roots of b are $\cos\theta_0$, $\cos(\theta_0 + 4\pi/3)$ and $\cos(\theta_0 + 2\pi/3)$, respectively. It follows that the three roots of Eq. (3.79) are $7(\cos\theta_0 + 1/2)$, $a_2^+ = 7[\cos(\theta_0 + 4\pi/3) + 1/2] > 0$ and $a_2^- = 7[\cos(\theta_0 + 2\pi/3) + 1/2] < 0$, respectively, and that

$$(a_2^+)^2 - (a_2^-)^2 = 49\sqrt{3}\sin\theta_0(1 - \cos\theta_0) > 0. \tag{3.82}$$

Since θ_0 is nearly equal to zero, therefore both a_2^+ and a_2^- are nearly equal to zero. The solution a_2^+ leads to prolate ellipsoid while a_2^- leads to oblate ellipsoid. On the other hand, the solution $7(\cos\theta_0 + 1/2) \simeq 8.5$ is not a small quantity and does not lead to a small distortion function. Equations (3.80) and (3.81) imply that for fixed δV, i.e., fixed η, whenever

$$c_0 R < -1.2 \tag{3.83}$$

the oblate shape has lower curvature-elastic energy and is more stable than the prolate ones. This result agrees with the stability analysis of Peterson[13] and Milner and Safran.[14]

In case of $\delta A = 0$, i.e., let $a_0 = -(2/5)a_2^2$, up to $O(a_2^3)$, Eq. (3.76) gives the variation of volume

$$\delta V = \delta V_A = \pi R^3 \left(-\frac{4}{5}a_2^2 + \frac{8}{105}a_2^3 \right). \tag{3.84}$$

The variation of the curvature-elastic energy δF_c, Eq. (3.80), can be written in the form

$$\delta F_c = -32\pi k_c(c_0 R + 1.2)a_2^3/21 - 2k_c(6 - c_0 R)\delta V_A/R^3. \tag{3.85}$$

With Eq. (3.85), the variation of the total energy $\delta F = \delta F_c + \Delta p\delta V$ can be written in the form

$$\delta F = [\Delta p - 2k_c R^{-3}(6 - c_0 R)]\delta V_A - 32\pi k_c(c_0 R + 1.2)a_2^3/21. \tag{3.86}$$

The leading term, i.e., the a_2^2 term in δV_A in Eq. (3.86) indicates that $\delta F \leq 0$ for $\Delta p \geq 2k_c R^{-3}(6 - c_0 R)$. In other words, the sphere begins to become unstable when $\Delta p \geq 2k_c R^{-3}(6 - c_0 R)$. The last term in Eq. (3.86) shows that for infinitesimal $\Delta p - 2k_c R^{-3}(6 - c_0 R)$, the oblate ellipsoids (a_2^-) are again stable for $c_0 R < 1.2$. Since δV_A contains a_2^2 term, the spontaneous curvature below which oblate ellipsoids are stable depends on the value of Δp.

3.5   Circular Cylindrical Vesicles

A circular cylinder may be described by

$$\mathbf{Y} = (R\cos\theta, R\sin\theta, z), \quad 0 \leq \theta \leq 2\pi, \ 0 \leq z \leq L. \tag{3.87}$$

It follows that

$$\mathbf{n} = (\cos\theta, \sin\theta, 0), \tag{3.88}$$

and the various non-vanishing basic terms are

$$g_{\theta\theta} = R^2, \quad g_{zz} = 1, \quad g = R^2, \quad g^{\theta\theta} = 1/R^2, \quad g^{zz} = 1,$$
$$L_{\theta\theta} = -R. \quad L = 0, \quad H = -1/(2R), \tag{3.89}$$

together with $K = 0$, $\Gamma^k_{ij} = 0$, $g^{ij}\nabla_i\psi_{,j} = \nabla^2\psi = (\partial^2/(R^2\partial\theta^2) + \partial^2/\partial z^2)\psi$. The circular cylinder, Eq. (3.87), is a solution of the shape (3.46), if the following relation is satisfied

$$\Delta p R^3 + \lambda R^2 + \frac{1}{2}k_c(c_0^2 R^2 - 1) = 0. \tag{3.90}$$

The circular cylinder may also be regarded as a cylinder closed by two nearly hemispherical caps at both ends. For $L \gg R$, the influence of the two ends may be neglected, and the shape energy, Eq. (3.38), becomes

$$F = \left[\frac{1}{2}k_c\left(c_0 - \frac{1}{R}\right)^2 + \lambda\right]2\pi R L + \Delta p\pi R^2 L.$$

Demanding $\delta F/\delta L = 0$, one gets another equilibrium condition

$$\Delta p R^3 + 2\lambda R^2 + k_c(c_0 R - 1)^2 = 0. \tag{3.91}$$

Equations (3.90) and (3.91) show that, for circular cylindrical vesicles in equilibrium, both ΔP and λ are fixed at

$$\Delta p = 2k_c(1 - c_0 R)/R^3,$$
$$\lambda = k_c(3 - c_0 R)(c_0 R - 1)/(2R^2), \tag{3.92}$$

while one of them can be chosen freely for spherical vesicles.

In order to study the stability of cylindrical vesicles, the ψ function of the slightly distorted cylinder may be expressed by

$$\psi = \sum_{m,n} b_{mn}\exp[i(m\theta + 2n\pi z/L)] = \sum_{m,n} b^*_{mn}\exp[-i(m\theta + 2n\pi z/L)]. \tag{3.93}$$

Equation (3.89) gives that

$$\oint \psi g^{ij}\nabla_i\psi_{,j}g^{1/2}dudv = \oint \psi\nabla^2\psi g^{1/2}dudv$$
$$= -2\pi R L\sum_{m,n}(m^2 + n^2 q^2)|b_{mn}|^2/R^2, \tag{3.94}$$

with $q = 2\pi R/L$. The variations of A, V, F_c given by Eqs. (3.34), (3.37), (3.44), and (3.50) become

$$\delta A = 2\pi RL(b_{00}/R) + \pi RL \sum_{m,n}(m^2 + n^2q^2)|b_{mn}|^2 R^{-2} + O(b_{mn}^3),$$

$$\delta V = 2\pi R^2 L(b_{00}/R) + \pi R^2 L \sum_{m,n}|b_{mn}|^2 R^{-2} + O(b_{mn}^3),$$

$$\delta F_c = k_c \pi L R^{-1}(c_0^2 R^2 - 1)(b_{00}/R)$$
$$+ k_c \pi L R^{-1} \sum_{m,n}\{[(1/2)(c_0^2 R^2 - 1) - 2c_0 R](m^2 + n^2q^2)$$
$$+ 2c_0 R m^2 + (m^2 + n^2q^2)^2 + 1 - 2m^2\}|b_{mn}|^2/R^2 + O(b_{mn}^2),$$

$$\tag{3.95}$$

respectively. Accordingly, the variation of the total energy is given by

$$\delta F = k_c \pi R^{-1} L \sum_{m,n}[2(c_0 R - 1)(m^2 - 1) + (m^2 + n^2q^2 - 1)^2]|b_{mn}|^2/R^2$$

$$+ O(b_{mn}^2),$$

$$\tag{3.96}$$

where Δp and λ are eliminated with Eq. (3.92). In case the deformation has rotational symmetry, $m = 0$ and δF reduces to

$$\delta F = k_c \pi R^{-1} L \sum_{n}[-2(c_0 R - 1) + (n^2q^2 - 1)^2]|b_{0n}|^2/R^2.$$

The necessary condition for the distorted cylinder being stable is then $c_0 R > 1$, and the nth mode ($n \geq 1$) is unstable if $n^2q^2 = (2\pi nR/L)^2 \leq 1 + (2c_0 R - 2)^{1/2}$. For an infinitely long cylinder, the least stable mode is characterized by $c_0 R = 1$ and $n^2q^2 = 1$, i.e., the period of the distortion along the z-axis, T, is equal to $T = L/n = 2\pi R$. The four rotational symmetric myelin forms calculated by Deuling and Helfrich[15] agree well with the above prediction.[9] In case of $m = 1$, Eq. (3.96) reduces to

$$\delta F = k_c \pi R^{-1} L \sum_{n} n^4 q^4 |b_{1n}|^2/R^2.$$

Here, $n = 0$ refers to a simple sideways translation of the cylinder requiring no energy. For $n > 0$, the modes represent bending of the tube and the energy of tube bending is always positive since δF now does not depend on c_0. In case of $m > 1$, the circular cylinder can be destabilized by negative spontaneous curvature. The simplest case of $n = 0$ means uniform deformation along the tube. The circular cylinder becomes unstable at $c_0 R \leq -1/2$, its cross-section turning into an ellipsoid ($m = 2$). In the absence of reservoirs, this is accompanied by a decrease of its length. As c_0 becomes more and more negative, the cylinder may be expected to transform into a tape.

3.6   Noncircular Cylindrical Vesicles

Cylindrical structures are often present in cell membranes.[16] For example, mitochondrial cristae are often tubular and form more or less hexagonal packing. There are also prismatic cristae with triangular cross-sections.[17] On the theoretical side, Seifert has made calculations on noncircular cylindrical surfaces to study the adhesion of vesicles in two dimensions.[18] Recently, general cylindrical surface solutions to the Helfrich variation problem were also fully investigated.[19-21] A cylinder along the y-axis may be represented by the equation of the curve of the cross-section

$$z = f(x). \tag{3.97a}$$

With $\psi = \psi(x)$ as the angle between the x-axis and the tangent line of the curve at any point x on the curve, the line element of the curve, $\mathrm{d}l$, and the space line element on the cylindrical surface, $\mathrm{d}s$, satisfy the relations:

$$\mathrm{d}x = \cos\psi \mathrm{d}l, \quad \mathrm{d}z = \sin\psi \mathrm{d}l,$$
$$\mathrm{d}s^2 = \mathrm{d}x^2 + \mathrm{d}y^2 + \mathrm{d}z^2 = \mathrm{d}l^2 + \mathrm{d}y^2, \quad \psi = \psi(l). \tag{3.97b}$$

With l and y as the two independent parameters, it shows that $g_{ll} = g_{yy} = g = 1$, $\mathbf{n} = (-\sin\psi, 0, \cos\psi)$, $L_{ll} = \mathrm{d}\psi/\mathrm{d}l = \cos\psi \mathrm{d}\psi/\mathrm{d}x$, $L_{yy} = L_{ly} = L = 0$, $2H = -(c_1 + c_2) = \mathrm{d}\psi/\mathrm{d}l = \cos\psi \mathrm{d}\psi/\mathrm{d}l$ and $K = 0$. The two principal curvatures are $-c_1 = t = \mathrm{d}\psi/\mathrm{d}l$ and $-c_2 = 0$. Equation (3.43) gives that

$$2\nabla^2 H = \cos\psi \left[\cos^2\psi \frac{\mathrm{d}^3\psi}{\mathrm{d}x^3} - 2\sin 2\psi \frac{\mathrm{d}\psi}{\mathrm{d}x}\frac{\mathrm{d}^2\psi}{\mathrm{d}x^2} - \cos 2\psi \left(\frac{\mathrm{d}\psi}{\mathrm{d}x}\right)^3\right].$$

The general shape Eq. (3.46) now becomes

$$\frac{\mathrm{d}^3\psi}{\mathrm{d}l^3} + \frac{1}{2}\left(\frac{\mathrm{d}\psi}{\mathrm{d}l}\right)^3 - \bar{\lambda}\frac{\mathrm{d}\psi}{\mathrm{d}l} + \overline{\Delta p} = 0 \tag{3.98a}$$

or

$$\cos^3\psi\frac{d^3\psi}{dl^3} - 2\sin 2\psi\cos\psi\frac{d\psi}{dx}\frac{d^2\psi}{dx^2} - \cos\psi\left(\frac{1}{2}\cos^2\psi - \sin^2\psi\right)\left(\frac{d\psi}{dx}\right)^3$$

$$-\bar\lambda\cos\psi\frac{d\psi}{dx} + \overline{\Delta p} = 0, \qquad (3.98b)$$

where

$$\bar\lambda = \lambda/k_c + c_0^2/2 \quad \text{and} \quad \overline{\Delta p} = \Delta p/k_c. \qquad (3.98c)$$

The contour of the cylinder is described by $z = \int\tan\psi(x)dx$. In terms of the curvature

$$t = -c_1 = \frac{d\psi}{dl} = \cos\psi\frac{d\psi}{dx}, \qquad (3.99)$$

and with the transformation $d/dl = (d\psi/dl)d/d\psi = td/d\psi$ or $d/dx = (d\psi/dx)d/d\psi$, Eq. (3.98a) becomes

$$t^2\frac{d^2t}{d\psi^2} + t\left(\frac{dt}{d\psi}\right)^2 + \frac{1}{2}t^3 - \bar\lambda t - \overline{\Delta p} = 0.$$

The transformation $d/d\psi = (dt/d\psi)d/dt$ changes it into

$$\frac{d}{dt}\left(t\frac{dt}{d\psi}\right)^2 + t^3 - 2\bar\lambda t - 2\overline{\Delta p} = 0.$$

The first integral of this equation is

$$\frac{dt}{d\psi} = \left(\frac{-t^2}{4} + \bar\lambda + \frac{2\overline{\Delta p}}{t} + \frac{C}{4t^2}\right)^{1/2}, \qquad (3.100)$$

where $C/4$ is the constant of integration. With Eq. (3.99), $d\psi = tdl$, integration of Eq. (3.100) gives the elliptic integral[22,23]

$$\int_0^\Lambda (-t^4 + 4\bar\lambda t^2 + 8\overline{\Delta p}t + C)^{-1/2}dt = \int_0^l \frac{1}{2}dl. \qquad (3.101)$$

The curvature t of the curve can then be expressed by

$$t = d\psi/dl = \Lambda(l/2). \qquad (3.102)$$

Elliptic functions are doubly periodic (one real and one imaginary) and analytic except for poles. For vesicles, only real periodic functions need to be considered. Since the curvature t is a periodic function of the arc length l of the curve, it is only necessary to consider one period. The angle ψ_0 turned in one period is given by

$$\psi_0 = \int_0^T t\,dl, \tag{3.103}$$

where T is the period of the elliptic function $\Lambda(l/2)$. For given $\bar{\lambda}$ and $\overline{\Delta p}$, ψ_0 may be considered as a function of C. The expression of the cross sectional curve can be obtained from Eqs. (3.97b) and (3.102). The curve is a closed curve only for those values of C for which $\psi_0(C)$ satisfies certain conditions.

The quartic function

$$P(t) = C - t(t^3 - 4\bar{\lambda}t - 8\overline{\Delta p}) \tag{3.104}$$

in the integrand of Eq. (3.101) depends on the constants $\bar{\lambda}$ and $\overline{\Delta p}$. The constant of integration C corresponds to a shift of the origin of the $t - P$ coordinate system along the P-axis. The extreme values of $P(t)$ are determined by the condition $dP(t)/dt = 0$, i.e., by the equation

$$t^3 - 2\bar{\lambda}t - 2\overline{\Delta p} = 0, \tag{3.105}$$

which is independent of C. For vesicles, both λ and k_c are positive, which makes $\bar{\lambda} > 0$. The constant $\overline{\Delta p}$ may be either positive or negative. Both Eqs. (3.104) and (3.105) are invariant under simultaneous changes of the sign of t and $\overline{\Delta p}$. This shows that shapes of the curve $P(t)$ under $\overline{\Delta p}$ and under $-\overline{\Delta p}$ are symmetric. They are simply mirror images of each other. Equation (3.105) belongs to the standard type of cubic equation of the form $x^3 + px + g = 0$ with determinant $\triangle = q^2/4 + p^3/27$. The theory of equations shows that in the case of $\triangle < 0$, i.e., $q^2/4 < -p^3/27$, the equation has three distinct roots $x_1 = 2r^{1/3}\cos(\theta/3), x_2 = 2r^{1/3}\cos[(\theta + 2\pi)/3]$ and $x_3 = 2r^{1/3}\cos[(\theta + 4\pi)/3]$, where $r = (-p^3/27)^{1/2}$ and $\cos\theta = (-q/2)(-27/p^3)^{1/2}$. In the case of $\triangle = 0$, i.e., $q^2/4 = -p^3/27$, the standard equation has one real root $x_1 = 2(-q/2)^{1/3}$ and a double root $x_2 = x_3 = -(-q/2)^{1/3}$. In the case of $\triangle > 0$, i.e., $q^2/4 > -p^3/27$, the equation has one reel root $x_1 = (-q/2+\sqrt{\triangle})^{1/3}+(-q/2-\sqrt{\triangle})^{1/3}$ and a pair of complex conjugate roots with no physical significance in the problem of the shape of vesicles. For Eq. (3.105), the discriminant is equal to $\triangle_1 = \overline{\Delta p}^2 - 8\bar{\lambda}^3/27$. The different cases will be discussed separately in detail.

Case I: $\triangle_1 < 0, \overline{\Delta p}^2 < (2\bar{\lambda}/3)^3$

The corresponding expressions of r and θ are given by $r = (2\bar{\lambda}/3)^{3/2}$ and $\cos\theta = (3/2\bar{\lambda})^{3/2}\overline{\Delta p}$, respectively. With $\overline{\Delta p}^2 < (2\bar{\lambda}/3)^{1/3}$, for $\overline{\Delta p} > 0$, the value of $\cos\theta$ is in the range $(0,1)$. Correspondingly, θ is in the range of $(0, \pi/2)$. The three values $\cos(\theta/3), \cos[(\theta + 2\pi)/3]$ and $\cos[(\theta + 4\pi)/3]$ are in the ranges $(\sqrt{3}/2, 1), (-\sqrt{3}/2, -1/2)$ and $(-1/2, 0)$, respectively. Consequently, one of the roots is positive and the other two are negative. They may be denoted as t_a, t_b and t_c in the order of $t_a > 0 > t_b > t_c$ with $t_a > |t_c|$, where $|t_c|$ is the absolute value of t_c. Similarly, for $\overline{\Delta p} < 0$, $\cos\theta$ and θ are in the ranges $(-1,0)$ and $(\pi/2, \pi)$, respectively. It follows that the three quantities $\cos(\theta/3), \cos[(\theta + 2\pi)/3]$ and $\cos[(\theta + 4\pi)/3]$ are in the ranges $(1/2, \sqrt{3}/2), (-1, -\sqrt{3}/2)$ and $(0, 1/2)$, respectively. These facts lead to $t_a > t_b > 0 > t_c$ and $|t_c| > t_a$. With $\mathrm{d}^2 P(t)/\mathrm{d}t^2 = -4(3t^2 - 2\bar{\lambda})$, it is not difficult to show that $P(t_a)$ and $P(t_c)$ are the two maxima of $P(t)$ and the minimum $P(t_b)$. Furthermore, from Eqs. (3.104) and (3.105), it can be shown that $P(t_a) - P(t_c) = (t_a + |t_c|)[(t_a - |t_c|)^3 + \overline{\Delta p}]$. This relation shows that for $\overline{\Delta p} > 0$, $P(t_a) > P(t_c)$ and for $\overline{\Delta p} < 0$, $P(t_c) > P(t_a)$. These results confirm again the fact that curves $P(t)$ with $\overline{\Delta p} > 0$ and with $\overline{\Delta p} < 0$ are mirror images of each other. For simplicity, in the following discussions, unless stated otherwise, the condition $\overline{\Delta p} > 0$ will be used and the constant C_i $(i = a, b, c, j)$ represents

$$-C_i = t_i^4 - 4\bar{\lambda}t_i^2 - 8\overline{\Delta p}t_i, \quad (i = a, b, c, j). \qquad (3.106)$$

Figure 3.2 gives a schematic sketch of the curve $P(t)$ and Fig. 3.3 gives plots of $P(t) - C$ with $C = 0$ against t with $\bar{\lambda} = 1.0k_c$ and (a) $\overline{\Delta p} = 0.3k_c$, (b) $\overline{\Delta p} = -0.3k_c$, (c) $\overline{\Delta p} = p_c = (2/3)^{3/2}k_c$, (d) $\overline{\Delta p} = 0$. Figure 3.3 indicates that any line $P(t) = P(t_j) = \text{const}$, where t_j is a constant, intersects the $P(t)$ curve at most at four points α, β, γ and δ with t values $t_\alpha > t_\beta > t_\gamma > t_\delta$. In other words, the solutions of the equation

$$P(t) - P(t_j) = -(t^4 - t_j^4) + 4\bar{\lambda}(t^2 - t_j^2) + 8\overline{\Delta p}(t - t_j) = 0 \qquad (3.107)$$

are $t_\alpha, t_\beta, t_\gamma$ and t_δ. The intersections depend upon the value of C. Table 3.1 summarizes the different situations in case of $\overline{\Delta p} > 0$. For the case of $\overline{\Delta p} < 0$, it is simply a change of t with $-t$.

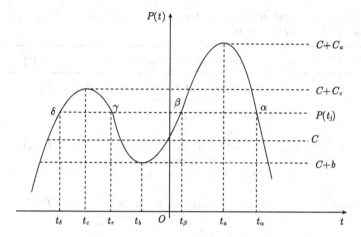

Fig. 3.2.   Schematic illustration of the $P(t)$ curve.

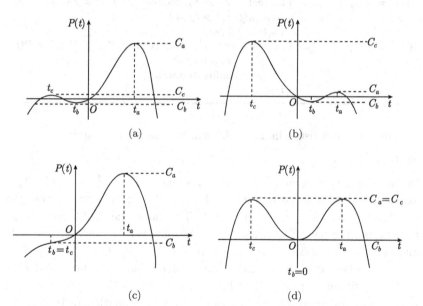

Fig. 3.3.   Schematic illustration of the $P(t) - C$ curve with $C = 0$ and $\bar{\lambda} = 1$, (a) $\overline{\Delta p} = 0.3k_c$, (b) $\overline{\Delta p} = -0.3k_c$, (c) $\overline{\Delta p} = p_c = (2/3)^{3/2}k_c$, (d) $\overline{\Delta p} = 0$.

Table 3.1.   The intersections of the line $P(t) = P(t_j)$ and the $P(t)$ curve (i.e., the roots of $P(t) - P(t_j) = 0$).

C	Intersections (Roots)	$P(t)$
(1) $C = -C_a$	One double root $t_\alpha = t_\beta(= t_a)$ and a pair of complex conjugate roots t_γ and $t_\delta = t_\gamma^*$	$P(t) = P(t_a)$
(2) $-C_c > C > -C_a$	Two real roots $t_\alpha(> t_a) > t_\beta$ and a pair of complex conjugate roots t_γ and $t_\delta = t_\gamma^*$	$P(t) = P(t_\alpha) = P(t_\beta)$
(3) $C = -C_c$	Two real roots and a double root $t_\alpha(> t_a) > t_\beta(> t_b) > t_\gamma = t_\delta(= t_c)$	$P(t) = P(t_c)$
(4) $-C_b > C > -C_c$	Four real roots $t_\alpha(> t_a) > t_\beta(> t_b) > t_\gamma(> t_c) > t_\delta$	$P(t) = P(t_\alpha) = P(t_\beta)$ $= P(t_\gamma) = P(t_\delta)$
(5) $C = -C_b$	Two real roots and a double root $t_\alpha(> t_a) > t_\beta = t_\gamma$ $(= t_b > t_c) > t_\delta$	$P(t) = P(t_b)$
(6) $C > -C_b$	Two real roots $t_\alpha(> t_a > t_b > t_c) > t_\delta$ and a pair of complex conjugate roots t_β and $t_\gamma = t_\beta^*$	$P(t) = P(t_\alpha) = P(t_\beta)$ $= P(t_\delta)$

The six states listed in Table 3.1 will be studied separately.

State (1): $C = -C_a$

In this state, $t_j = t_a$ and Eq. (3.107) becomes $-(t^4 - t_a^4) + 4\bar{\lambda}(t^2 - t_a^2) + 8\overline{\Delta p}(t - t_a) = 0$. Since t_a satisfies Eq. (3.105), it is easy to show that $(t - t_a)(t_a - t)[(t + t_a)^2 + 4\overline{\Delta p}/t_a] = 0$. Therefore, the double root is $t_\alpha = t_\beta = t_a$ and the pair of complex conjugate roots is $t_\gamma = -t_a + 2i(\overline{\Delta p}/t_a)^{1/2}$ and $t_\delta = -t_a - 2i(\overline{\Delta p}/t_a)^{1/2}$. The real root $t_a = 2(2\bar{\lambda}/3)^{1/2}\cos(\theta/3)$ is a constant, where $\cos\theta = [3/(2\bar{\lambda})]^{3/2}\overline{\Delta p}$. Correspondingly, the cylinder is a circular cylinder of radius $R = 1/t_a$. Integration of Eq. (3.99) gives simply the circumference of the cylinder $L = 2\pi R$. Equation (3.105) is simply Eq. (3.90), the constraint for the circular cylinder. If t_γ and t_δ are taken into consideration, Eq. (3.101) does not lead to an elliptic integral and the angle ψ_0 turned in one period is not necessarily periodic. However, in case of $\overline{\Delta p} = 0$, Eq. (3.105) shows that $2\bar{\lambda} = t_\alpha^2$. It follows that Eq. (3.106) becomes $-C_a = t_a^4 - 4\bar{\lambda}t_a^2 = -t_a^4$ and Eq. (3.104) becomes $P(t) = -(t_a^2 - t^2)^2$. Equation (3.101) now gives $\Lambda = it_a\tan(t_aL/2)$ and

Eq. (3.103) leads to $\psi_0 = -2\mathrm{i} \ln \cos(t_a L/2)$, where L is the circumference of the cylindrical cross-section.[24] The expression of ψ_0 may be written as $\cos(t_a L/2) = \cos(\psi_0/2) + \mathrm{i} \sin(\psi_0/2)$. The requirement that $\cos(t_a L/2)$ should be real shows that ψ_0 should be equal to $2n\pi$, where n is any integer. It follows that $t_a L = 2n\pi$ as it should be.

State (2): $-C_c > C > -C_a$

In this state, Eq. (3.107) becomes $(t_\alpha - t)(t - t_\beta)(t - t_\gamma)(t - t_\gamma^*) = 0$, where $t_\alpha > t > t_\beta$. Under the transformation[25] $\tau = [(t_\alpha - t)B - (t - t_\beta)A]/[(t_\alpha - t)B + (t - t_\beta)A]$, Eq. (3.101) becomes

$$\int_{t_\beta}^t [(t_\alpha - t)(t - t_\beta)(t - t_\gamma)(t - t_\gamma^*)]^{-1/2}\mathrm{d}t$$

$$= g \int_\tau^1 [(1 - \tau^2)(k'^2 + k^2\tau^2)]^{-1/2}\mathrm{d}\tau = \frac{l}{2}, \qquad (3.108)$$

where

$$g = (AB)^{-1/2}, \quad k^2 = [(t_\alpha - t_\beta)^2 - (A - B)^2]/4AB, \quad k'^2 = 1 - k^2,$$

$$A^2 = (t_\alpha - b_1)^2 + a_1^2, \quad B^2 = (t_\beta - b_1)^2 + a_1^2,$$

$$a_1^2 = -(t_\gamma - t_\gamma^*)^2/4, \quad b_1 = (t_\gamma + t_\gamma^*)^2/2.$$

By definition, $\tau = \mathrm{cn}(l/(2g), k)$. It follows that

$$t = [t_\alpha B + t_\beta A - (t_\alpha B - t_\beta A)\mathrm{cn}(l/(2g), k)]/$$

$$[(A + B) + (A - B)\mathrm{cn}(l/(2g), k)]. \qquad (3.109)$$

The function $\mathrm{cn}(l/(2g), k)$ has a period of $4K(k)$, where $K(k)$ is the complete elliptic integral of the first kind. It follows that the period of t is also equal to $4K(k)$ and the angle $\psi_0(C)$ turned in one period is now given by

$$\psi_0(C) = \int_0^{4K(k)} 2gt\mathrm{d}L$$

$$= 2g \int_0^{4K(k)} \{-(t_\alpha B - t_\beta A)/(A - B)$$

$$+ [2AB/(t_\alpha + t_\beta - 2b_1)]/[1 + (A - B)\mathrm{cn}(L, k)/(A + B)]\}\mathrm{d}L,$$

where $L = l/(2g)$. The second integral may be converted into elliptic integral of the third kind $\Pi(\phi, \alpha, k)$ by the relation[25]

$$\int \frac{\mathrm{d}u}{1 + \beta\mathrm{cn}u}$$

$$= (1 - \beta^2)^{-1} \left[\Pi\left(\phi, \frac{\beta^2}{(\beta^2 - 1)}, k \right) - \beta f_1 \right], \quad \phi = \mathrm{am}u, \quad \beta^2 \neq 1,$$

where

$$f_1 = (1 - \beta^2)^{1/2}[\beta^2 + k^2(1 - \beta^2)]^{-1/2}$$

$$\cdot \arctan\{(1 - \beta^2)^{-1/2}[\beta^2 + k^2(1 - \beta^2)^{1/2}]sdu\},$$

if $\beta^2/(\beta^2 - 1) < k^2$. In the present case, $1 > \beta = (A - B)/(A + B) \geq 0$. With $sd4K(k) = sd0 = 0$, f_1 is equal to $(1 - \beta^2)^{1/2}[\beta^2 + k^2(1 - \beta^2)]^{-1/2}p\pi$, where p is any integer. With Eq. (3.108), it can be shown that the angle $\psi_0(C)$ turned in one period is now given by

$$\psi_0(C) = 2g\left[\frac{4K(k)(t_\beta A - t_\alpha B)}{(A - B)} + \frac{2AB(1 - \alpha)}{t_\alpha + t_\beta - 2b_1}\Pi(2\pi, \alpha, k)\right] - 2p\pi,$$

$$(3.110)$$

where $\alpha = \beta^2/(\beta^2 - 1) = -(A - B)^2/4AB$. By definition,

$$\Pi(2\pi, \alpha, k) = \int_0^{2\pi} \frac{d\theta}{[(1 - \alpha\sin^2\theta)(1 - k^2\sin^2\theta)^{1/2}]}$$

$$= 2(1 - \alpha)^{-1}(1 - k^2)^{-1/2}\Pi[\pi/2, \alpha/(1 - \alpha), \alpha^2/(1 - k^2)]$$

$$+ 2\Pi(\pi/2, \alpha, k). \qquad (3.111)$$

The constant $2p\pi$ in Eq. (3.110) may be neglected simply by properly shifting the zero point of $\psi_0(C)$ and $\psi_0(C) \geq -\infty$.

When $C \to -C_a$, $t_\alpha = t_\beta = t_a$, $t_\gamma = -t_a + 2i(\overline{\Delta p}/t_a)^{1/2}$, it follows that $a_1^2 = 4\overline{\Delta p}/t_a$, $b_1 = -t_a$, $A^2 = B^2 = 4(t_a^2 + \overline{\Delta p}/t_a)$, $\alpha = \beta = k = 0$, $g = (1/2)(t_a^2 + \overline{\Delta p}/t_a)^{-1/2}$, and $K(0) = \pi/2$. With $t_\alpha = t_\beta$ and $A = B$, Eq. (3.109) shows that t is independent of $cn(l/(2g))$ and straightforward integration gives that

$$\psi_0 = 4gt_a\pi, \qquad (3.112)$$

the case of a circular cylinder as stated in state (1). In this range $-\infty < \psi_0 \leq 4\pi gt_a\pi$, only contours corresponding to values of C for which $\psi_0(C) = 2\pi - 2\pi/n$ are closed curves, where n is a positive integer. Since $t_a > 0$, the corresponding cross-section has n outward nodules as shown in Fig. 3.4. For each n, the curve has a n-fold axis of symmetry.

With $C \to -C_c$, $t_\gamma = t_\gamma^* = t_c$, $t_\alpha = 2(-\overline{\Delta p}/t_c)^{1/2} - t_c$, $t_\beta = -2(-\overline{\Delta p}/t_c)^{1/2} - t_c$, $a_1^2 = 0$, $b_1 = t_c$, $A = t_\alpha - t_c$, $B = t_\beta - t_c$, $g = [(t_\alpha - t_c)(t_\beta - t_c)]^{-1/2}$, $\alpha = -(t_\alpha - t_\beta)^2/[4(t_\alpha - t_c)(t_\beta - t_c)] < 1$, and $k = 0$. It follows that $K(k) = \pi/2$. With $\Pi(\phi, \alpha, 0) = (1 - \alpha)^{-1/2}\tan^{-1}[(1 - \alpha)^{1/2}\tan\phi]$ for

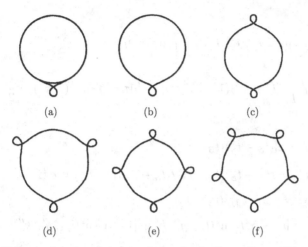

Fig. 3.4.   Shapes of cross section in the region $-C_b > C > -C_a$, $t > t_b$ with $\bar{\lambda} = 1$, $\overline{\Delta p} = 0.2$, (a) $C = -C_a$, (b) $n = 1$, (c) $n = 2$, (d) $n = 3$, (e) $n = 4$, (f) $n = 5$.

$\alpha < 1$, it gives that $\Pi(2\pi, \alpha, 0) = 0$. Aside from a constant $2p\pi$, Eq. (3.110) gives

$$\psi_0 = 4gt_c\pi, \tag{3.113}$$

the case of a circular cylinder of radius $R = 1/t_c$.

State (3): $C = -C_c$

In this state, Eq. (3.107) becomes $-(t^4 - t_c^4) + 4\bar{\lambda}(t^2 - t_c^2) + 8\overline{\Delta p}(t - t_c) = 0$, where $t_c < 0$ and $t_c^3 - 2\bar{\lambda}t_c - 2\overline{\Delta p} = 0$. The equation may then be written as $-(t - t_c)^2(t^2 + 2t_ct + 3t_c^2 - 4\bar{\lambda}) = 0$. It follows that $t_\alpha = -t_c + 2(-\overline{\Delta p}/t_c)^{1/2}$ and $t_\beta = -t_c - 2(-\overline{\Delta p}/t_c)^{1/2}$. The real double root $t_\gamma = t_\delta = t_c$ is equal to $2(2\bar{\lambda}/3)^{1/2}\cos[(\theta + 2\pi)/3]$ where $\cos\theta = [3/(2\bar{\lambda})]^{3/2}\overline{\Delta p}$. Corresponding to t_c, the cylinder is circular with radius $R = 1/t_c$ and circumference $L = 2\pi R$. When t_α and t_β are taken into consideration, similar to State (1), the expression of Λ is rather complicated. However, in case of $\overline{\Delta p} = 0$, Eq. (3.101) leads to $\Lambda = it_c\tan(Lt_c/2)$ and Eq. (3.103) gives $\psi_0 = -2i\ln\cos(Lt_c/2)$. The requirement that ψ_0 should be real leads to the obvious result $t_cL = 2\pi$.

State (4): $-C_b > C > -C_c$

In this state, there are two cases, i.e., $t_\alpha \geq t \geq t_\beta$ and $t_\gamma \geq t \geq t_\delta$. For $t_\alpha \geq t \geq t_\beta$, the transformation $\tau^2 = (t_\alpha - t_\gamma)(t - t_\beta)/(t_\alpha - t_\beta)(t - t_\gamma)$

leads to

$$\int_{t_\beta}^{t} [(t_\alpha - t)(t - t_\beta)(t - t_\gamma)(t - t_\delta)]^{-1/2}dt$$

$$= g' \int_0^\tau [(1 - \tau^2)(1 - k''^2\tau^2)]^{-1/2}d\tau = g'\mathrm{sn}^{-1}(\tau, k'') = \frac{l}{2}, \quad (3.114)$$

where

$$g' = 2[(t_\alpha - t_\gamma)(t_\beta - t_\delta)]^{-1/2},$$

$$k''^2 = (t_\alpha - t_\beta)(t_\gamma - t_\delta)/(t_\alpha - t_\gamma)(t_\beta - t_\delta) < \alpha^2,$$

$$\alpha^2 = (t_\alpha - t_\beta)/(t_\alpha - t_\gamma),$$

$$t = [t_\beta - \alpha^2 t_\gamma \mathrm{sn}^2(l/(2g'), k'')]/[1 - \alpha^2\mathrm{sn}^2(l/(2g'), k'')].$$

Since $\mathrm{sn}(u + 2K) = -\mathrm{sn}u$ and t depends only on $\mathrm{sn}^2 u$, the period of t may be taken as $2K(k)$ instead of $4K(k)$. With $L = l/(2g')$, the angle $\psi_0(C)$ turned in one period is now given by

$$\psi_0(C) = 2g' \int_0^{2K(k'')} \left[t_\gamma + \frac{t_\beta - t_\gamma}{1 - \alpha^2\mathrm{sn}^2(L, k'')} \right] dL$$

$$= 2g'[2K(k'')t_\gamma + (t_\beta - t_\gamma)\Pi(\pi, \alpha^2, k'')], \quad (3.115)$$

where $\Pi(\pi, \alpha^2, k'')$ may be expressed in terms of two $\Pi(\pi/2, \beta, k'')$ functions as shown in Eq. (3.111). When $C \to C_c$, $t_\gamma = t_\delta = t_c$, $\alpha^2 = (t_\alpha - t_\beta)/(t_\alpha - t_c)$, $k'' = 0$, it follows that $\Pi(\pi, \alpha^2, 0) = (1 - \alpha^2)^{-1/2}\tan^{-1}[(1 - \alpha^2)^{1/2}\tan\pi] = 0$ and $\psi_0 = 2g't_c\pi$ which is simply Eq. (3.113) since $g' = 2g$. As $t_c < 0$, it is only for those values of C for which $\psi_0(C) = -2\pi - 2\pi/n$, where n is a positive integer, that the curve is a closed curve with n inward nodules. When $C \to C_b$, the above transformation is inapplicable on account of $t_\beta = t_\gamma = t_b$. Another transformation $\tau^2 = (t_\beta - t_\delta)(t_\alpha - t)/(t_\alpha - t_\beta)(t - t_\delta)$ leads to

$$\int_t^{t_\alpha} [(t_\alpha - t)(t - t_\beta)(t - t_\gamma)(t - t_\delta)]^{-1/2}dt = 2g'\mathrm{sn}^{-1}\left(\frac{l}{2}g', k''\right) = \frac{l}{2}, \quad (3.116)$$

where $t = t_\delta + (t_\alpha - t_\delta)/[1 + \alpha'^2\mathrm{sn}^2(L, k'')]$, $\alpha'^2 = (t_\alpha - t_\beta)/(t_\beta - t_\delta)$, $L = l/(2g)'$. The angle ψ_0 turned in one period is equal to

$$\psi_0 = 2g' \int_0^{2K(k'')} t dL = 2g'[2K(k'')t_\delta + (t_\alpha - t_\delta)\Pi(\pi, -\alpha'^2, k'')].$$

In this case, since $k'' = 1$, $K(1) = \infty$, $\Pi(\pi, -\alpha'^2, 1) = \infty$ and $\text{sn}(L, 1) = \tanh L$, t is nonperiodic. As $L \to \infty$, $t \to t_\beta = t_b$, and the cylinder becomes circular with radius $R = 1/t_b$. The angle ψ_0 turned in one period becomes negative infinite since $t_b < 0$.

For $t_\gamma \geq t \geq t_\delta$, the transformation $\tau^2 = (t_\alpha - t_\gamma)(t - t_\delta)/(t_\gamma - t_\delta)$ $(t_\alpha - t)$ leads to

$$\int_{t_\delta}^{t} [(t_\alpha - t)(t_\beta - t)(t_\gamma - t)(t - t_\delta)]^{-1/2} dt = g' \text{sn}^{-1}\left(\frac{l}{2}g', k''\right) = \frac{l}{2} = g'L,$$

(3.117)

where $t = t_\alpha - (t_\alpha - t_\delta)/[1 + \alpha''^2 \text{sn}^2(L, k'')]$, $\alpha''^2 = (t_\gamma - t_\delta)/(t_\alpha - t_\gamma) < 1$. The angle ψ_0 turned in one period is now equal to[25]

$$\psi_0 = 2g' \int_0^{2K(k'')} t\, dL = 2g'[2K(k'')t_\alpha - (t_\alpha - t_\delta)\Pi(\pi, -\alpha''^2, k'')].$$

(3.118)

When $C \to -C_c$, $t_\gamma = t_\delta = t_c$, $k'' = 0$, $\alpha'' = 0$, $g' = (t_c^2 + \overline{\Delta p}/t_c)^{-1/2}$, it follows that $\psi_0 = 2\pi t_c/(t_c^2 + \overline{\Delta p}/t_c)^{1/2} = 2g't_c\pi$, where $t_c < 0$. Again, it agrees with Eq. (3.113). The cross-sections in this region have inward nodules as shown in Fig. 3.5. When $C \to -C_b$, $t_\beta = t_\gamma = t_b$, $k'' = 1$ and the transformation $\tau^2 = (t_\beta - t_\delta)(t_\gamma - t)/[(t_\gamma - t_\delta)(t_\beta - t)]$ leads to[25]

$$\int_{t}^{t_c} [(t_\alpha - t)(t_\beta - t)(t_\gamma - t)(t - t_\delta)]^{-1/2} dt = g' \text{sn}^{-1}\left(\frac{l}{2}g', k''\right) = \frac{l}{2} = g'L$$

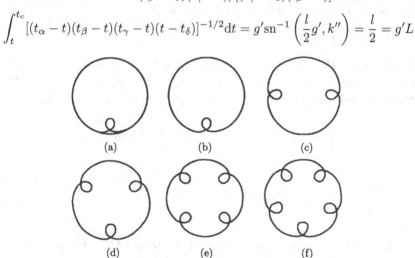

(a)   (b)   (c)

(d)   (e)   (f)

Fig. 3.5. Shapes of cross sections in the region $-C_b > C > -C_c$, $t < t_b$ with $\bar{\lambda} = 1$, $\overline{\Delta p} = 0.2$, (a) $C = -C_b$, (b) $n = 1$, (c) $n = 2$, (d) $n = 3$, (e) $n = 4$, (f) $n = 5$.

where $t = t_\beta - (t_\beta - t_\gamma)/[1 - \alpha'''^2 \mathrm{sn}^2(L, k'')]$, $\alpha'''^2 = (t_\beta - t_\gamma)/(t_\beta - t_\delta)$. In this case, $\alpha'''^2 = 0$ and $t = t_b$, the cylinder becomes circular with radius $R = 1/t_b$. The angle ψ_0 turned in one period is equal to

$$\psi_0 = 4g'K(k'')t_b.$$

Since $k'' = 1$, the angle ψ_0 becomes negative infinite.

State (5): $C = -C_b$

In this state, $t_\beta = t_\gamma = t_b$ and $-(t^4 - t_b^4) + 4\bar\lambda(t^2 - t_b^2) + 8\overline{\Delta p}(t - t_b) = 0$, where $t_b < 0$ and $t_b^3 - 2\bar\lambda t_b - 2\overline{\Delta p} = 0$. In case of $\overline{\Delta p} \neq 0$, the above equation of t may be written as $-(t - t_b)^2[t^2 + 2t_bt + (3t_b^2 - 4\bar\lambda)] = 0$. This gives that $t_\alpha = -t_b + 2\sqrt{-\overline{\Delta p}/t_b}$ and $t_\delta = -t_b - 2\sqrt{-\overline{\Delta p}/t_b}$. The double root t_b is equal to $2(2\bar\lambda/3)^{1/2}\cos[(\theta + 4\pi)/3]$, where $\cos\theta = [3/(2\bar\lambda)]^{3/2}\overline{\Delta p}$. Corresponding to t_b, the cylinder is circular with radius $R = 1/t_b$ and circumference $2\pi R$. When t_α and t_δ are taken into consideration, similar to States (1) and (3), the expression of Λ is rather complicated. It should be noticed that the above expressions for t_α and t_δ cannot be used in the case of $\overline{\Delta p} = 0$. When $\overline{\Delta p} = 0$, Eq. (3.105) shows that $t_b = 0$, $t_a = 2\sqrt{\lambda}$, consequently $t_\alpha = 2\sqrt{\lambda}$ and $t_\delta = -2\sqrt{\lambda}$. Here, $P(t) = t^2(4\bar\lambda - t^2)$ and Eq. (3.101) indicates that the curvature t is nonperiodic. More detailed discussions on this case will be given in Case IV.

State (6): $C > -C_b$

In this state, $P(t)$ has two real roots $t_\alpha(> t_a > t_b > t_c) > t_\delta$ and a pair of complex conjugate roots t_β and $t_\gamma = t_\beta^*$ with $t_\alpha \geq t \geq t_\delta$. The transformation $\tau = [(t_\alpha - t)B - (t - t_\delta)A]/[(t_\alpha - t)B + (t - t_\delta)A]$ leads to[25]

$$\int_{t_\delta}^{t} [(t_\alpha - t)(t - t_\beta)(t - t_\gamma)(t - t_\delta)]^{-1/2}\mathrm{d}t$$

$$= g\int_0^1 [(1 - \tau^2)(k'^2 + k^2\tau^2)]^{-1/2}\mathrm{d}\tau = \frac{l}{2}, \qquad (3.119)$$

where

$$g = (AB)^{-1/2}, \quad k^2 = [(t_\alpha - t_\delta)^2 - (A - B)^2]/4AB, \quad k'^2 = 1 - k^2,$$

$$A^2 = (t_\alpha - b_1)^2 + a_1^2, \quad B^2 = (t_\delta - b_1)^2 + a_1^2,$$

$$a_1^2 = -(t_\beta - t_\beta^*)^2/4, \quad b_1 = (t_\beta + t_\beta^*)/2.$$

It follows that $t = (t_\delta A - t_\alpha B)/(A - B) + [2AB/(t_\alpha + t_\delta - 2b_1)]/[1 + (A - B)\text{cn}(l/(2g), k)/(A + B)]$ and aside from a constant $2p\pi$, where p is an integer,

$$\psi_0 = 2g\{4K(k)(t_\delta A - t_\alpha B)/(A - B)$$
$$+ [(A + B)^2/(2t_\alpha + 2t_\delta - 4b_1)]\Pi(2\pi, \alpha, k)\}, \qquad (3.120)$$

where $L = l/(2g)$ and $\alpha = -(A - B)^2/(4AB)$. When $C \to -C_b$, $a_1^2 = 0$, $b_1 = t_b$, $A^2 = (t_\alpha - t_b)^2$, $B^2 = (t_b - t_\delta)^2$, $k^2 = 1$ and $\text{cn}(L, 1) = \text{sech}L$, t is non-periodic. With $L \to \infty$, $\text{cn}(L, 1) \to 0$ and $t = t_b$, the cylinder is circular with radius $R = 1/t_b$. Figure 3.6 shows some cross-sections in this state with $\bar\lambda = 1$ and $\overline{\Delta p} = 0.2$. Since $t_b < 0$, Eq. (3.103) shows that ψ_0 becomes negative infinite as $C \to -C_b$. On the other hand, as $C \to \infty$, $P(t)$ becomes infinite and $\psi_0 \to 0$. Numerical calculations indicate that somewhere between $C = -C_b$ and $C \to \infty$, there exists a value of $C = C_m$ for which ψ_0 reaches its maximum value ψ_m.

As a brief summary, for $\overline{\Delta p} > 0$, the space under the $P(t) - t$ curve may be divided into three different regions: (I) the region $-C_b \geq C \geq -C_a$ with $t \geq t_b$ (which is negative), (II) the region $-C_b \geq C \geq -C_c$ with $t_b \geq t$ and (III) the region $C > -C_b$. In case of $\bar\lambda = 1$ and $\overline{\Delta p} = 0.2$, the $\psi_0 - C$ curves in the three different regions are plotted in Fig. 3.7.

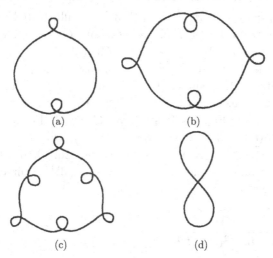

Fig. 3.6.   Shapes of cross-sections in the range $C > -C_b$ with $\bar\lambda = 1$, $\overline{\Delta p} = 0.2$, (a) $n = 1$, (b) $n = 2$, (c) $n = 3$, (d) $n = \infty$.

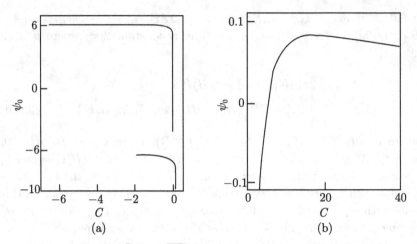

Fig. 3.7. $\psi_0(C)$-C for $\bar{\lambda} = 1$, $\overline{\Delta p} = 0.2$. (a) The upper curve corresponds to region (I) which tends to $4gt_a\pi(< 2\pi)$ as $C \to -C_a$, and $-\infty$ as $C \to -C_b$. The lower curve corresponds to region (II) which tends to $4gt_c\pi(< -2\pi)$ as $C \to -C_c$ and $-\infty$ as $C \to -C_b$. (b) The curve corresponds to region (III) which tends to $-\infty$ as $C \to -C_b$ and 0 as $C \to -\infty$.

Case II: $\triangle_1 = 0, \overline{\Delta p}^2 = (2\bar{\lambda}/3)^3$

In this case, Eq. (3.105) has a real root $t_a = 2(\overline{\Delta p})^{1/3} = 2(2\bar{\lambda}/3)^{1/2}$ and a double root $t_b = t_c = -\overline{\Delta p}^{1/3} = -(2\bar{\lambda}/3)^{1/2}$ as illustrated in Fig. 3.3(c). There are two regions $-C_b > C > -C_a$ and $C > -C_b$ with different states: (1) $C = -C_a$, (2) $-C_b > C > -C_a$, (3) $C = -C_b$ and (4) $C > -C_b$. State (1) is similar to the state (1) of case I, except with a different value of t_a. State (2) is similar to the state (2) of case I in the region $-C_b \geq C \geq -C_a$, except in the case of $C > -C_c$. The shapes of the cross-sections are shown in Fig. 3.8. In state (3), $C = -C_b$, Eq. (3.107) becomes
$$-(t^4 - t_b^4) + 4\bar{\lambda}(t^2 - t_b^2) + 8\overline{\Delta p}(t - t_b) = -(t - t_b)^2(t^2 + 2t_bt + 3t_b^2 - 4\bar{\lambda}) = (t + \sqrt{2\bar{\lambda}})^3(3\sqrt{2\bar{\lambda}/3} - t) = 0.$$
In other words, Eq. (3.107) has one root $t_\alpha = 3(2\bar{\lambda}/3)^{1/2}$ and a triple root $t = t_b = -(2\bar{\lambda}/3)^{1/2}$. Direct integration of Eq. (3.101) gives[24]

$$\int_0^\Lambda \frac{dt}{(t - t_b)^{3/2}(t_\alpha - t)^{1/2}}$$

$$= \frac{2}{t_\alpha - t_b}[(-t_\alpha/t_b)^{1/2} - (t_\alpha - \Lambda)^{1/2}(\Lambda - t_b)^{-1/2}] = \frac{l}{2},$$

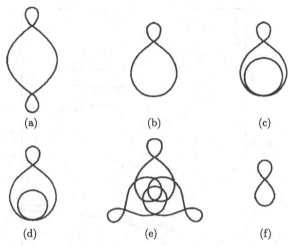

Fig. 3.8.   Shapes of cross-sections with $\bar{\lambda} = 1$ and $\overline{\Delta p} = p_c = (2/3)^{3/2}$. (a), (b) are shapes in region $-C_b > C > -C_a$, $t > t_b$, (c) on the boundary $C = -C_b$, and (d), (e), (f) in region $C > -C_b$.

i.e., $\Lambda = -(t_b^2 l/2)[l(t_\alpha - t_b)/8 - (-t_b/t_\alpha)^{1/2}]/\{1 - (l t_b/2)[l(t_\alpha - t_b)/8 - (-t_b/t_\alpha)^{1/2}]\}$. The curvature is non-periodic. As $l \to \infty, t = \Lambda \to t_b$. State (4) is similar to state (6) of case I.

Case III: $\triangle_1 > 0, \overline{\Delta p}^2 > (2\bar{\lambda}/3)^3$

In this case, Eq. (3.105) has only one real root. In other words, the $P(t)$ curve has only one maximum at $t_a = [\overline{\Delta p} + (\overline{\Delta p}^2 - 8\bar{\lambda}^3/27)^{1/2}]^{1/3} + [\overline{\Delta p} - (\overline{\Delta p}^2 - 8\bar{\lambda}^3/27)^{1/2}]^{1/3}$. When $C = -C_a$, the situation is just the same as stated in state (1) of case I. When $C > -C_a$, Eq. (3.107) has two real roots t_α and t_δ and a pair of complex conjugate roots t_β and $t_\gamma = t_\beta^*$. The expressions of the curvature t and the angle ψ_0 are the same as those given in Eq. (3.120). In order to compare with case II, the ψ_0 vs C curves in case of $\bar{\lambda} = 1$ and different values of $\overline{\Delta p}$ are given in Fig. 3.9.

For $\triangle_1 = 0, \psi_0(C)$ approaches $-\infty$ at $C_2 \approx -4/3$. $\psi_0(C)$ has two real roots C_{01} and C_{02} at different sides of $C = -C_b$. The corresponding shapes are Figs. 3.8(b) and 3.8(f), respectively. For $\overline{\Delta p} > (2/3)^{3/2}, \psi_0$ is a continuous function of C. When $\overline{\Delta p} = p_0 \approx 0.69, \psi_0(C) = 0$ has a double root C_0 such that Figs. 3.8(b) and 3.8(f) degenerate and all the five shapes between them vanish. For $\overline{\Delta p} > p_0, \psi_0$ is always positive (or equivalently negative). Only for those values of C for which $\psi_0 = 2\pi/n$, where n is a positive integer, the shape is a closed curve with a n-fold axis of symmetry.

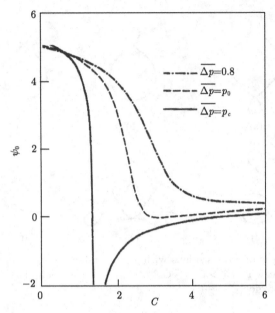

Fig. 3.9. $\psi_0(C)$-C for $\bar{\lambda} = 1$, $\overline{\Delta p} = 0.8$, $\overline{\Delta p} = p_0 \simeq 0.6987$ and $\overline{\Delta p} = p_c = (2/3)^{3/2}$.

However, for different values of $\overline{\Delta p}$, the shapes with the same $\bar{\lambda}$ and n are different. Figure 3.10 shows that only for large values of $\overline{\Delta p}$ the shapes are free from intersecting. Analysis of the values of t_α and t_δ together with numerical calculations of the corresponding ψ_0 will show the aforementioned results.

Case IV: $\overline{\Delta p} = 0$

When $\overline{\Delta p} = 0$, the $P(t)$ curve is symmetric and $t_a = \sqrt{2\bar{\lambda}}, t_b = 0, t_c = -\sqrt{2\bar{\lambda}}$ as shown in Fig. 3.3(d). For $0 = -C_b > C > -C_c = -C_a$, Eq. (3.107) has four real roots t_α, t_β, $t_\gamma = -t_\beta$ and $t_\delta = -t_\alpha$. In the range $t_\alpha > t > t_b = 0$, with $t_\alpha = -t_\delta$, $t_\beta = -t_\gamma$, the transformation $\tau^2 = (t_\alpha + t_\beta)(t - t_\beta)/[(t_\alpha - t_\beta)(t + t_\beta)]$ gives

$$\int_{t_\beta}^{t} [(t_\alpha - t)(t - t_\beta)(t - t_\gamma)(t - t_\delta)]^{-1/2} dt$$

$$= g_1 \text{sn}^{-1} \left(\frac{l}{2g}, k_1 \right) = \frac{l}{2} = g_1 L, \tag{3.121}$$

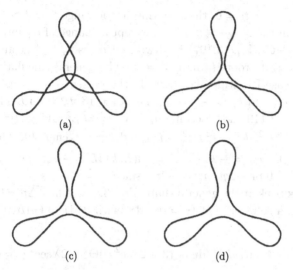

Fig. 3.10.   Triangular shapes for $\bar{\lambda} = 1$, (a) $\overline{\Delta p} = 2$, (b) $\overline{\Delta p} = 6$, (c) $\overline{\Delta p} = 10$, (d) $\overline{\Delta p} = 30$.

where

$$g_1 = 2/(t_\alpha + t_\beta),$$

$$k_1^2 = (t_\alpha - t_\beta)^2/(t_\alpha + t_\beta)^2 < \alpha_1^2 = (t_\alpha - t_\beta)/(t_\alpha + t_\beta) < 1,$$

$$t = t_\beta\{-1 + 2/[1 - \alpha^2\text{sn}^2(L, k_1)]\}.$$

The angle turned in one period, $\psi_0(C)$, now becomes[25]

$$\psi_0(C) = -4g_1K(k_1)t_\beta + 4g_1t_\beta\Pi(\pi, \alpha_1^2, k_1). \tag{3.122}$$

When $C \to -C_a$, $t_\alpha = t_\beta = t_a$, $t_\gamma = t_\delta = -t_a$, $k_1^2 = \alpha_1^2 = 0$, $g_1 = 1/t_a$, $K(k_1) = \pi/2$, $\Pi(\pi, 0, 0) = \pi$, it follows that $\psi_0 = 2\pi$, when $C \to -C_b = 0$, $t_\alpha = -t_\delta = t_a$, $t_\beta = t_\gamma = t_b = 0$, $k_1^2 = \alpha_1^2 = 1$ and the above transformation is inapplicable. The transformation $\tau^2 = (t_\beta - t_\delta)(t_\alpha - t)/[(t_\alpha - t_\beta)(t - t_\delta)] = (t_a - t)/(t + t_a)$ gives

$$\int_t^{t_a} [(t_\alpha - t)(t - t_\beta)(t - t_\gamma)(t - t_\delta)]^{-1/2}\text{d}t = \frac{2}{t_a}\text{sn}^{-1}\left(\frac{lt_a}{4}, 1\right) = \frac{l}{2}.$$

Since $\text{sn}(u, 1) = \tanh u$, it follows that

$$t = t_a[1 - \tanh^2(lt_a/4)]/[1 + \tanh^2(lt_a/4)], \tag{3.123}$$

which is non-periodic. As $l \to \infty$, the curvature becomes zero and the cylinder becomes a plane. Since the $P(t)$ vs t curve is symmetric with respect to $t = t_\beta = t_b = 0$, the situation in the range $t_\gamma > t > t_\delta$ is exactly the same as in the range $t_\alpha > t > t_\beta$ except a change of α_1^2 into $-\alpha_1^2$.

For $C > -C_b$, Eq. (3.107) has two real roots t_α and t_δ and a pair of complex conjugate roots t_β and $t_\gamma = t_\beta^*$. The general calculation given in State (6) of case I is still valid. Only in the present case when $C \to -C_b$, $t_\alpha = -t_\delta = t_a$, and $t_\gamma = -t_\beta = t_\beta^*$, i.e., $t_\beta = i\beta$, $t_\gamma = -i\beta$. The different terms in Eq. (3.119) now become $b_1 = 0$, $a_1^2 = \beta^2$, $A^2 = B^2 = t_\alpha^2 + t_\beta^2$, $g = (t_\alpha^2 + t_\beta^2)^{-1/2}$, $k^2 = t_\alpha^2/(t_\alpha^2 + t_\beta^2)$ and $t = -t_\alpha \mathrm{cn}(L, k)$. The angle ψ_0 is now given by $\psi_0 = -2g \int_0^{4K(k)} t_\alpha \mathrm{cn}(L, k) \mathrm{d}L = -2g t_\alpha[\sin^{-1}(k\mathrm{cn}4K) - \sin^{-1}(k\mathrm{sn}0)] = 0$ or $-2g t_\alpha n\pi = -2n\pi$ since $t_\beta = 0$.

Now look back at the general shape Eq. (3.98). When $\overline{\Delta p} = 0$, the plane $\psi = \mathrm{const}$ is a solution. Let $\Theta = \mathrm{d}\psi/\mathrm{d}x$, with $\mathrm{d}/\mathrm{d}x = \Theta \mathrm{d}/\mathrm{d}\psi$, Eq. (3.98) becomes

$$\mathrm{d}^2\Theta^2/\mathrm{d}\psi^2 - 4\tan\psi \mathrm{d}\Theta^2/\mathrm{d}\psi - (3 - 2\sec^2\psi)\Theta^2 - 2\bar{\lambda}\sec^2\psi = 0. \quad (3.124)$$

In Sec. 3.4, it has been shown that the general shape equation has a circular cylinder solution with radius R satisfying the condition Eq. (3.90) (note that $\bar{\lambda} = \lambda/k_c + c_0^2/2$). In the case of $\overline{\Delta p} = 0$, Eq. (3.90) becomes

$$2\bar{\lambda}R^2 = 1. \quad (3.125)$$

The equation of the circular cross section of the cylinder may be taken as

$$X = R\sin\psi, \quad (3.126)$$

and Eq. (3.124) gives directly Eq. (3.125). It can be shown easily that $\sec\psi$ and $\sec\psi\tan\psi$ are two solutions of the homogeneous part of Eq. (3.124). The general solution of Eq. (3.124) may now be written as

$$\Theta^2 = \sec^2\psi/R^2 + A\sec\psi/R^2 + B\sec\psi\tan\psi/R^2, \quad (3.127)$$

where A and B are two arbitrary constants. Therefore, the angle ψ is determined by

$$\int \frac{\cos\psi \mathrm{d}\psi}{(1 + A\cos\psi + B\sin\psi)^{1/2}} = \int_{X_0}^{X} \frac{\mathrm{d}X}{R} = \frac{X - X_0}{R}, \quad (3.128)$$

where X_0 is the constant of integration. For convenience, let $C^2 = A^2 + B^2$ ($C > 0$), $\sin\psi_0 = A/C$ and $\cos\psi_0 = B/C$. The integral in Eq. (3.128) now becomes

$$\int \frac{\cos\psi \mathrm{d}\psi}{(1 + A\cos\psi + B\sin\psi)^{1/2}} = \int \frac{\cos\psi_0 \cos\theta \mathrm{d}\theta}{(1 + C\sin\theta)^{1/2}} + \int \frac{\sin\psi_0 \sin\theta \mathrm{d}\theta}{(1 + C\sin\theta)^{1/2}},$$

where $\theta = \psi + \psi_0$. Obviously,

$$\int \frac{\cos \psi_0 \cos \theta d\theta}{(1 + C \sin \theta)^{1/2}} = \frac{2}{C} \cos \psi_0 (1 + C \sin \theta)^{1/2}.$$

However, the integration of the second integral is not so simple. It depends on the value of C. For $C < 1$, the transformation $\mathrm{sn}^2 u = C(1 - \sin\theta)/(1 + C) = (1 - \sin\theta)/(2k^2)$, where $k^2 = (1 + C)/(2C)$, gives[25]

$$\int \frac{\sin\theta}{(1 + C\sin\theta)^{1/2}} d\theta = -g \int (1 - 2k^2 \mathrm{sn}^2 u) du = g[u - 2E(u, k)],$$

where $g = \sqrt{2/C}$ and

$$E(u, k) = \int_0^y (1 - 2k^2 \sin^2\theta)^{1/2} d\theta = \int_0^u \mathrm{dn}^2 u \, du$$

is the normal elliptic integral of the second kind. For $C > 1$, the transformation $\mathrm{sn}^2 u = (1 - \sin\theta)/2$ with $k^2 = 2C/(1 + C)$ gives[25]

$$\int \frac{\sin\theta d\theta}{(1 + C\sin\theta)^{1/2}} = -g' \int (1 - 2\mathrm{sn}^2 u) du = g' \left[\left(\frac{2}{k^2} - 1 \right) u + \frac{2}{k^2} E(u, k) \right],$$

where $g' = 2/(1 + C)^{1/2}$. For $C = 1$, under the transformation $1 + \sin\theta = 2y^2 = 2z$, it is easy to show that

$$\int \frac{\sin\theta}{(1 + C\sin\theta)^{1/2}} d\theta = \int \frac{(2z - 1)}{z\sqrt{2(1 - z)}} dz,$$

which is non-periodic. Therefore, the periodic solutions for ψ under $C < 1$ and $C > 1$, respectively, are

$$\frac{2\cos\psi_0 (1 + C\sin\theta)^{1/2}}{C} + \sqrt{\frac{2}{C}} \sin\psi_0 [u - 2E(u)] = \frac{X - X_0}{R}, \quad (3.129a)$$

where

$$E(u) = E[u, (1 + C)^{1/2}/(2C)^{1/2}], \quad \text{for } C < 1,$$

and

$$\frac{2\cos\psi_0 (1 + C\sin\theta)^{1/2}}{C} + \frac{2\sin\psi_0}{(1 + C)^{1/2}} \left[\frac{u}{C} - \left(\frac{1}{C} + 1 \right) E(u) \right] = \frac{X - X_0}{R},$$

$$(3.129b)$$

where

$$E(u) = E[u, (2C)^{1/2}/(1+C)^{1/2}], \quad \text{for } C > 1,$$

with

$$\theta = \psi + \psi_0, \quad C^2 = A^2 + B^2, \quad C < 0, \quad \sin\psi_0 = A/C \quad \text{and} \quad \cos\psi_0 = B/C.$$
$$\text{(3.129c)}$$

The solutions (3.129a) and (3.129b) are difficult to solve for ψ. However, there are some simple cases worth noting. In the case of $A = 0$, it follows that $C = B$, $\sin\psi_0 = 0, \cos\psi_0 = 1$ and Eq. (3.128) gives

$$\sin\psi = B(X - X_0)^2/(4R^2) - 1/B. \tag{3.130}$$

This result holds even for $B < 0$. With proper choice of the origin on the x-axis, the constant of integration X_0 may be taken as zero. In the case of $1 > 1/B > 0$, with $X_0 = 0$, Eq. (3.130) represents undulatory surfaces. Let $X_m = 2R(1/B + 1/B^2)^{1/2} = 2R(1+B)^{1/2}/B$ and $\chi = X/X_m$. With $X_0 = 0$, Eq. (3.130) becomes

$$\sin\psi = (1 + 1/B)\chi^2 - 1/B, \quad (-1 \le \chi \le 1), \tag{3.131}$$

and the cross-section of the contour given by Eq. (3.190) becomes

$$Z(\chi) - Z(\chi') = -\int_{\chi'}^{\chi} \tan\psi d\chi, \tag{3.132}$$

where $Z = z/X_m$. The integral on the right-hand side leads to elliptic functions. Figure 3.11 shows one period of the contour of the cross-section for different values of $1/B$ calculated numerically.[19] It indicates that the period of the cross-section reaches its maximum value as $1/B \to 0$, at which $\sin\psi = \chi^2$ and the maximum period T is given by

$$T = 2\int_{-1}^{1} \chi^2 \frac{d\chi}{(1 - \chi^4)^{1/2}} \approx 2.4.$$

The shape shown in the inset of Fig. 3.11 had been observed by Harbich and Helfrich in the experiment of the swelling of egg lecithin in excess water.[26] They give no explanation on the wavy surface. Actually, it is a solution of the general shape equation.

In the case of $1/B > 1$, Eq. (3.130) represents another kind of periodic structure, the nodoidlike cylinder, as shown in Fig. 3.12.[19] In this case, if

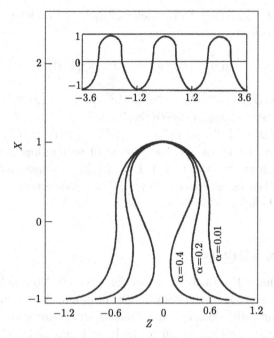

Fig. 3.11.   One period of the cross-sections of the undulatory cylindrical surfaces with different values of $1/B$ ($= \alpha$). The inset shows three periods for $B = 100$.

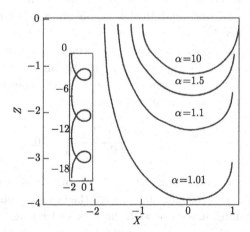

Fig. 3.12.   Half period of the cross-sections of the nodoidlike cylindrical surfaces for different values of $1/B$ ($= \alpha$). The inset shows three periods of the surface for $B = 1/1.01$.

the constant of integration X_0 is taken as $2R/B$ and we let $\chi = X/R$, then Eq. (3.130) becomes

$$\sin \psi = (B/4)(\chi + 2/B)^2 - (1/B), \tag{3.133}$$

with $2(1/B+1/B^2)^{1/2}-2/B \geq \chi \geq 2(1/B^2-1/B)^{1/2}-2/B$. In the limiting case of $B \to 0$, since $\lim_{B\to 0}[(1/B)(B\chi/2+1)^2-1/B] = \chi$, $\lim_{B\to 0}[(1/B+1/B^2)^{1/2} - 1/B] = 1/2$ and $\lim_{B\to 0}[(1/B^2 - 1/B) - 1/B] = -1/2$, the surface degenerates into a circular cylinder of unit radius. As to the case of $B = 1$, which amounts to $C = 1$, Eq. (3.130) becomes $\sin \psi = 2\chi^2 - 1$, where X_0 is taken as zero and $\chi = X/R$. The cross-section is non-periodic as mentioned before.

3.7   Clifford Torus

The general shape Eq. (3.46) is rather complicated. No one had ever discussed how to solve it analytically. There does not exist helpful guidance to find any particular solution. In problems with axisymmetry, the complexity of the shape equation will certainly be highly reduced. So far, only equilibrium shapes of axisymmetric vesicles have been extensively calculated numerically.[27-39] The finding of any particular analytical axisymmetric solution still depends on heuristic procedure. Besides the genus zero spherical surface, the first analytical solution obtained is the genus one Clifford torus surface.[40]

The anchor ring, i.e., Clifford torus, is formed by revolving a circle of radius r around an axis, e.g., the z-axis, in the plane of the circle at a distance R ($R > r$) from the center of the r circle. The equation of the family of tori in Cartesian coordinates is given by[41]

$$\frac{(x^2 + y^2 + r^2)^2}{(R/r)^2} - \frac{4r^2(x^2 + y^2)}{(R/r)^2 - 1} = 0. \tag{3.134}$$

There is another way of defining the family of tori. As shown in Fig. 3.13, A and B are points on a straight line through the origin O, perpendicular to the z-axis, and making an angle ϕ with the x-axis with length $AB = 2c$. Let P be a point in the plane $\phi = $ const. and $\theta = \angle APB$. Let $\eta = \ln(AP/BP)$, the surfaces $\eta = $ const. will be the family of tori generated by the revolution around the z-axis of the circles of the family of coaxial circles of which A and B are the limiting points. The parametric equation for the family of

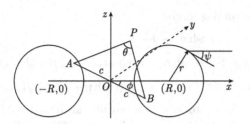

Fig. 3.13.   Generation of torus.

tori is given by[42]

$$\mathbf{Y} = \left(\frac{c \sinh \eta \cos \phi}{\cosh \eta - \cos \theta}, \frac{c \sinh \eta \sin \phi}{\cosh \eta - \cos \theta}, \frac{c \sin \theta}{\cosh \eta - \cos \theta} \right), \qquad (3.135)$$

where ϕ varies from 0 to 2π and θ varies from $-\pi$ to π. Equation 3.135 gives that

$$x^2 + y^2 = c^2 \sinh^2 \eta / (\cosh \eta - \cos \theta)^2,$$

$$x^2 + y^2 + z^2 = c^2 (\sinh^2 \eta + \sin^2 \theta) / (\cosh \eta - \cos \theta)^2.$$

Elimination of θ yields

$$(x^2 + y^2 + z^2 + c^2)^2 / \cosh^2 \eta - 4c^2 (x^2 + y^2) / (\cosh^2 \eta - 1) = 0. \qquad (3.136)$$

Comparison of Eqs. (3.134) and (3.136) shows that they are the same equation with

$$c^2 = r^2, \quad \cosh^2 \eta = (R/r)^2, \quad \sinh^2 \eta = (R^2 - r^2)/r^2. \qquad (3.137)$$

From Eqs. (3.135) and (3.137), the following non-vanishing components can be obtained:

$$g_{\theta\theta} = 1/g^{\theta\theta} = c^2 / (\cosh \eta - \cos \theta)^2,$$

$$g_{\phi\phi} = 1/g^{\phi\phi} = c^2 \sinh^2 \eta / (\cosh \eta - \cos \theta)^2,$$

$$L_{\theta\theta} = c \sinh \eta / (\cosh \eta - \cos \theta)^2,$$

$$L_{\phi\phi} = c \sinh \eta (\cosh \eta \cos \theta - 1) / (\cosh \eta - \cos \theta)^2, \qquad (3.138)$$

$$H = (1/2) g^{ij} L_{ij} = [1/(2c)][\sinh \eta + (\cosh \eta \cos \theta - 1)/\sinh \eta],$$

$$K = (\cosh \eta \cos \theta - 1)/c^2,$$

$$\nabla^2 H = -(\cosh \eta - \cos \theta)^2 \cosh \eta \cos \theta / (2c^3 \sinh \eta).$$

The shape equation now becomes

$$\Delta p - 2k_c c^{-2} c_0 (\cosh \eta \cos \theta - 1)$$
$$- (2\lambda + k_c c_0^2)[\sinh \eta + (\cosh \eta \cos \theta - 1)/\sinh \eta]/(2c) \qquad (3.139)$$
$$- k_c c^{-3} \cosh \eta (\cosh^2 \eta - 2)(\cosh \eta - \cos \theta)^3 / (2 \sinh^3 \eta) = 0.$$

Since θ varies from $-\pi$ to π, this equation can be satisfied only when the coefficients of $\cos^m \theta$ for $m = 0, 1, 3$ all vanish. Thus, it leads to

$$\cosh^2 \eta = (R/r)^2 = 2, \quad \Delta p = -2k_c c_0/c^2, \quad \lambda = -2k_c(\pm c_0/c + c_0^2/4).$$
$$(3.140)$$

The first condition implies that only those tori with the ratio of the radii of their generating circles equal to $\sqrt{2}$ can satisfy the shape equation. Experimental observations on vesicles of diacetylenic phospholipids are in quantitative agreement with this prediction.[43-45] Since $\cosh \eta$ is always positive and $\cosh \eta = \pm R/r$, if R and r are in the same direction then $\cosh \eta = R/r$, otherwise $\cosh \eta = -R/r$. Both c and $\sinh \eta$ may be either positive or negative. The second term of Eq. (3.139) is independent of the sign of c, but the third term may be either positive or negative depending on whether c and $\sinh \eta$ have the same sign or not. When c and $\sinh \eta$ have the same sign $\lambda = -2k_c(c_0/r + c_0^2/4)$; otherwise, $\lambda = -2k_c(-c_0/r + c_0^2/4)$. Although there is a difference of sign for c_0 in the two cases, actually they represent the same asymmetry property. In one case it is looking from inside to outside of the vesicle membrane, while the other case is looking from outside to inside of the vesicle membrane.

It is interesting to investigate the limiting case of an infinitely large torus with $R \to \infty$ and $r \to \infty$, but keeping $R - r = \rho_0$. In this case, Eq. (3.137) shows that for $c \to -\infty$ (outward normal is in the positive direction), $c/\cosh \eta \to -\rho_0$ and $c/\sinh \eta \to -\rho_0$. From Eq. (3.138), it follows that $H \to -1/(2\rho_0)$, $K \to 0$ and $\Delta^2 H \to 0$. The general shape equation (3.46) now gives

$$\Delta p \rho_0^3 + \lambda \rho_0^2 + (1/2)k_c(c_0^2 \rho_0^2 - 1) = 0,$$

which is exactly the constraint Eq. (3.90) that a circular cylinder of radius ρ_0 should satisfy. This shows that the limiting torus is nothing but a straight cylinder.

The condition for the stability of the torus is not easy to find. A trial function ψ may be taken as

$$\psi = \sum_{m} [a_m(\theta) \sin m\phi + b_m(\theta) \cos m\phi], \qquad (3.141)$$

where $a_m(\theta) = a_m(-\theta)$ and $b_m(\theta) = b_m(-\theta)$. With $R^2 = 2r^2$, the following non-zero terms can be obtained:

$$g_{\theta\theta} = g_{\phi\phi} = 1/g^{\theta\theta} = 1/g^{\phi\phi} = g^{1/2} = rL_{\theta\theta} = r/L^{\theta\theta} = r^2/(\sqrt{2} - \cos\theta)^2,$$

$$L_{\phi\phi} = 1/L^{\phi\phi} = r(\sqrt{2}\cos\theta - 1)/(\sqrt{2} - \cos\theta)^2,$$

$$L = r^2(\sqrt{2}\cos\theta - 1)/(\sqrt{2} - \cos\theta)^4,$$

$$H = \sqrt{2}\cos\theta/2r, \quad K = (\sqrt{2}\cos\theta - 1)/r^2,$$

$$\Gamma^\theta_{\theta\theta} = -\Gamma^\theta_{\phi\phi} = \Gamma^\phi_{\theta\phi} = -\sin\theta/(\sqrt{2} - \cos\theta). \tag{3.142}$$

The second variation of the shape energy, Eq. (3.54), now becomes

$$\delta^2 F = k_c \oint \{(1/2)g^{1/2}(g^{ij}\nabla_i\psi_{,j})^2 - 2(K + c_0H)g^{1/2}g^{ij}\psi\nabla_i\psi_{,j}$$

$$+ g^{1/2}[-(5H^2 + 2c_0H + c_0/r)g^{ij} + (2H + c_0)KL^{ij}]\psi_{,i}\psi_{,j}$$

$$+ [g^{1/2}(2Hg^{ij} - KL^{ij})(2H + c_0)]_{,ij}\psi^2$$

$$- \{g^{1/2}(H + c_0/2)[g^{ij}(g^{kl}L_{ik})_{,j}$$

$$- \Gamma^m_{ij}g^{ij}g^{kl}L_{km} - \Gamma^l_{ij}(2Hg^{ij} - KL^{ij})]\}_{,l}\psi^2$$

$$+ 2g^{1/2}[(H^2 - K)(4H^2 - K) - (c_0/r)(K - H/r)]\psi^2\}d\theta d\phi. \tag{3.143}$$

It is not difficult to verify the following relations:

$$\nabla_\theta\psi_{,\theta} = \sum_m \left[\left(a_{m,\theta\theta} + \frac{\sin\theta a_{m,\theta}}{\sqrt{2} - \cos\theta}\right)\sin m\phi\right.$$

$$\left. + \left(b_{m,\theta\theta} + \frac{\sin\theta b_{m,\theta}}{\sqrt{2} - \cos\theta}\right)\cos m\phi\right],$$

$$\nabla_\phi\psi_{,\phi} = -\sum_m \left[\left(\frac{\sin\theta a_{m,\theta}}{\sqrt{2} - \cos\theta} + m^2a_m\right)\sin m\phi\right.$$

$$\left. + \left(\frac{\sin\theta b_{m,\theta}}{\sqrt{2} - \cos\theta} + m^2b_m\right)\cos m\phi\right],$$

$$g^{ij}\nabla_i\psi_{,j} = \frac{(\sqrt{2} - \cos\theta)^2}{r^2}\sum_m[(a_{m,\theta\theta} - m^2a_m)\sin m\phi$$

$$+ (b_{m,\theta\theta} - m^2b_m)\cos m\phi],$$

$$\int_0^{2\pi} \psi_{,\theta}^2 d\phi = \pi \sum_m (a_{m,\theta}^2 + b_{m,\theta}^2), \quad \int_0^{2\pi} \psi_{,\phi}^2 d\phi = \pi \sum_m m^2(a_m^2 + b_m^2),$$

$$\int_0^{2\pi} \psi_{,\theta\theta}^2 d\phi = \pi \sum_m (a_{m,\theta\theta}^2 + b_{m,\theta\theta}^2), \quad \int_0^{2\pi} \psi_{,\phi\phi}^2 d\phi = \pi \sum_m m^4(a_m^2 + b_m^2),$$

$$\int_0^{2\pi} \psi\psi_{,\phi\phi} d\phi = -\pi \sum_m m^2(a_m^2 + b_m^2), \quad \int_0^{2\pi} \psi^2 d\phi = \pi \sum_m (a_m^2 + b_m^2),$$

$$\int_0^{2\pi} \psi\psi_{,\phi\phi} d\phi = \pi \sum_m (a_m a_{m,\theta\theta} + b_m b_{m,\theta\theta})$$

$$= \frac{\pi}{2} \sum_m (a_m^2 + b_m^2)_{,\theta\theta} - \pi \sum_m (a_{m,\theta}^2 + b_{m,\theta}^2),$$

$$\int_0^{2\pi} \psi_{,\theta\theta}\psi_{,\phi\phi} d\phi = \pi \sum_m m^2(a_{m,\theta}^2 + b_{m,\theta}^2) - \frac{\pi}{2} \sum_m m^2(a_m^2 + b_m^2)_{,\theta\theta},$$

$$\int_{-\pi}^{\pi} f(\theta)(a_m a_{m,\theta\theta} + b_m b_{m,\theta\theta}) d\theta$$

$$= \frac{1}{2}\int_{-\pi}^{\pi} f(\theta)(a_m^2 + b_m^2)_{,\theta\theta} d\theta - \int_{-\pi}^{\pi} f(\theta)(a_{m,\theta}^2 + b_{m,\theta}^2) d\theta,$$

$$\int_{-\pi}^{\pi} f(\theta)(a_m^2 + b_m^2)_{,\theta\theta} d\theta$$

$$= \left[f(\theta)(a_m^2 + b_m^2)_{,\theta} - \frac{df}{d\theta}(a_m^2 + b_m^2) \right]_{-\pi}^{\pi} + \int_{-\pi}^{\pi} \frac{d^2 f}{d\theta^2}(a_m^2 + b_m^2) d\theta$$

$$= \int_{-\pi}^{\pi} \frac{d^2 f}{d\theta^2}(a_m^2 + b_m^2) d\theta, \tag{3.144}$$

where the subscript θ or ϕ after the comma sign means differentiation with respect to θ or ϕ and $f(\theta)$ is a function of $\sin\theta$ and $\cos\theta$. With Eq. (3.144), it is not difficult to show that

$$\oint \frac{1}{2}g^{1/2}(g^{ij}\nabla_i\psi_{,j})^2 d\theta d\phi = \frac{\pi}{2}\int_{-\pi}^{\pi} \left\{ \left(\frac{\sqrt{2} - \cos\theta}{r} \right)^2 \sum_m [(a_{m,\theta\theta}^2 + b_{m,\theta\theta}^2) \right.$$

$$+ 2m^2(a_{m,\theta}^2 + b_{m,\theta}^2) + m^4(a_m^2 + b_m^2)]$$

$$\left. - \frac{2(1 + \sqrt{2}\cos\theta - 2\cos^2\theta)}{r^2} \sum_m m^2(a_m^2 + b_m^2) \right\} d\theta,$$

$$\oint 2(K + c_0 H) g^{1/2} g^{ij} \psi \nabla_i \psi_{,j} \mathrm{d}\theta \mathrm{d}\phi$$

$$= -\pi \int_{-\pi}^{\pi} \sum_m \left\{ \frac{2\sqrt{2}\cos\theta - 2 + c_0 r \sqrt{2}\cos\theta}{r^2} [(a_{m,\theta}^2 + b_{m,\theta}^2) \right.$$

$$\left. + m^2(a_m^2 + b_m^2)] + \frac{\sqrt{2}(2 + c_0 r)\cos\theta}{2r^2} (a_m^2 + b_m^2) \right\} \mathrm{d}\theta,$$

$$\oint g^{1/2}[-(5H^2 + 2c_0 H + c_0/r)g^{ij} + 2K(H + c_0/r)L^{ij}]\psi_{,i}\psi_{,j}\mathrm{d}\theta\mathrm{d}\phi$$

$$= \frac{\pi}{r^2} \int_{-\pi}^{\pi} \sum_m \left[-\left(\frac{1}{2}\cos^2\theta + \sqrt{2}\cos\theta + 2c_0 r \right) (a_{m,\theta}^2 + b_{m,\theta}^2) \right.$$

$$\left. + \left(-\frac{5}{2}\cos^2\theta + \sqrt{2}\cos\theta - \sqrt{2}c_0 r \cos\theta \right) m^2(a_m^2 + b_m^2) \right] \mathrm{d}\theta,$$

$$\oint [g^{1/2}(2Hg^{ij} - KL^{ij})(2H + c_0)]_{,ij}\, \psi^2 \mathrm{d}\theta\mathrm{d}\phi$$

$$= -\pi \int_{-\pi}^{\pi} \frac{\sqrt{2}\cos\theta}{r^2} \sum_m (a_m^2 + b_m^2)\mathrm{d}\theta,$$

$$\oint \left\{ g^{1/2} \left(H + \frac{c_0}{2} \right) [g^{ij}(g^{kl}L_{ik})_{,j} \right.$$

$$\left. - \Gamma_{ij}^m g^{ij} g^{kl} L_{km} - \Gamma_{ij}^l (2Hg^{ij} - KL^{ij})] \right\}_{,l} \psi^2 \mathrm{d}\theta\mathrm{d}\phi$$

$$= \frac{\pi}{2r^2} \int_{-\pi}^{\pi} (4\cos^2\theta - 2 + c_0 r \sqrt{2}\cos\theta) \sum_m (a_m^2 + b_m^2)\mathrm{d}\theta,$$

$$\oint 2g^{1/2} \left[(H^2 - K)(4H^2 - K) - \frac{c_0}{r} \left(K - \frac{H}{r} \right) \right] \psi^2 \mathrm{d}\theta\mathrm{d}\phi$$

$$= \pi \int_{-\pi}^{\pi} \left[\frac{2\cos^2\theta - \sqrt{2}\cos\theta + 1}{r^2} + \frac{\sqrt{2}c_0}{r(\sqrt{2} - \cos\theta)} \right] \sum_m (a_m^2 + b_m^2)\mathrm{d}\theta.$$

$$(3.145)$$

Finally, the substitution of Eq. (3.145) into (3.143) gives

$$\delta^2 F = \frac{\pi k_c}{r^2} \int_{-\pi}^{\pi} \sum_m \left\{ \frac{1}{2}(\sqrt{2} - \cos\theta)^2 (a_{m,\theta\theta}^2 + b_{m,\theta\theta}^2) + \left[(\sqrt{2} - \cos\theta)^2 m^2 \right. \right.$$

$$\left. - \frac{1}{2}\cos^2\theta + \sqrt{2}\cos\theta + 2 + c_0 r(2 - \sqrt{2}\cos\theta) \right] (a_{m,\theta}^2 + b_{m,\theta}^2)$$

$$+ \left[\frac{1}{2}(\sqrt{2} - \cos\theta)^2 m^4 + \left(-\frac{1}{2}\cos^2\theta + 2\sqrt{2}\cos\theta - 3 \right) m^2 \right.$$

$$\left. + 2 - \sqrt{2}\cos\theta + \sqrt{2}c_0 r(\sqrt{2} - \cos\theta)^{-1} \right] (a_m^2 + b_m^2) \Bigg\} d\theta. \qquad (3.146)$$

The stability analysis requires that $\delta^2 F$ should be positive definite for $\psi \neq 0$. In Eq. (3.146), the necessary condition for the coefficient of $(a_{m,\theta}^2 + b_{m,\theta}^2)$ being positive is that

$$c_0 r < -1 - \cos^2\theta/[2\sqrt{2}(\sqrt{2} - \cos\theta)] < -\frac{3}{2} - \frac{\sqrt{2}}{4} \simeq -1.8536.$$

On the other hand, the requirement that the coefficient of $(a_m^2 + b_m^2)$ being positive is that

$$c_0 r > -(\sqrt{2} - \cos\theta)^2,$$

which gives a lower limit of $c_0 r > -5.8284$, not inconsistent with $c_0 r < -1.8536$. Therefore, the Clifford torus is stable when $c_0 r$ is in the range $-5.8284 < c_0 r < -1.8536$.

3.8   Dupin Cyclide

Observations on toroidal vesicles of diacetylenic phospholipids by Fourcade, Mutz and Bensimon[43–45] show the existence of a few non-axisymmetric toroidal vesicles. The geometrical similarities between amphiphilic membranes and smectic liquid crystals led Ou-Yang to believe that the shape of toroidal vesicles is closely related to the texture of smectic liquid crystals and to study the Dupin cyclide solution[46] of the general shape Eq. (3.46).

The Dupin cyclide is directly related to the pair of conic sections

$$(X/a)^2 + (Y/b)^2 = 1, \quad Z = 0, \qquad (3.147a)$$

and

$$(X/c)^2 - (Z/b)^2 = 1, \quad Y = 0, \qquad (3.147b)$$

with

$$a^2 = b^2 + c^2, \qquad (3.147c)$$

as shown in Fig. 3.14. Let the family of spheres of radii $-cX/a + \mu$ with arbitrary constants μ, where $a > \mu > c$, be drawn with centers at points

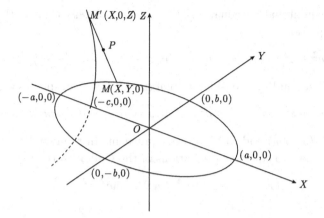

Fig. 3.14. Generating conic sections of Dupin cyclide.

$M(X, Y, 0)$ on the ellipse (3.147a) and another family of spheres of radii $-aX'/c + \mu$ with centers at $M'(X', 0, Z')$ on the hyperbola (3.147b). The radii of the two families of circles differ by $cX/a - aX'/c$, which is exactly the distance between M and M'. This means that the two families of spheres are tangent to each other at point P and have the same enveloping surfaces, the Dupin cyclides. Spheres of the M family can be represented by

$$
\begin{aligned}
F(x, y, z, X) &= (x - X)^2 + (y - Y)^2 + z^2 - (-cX/a + \mu)^2 \\
&= x^2 + y^2 + z^2 - 2X(x - c\mu/a) \\
&\quad - 2b(1 - X^2/a^2)^{1/2}y + b^2 - \mu^2 \\
&= 0,
\end{aligned}
\tag{3.148}
$$

where X is taken as a parameter and Y is related to X by Eq. (3.147a). The equation of the Dupin cyclide can be obtained by solving Eq. (3.148) and $\partial F/\partial X = 0$ simultaneously. From Eq. (3.148), the equation $\partial F/\partial X = 0$ is in the form

$$
(x - X) + (y - Y)\frac{\partial Y}{\partial X} - \frac{c}{a}\left(-\frac{c}{a}X + \mu\right) = 0,
$$

which may be converted into

$$
\frac{X^2}{a^2} = \frac{(x - c\mu/a)^2}{(x - c\mu/a)^2 + b^2y^2/a^2}.
\tag{3.149}
$$

The two simultaneous equations (3.148) and (3.149) lead to the Dupin cyclide equation

$$(x^2 + y^2 + z^2 + a^2 - c^2 - \mu^2)^2 = 4(ax - c\mu)^2 + 4(a^2 - c^2)y^2. \quad (3.150a)$$

On the other hand, the consideration of the M' family of spheres will lead to

$$(x^2 + y^2 + z^2 - a^2 + c^2 - \mu^2)^2 = 4(cx - a\mu)^2 + 4(c^2 - a^2)z^2. \quad (3.150b)$$

Equations (3.150a) and (3.150b) are equivalent. In the case of $c = 0$, they reduce to the two equivalent equations of the Clifford torus

$$(x^2 + y^2 + z^2 + a^2 - \mu^2)^2 = 4a^2(x^2 + y^2)$$

and

$$(x^2 + y^2 + z^2 - a^2 - \mu^2)^2 = 4a^2(\mu^2 - z^2),$$

respectively. The projections of the Dupin cyclide, Eq. (3.150), and that of the torus, Eq. (3.134), are shown schematically in Fig. 3.15 for comparison.

The parametric equations of Dupin cyclide are given by[47]:

$$x = [\mu(c - a\cos\theta\cos\phi) + b^2\cos\theta]/(a - c\cos\theta\cos\phi),$$

$$y = b\sin\theta(a - \mu\cos\phi)/(a - c\cos\theta\cos\phi), \qquad (3.151)$$

$$z = b\sin\phi(c\cos\theta - \mu)/(a - c\cos\theta\cos\phi),$$

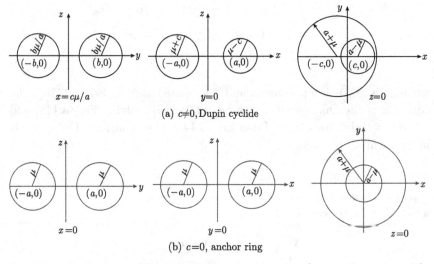

(a) $c \neq 0$, Dupin cyclide

(b) $c=0$, anchor ring

Fig. 3.15.   Projections of Dupin cyclide and anchor ring (not in the same scale).

where both θ and ϕ range from 0 to 2π. The following non-vanishing terms can be obtained from Eq. (3.151):

$$g_{\theta\theta} = b^2(a - \mu\cos\phi)^2/(a - c\cos\theta\cos\phi)^2,$$

$$g_{\phi\phi} = b^2(\mu - c\cos\theta)^2/(a - c\cos\theta\cos\phi)^2,$$

$$L_{\theta\theta} = b^2(a - \mu\cos\phi)\cos\phi/(a - c\cos\theta\cos\phi)^2,$$

$$L_{\phi\phi} = -b^2(\mu - c\cos\theta)/(a - c\cos\theta\cos\phi)^2,$$

$$n_x = (c - a\cos\theta\cos\phi)/(a - c\cos\theta\cos\phi), \qquad (3.152)$$

$$n_y = -b\sin\theta\cos\phi/(a - c\cos\theta\cos\phi),$$

$$n_z = -b\sin\phi/(a - c\cos\theta\cos\phi),$$

$$H = (1/2)[\cos\phi/(a - \mu\cos\phi) - 1/(\mu - c\cos\theta)],$$

$$K = -\cos\phi/[(a - \mu\cos\phi)(\mu - c\cos\theta)].$$

In order to find the conditions that the Dupin cyclide, Eq. (3.151), satisfies the general shape Eq. (3.46), for convenience, let[48]:

$$\Delta p' = \Delta p/k_c, \quad \gamma = \lambda/k_c + c_0^2/2, \quad A = \cos\theta,$$

$$B = \cos\phi, \quad D = (a - \mu\cos\phi)^3(\mu - c\cos\theta)^3. \qquad (3.153)$$

The Laplacian operator ∇^2, Eq. (3.48), now becomes

$$\nabla^2 = \frac{(a - cAB)^2}{b^2(a - \mu B)(\mu - cA)}\left[\frac{\partial}{\partial\theta}\left[\frac{\mu - cA}{a - \mu B}\frac{\partial}{\partial\theta}\right] + \frac{\partial}{\partial\phi}\left(\frac{a - \mu B}{\mu - cA}\frac{\partial}{\partial\phi}\right)\right].$$

$$(3.154)$$

Detailed manipulation gives that (note, there are some typographic errors in Ref. 48):

$$\Delta p' - (2\gamma - c_0^2)H$$

$$= D^{-1}[(\mu\Delta p' + 2\gamma')\mu^2 c^3 A^3 B^3 - (3\mu\Delta p' + 2\gamma')\mu a c^3 A^3 B^2$$

$$+ (3\mu\Delta p' + \gamma')a^2 c^3 A^3 B^2 - \Delta p' a^3 c^3 A^3 - (3\mu\Delta p' + 4\gamma')\mu^3 c^2 A^2 B^3$$

$$+ 9(\mu\Delta p' + \gamma')\mu^2 a c^2 A^2 B^2 - (9\mu\Delta p' + 6\gamma')\mu a^2 c^2 A^2 B$$

$$+ (3\mu\Delta p' + \gamma')a^3 c^2 A^2 + (3\mu\Delta p' + 5\gamma')\mu^4 c A B^3$$

$$- (9\mu\Delta p' + 12\gamma')\mu^3 a c A B^2 + 9(\mu\Delta p' + \gamma')\mu^2 a^2 c A B$$

$$- (3\mu\Delta p' + 2\gamma')\mu a^3 c A - (\mu\Delta p' + 2\gamma')\mu^5 B^3 + (3\mu\Delta p' + 5\gamma')\mu^4 a B^2$$

$$- (3\mu\Delta p' + 4\gamma')\mu^3 a^2 B + (\mu\Delta p' + \gamma')\mu^2 a^3], \qquad (3.155a)$$

where $\gamma' = \gamma - (1/2)c_0^2$, and

$$(2H + c_0)(2H^2 - 2K - c_0H)$$
$$= (1/2)D^{-1}[(\mu^2 c_0^2 - 1)c^3 A^3 B^3 - 2\mu c_0^2 ac^3 A^3 B^2 + c_0^2 a^2 c^3 A^3 B$$
$$- (4\mu^2 c_0^2 - 4\mu c_0 - 2)\mu c^2 A^2 B^3 + (9\mu^2 c_0^2 - 8\mu c_0 + 1)ac^2 A^2 B^2$$
$$- 2(3\mu c_0 - 2)c_0 a^2 c^2 A^2 B + c_0^2 a^3 c^2 A^2 + (5\mu c_0 - 8)\mu^3 c_0 cAB^3$$
$$- (12\mu^2 c_0^2 - 16\mu c_0 + 4)\mu acAB^2 + (9\mu^2 c_0^2 - 8\mu c_0 + 1)a^2 cAB$$
$$- 2\mu c_0^2 a^3 cA - 2(\mu c_0 - 2)\mu^4 c_0 B^3 + (5\mu c_0 - 8)\mu^3 c_0 aB^2$$
$$- (4\mu^2 c_0^2 - 4\mu c_0 - 2)\mu a^2 B + (\mu^2 c_0^2 - 1)a^3]. \tag{3.155b}$$

$$2\nabla^2 H = D^{-1}[(a^2 - \mu^2)b^{-2}c^3 A^3 B^3 - \mu c^2 A^2 B^3$$
$$+ (3\mu^2 - 2a^2 - c^2)b^{-2}ac^2 A^2 B^2 + 2\mu acAB^2$$
$$- (3\mu^2 - a^2 - 2c^2)a^2 b^{-2}cAB - \mu a^2 B + (\mu^2 - c^2)a^3 b^{-2}]. \tag{3.155c}$$

Since the shape equation must be satisfied for any value of θ and ϕ in the range $(0, 2\pi)$, b is non-zero and D is always positive, therefore the coefficient of the term $A^n B^m$ must vanish for any integers n and m. In this way, with Eq. (3.147c), the following conditions must be satisfied for the existence of Dupin cyclide toroidal vesicles:

(n, m)

$(3, 3)$ $c^3[\mu^2(\mu\Delta p' + \gamma) - (2\mu^2 - a^2 - c^2)/(2b^2)] = 0,$ (3.156a)

$(3, 2)$ $\mu ac^3(3\mu\Delta p' + 2\gamma) = 0,$ (3.156b)

$(3, 1)$ $a^2 c^3(3\mu\Delta p' + \gamma) = 0,$ (3.156c)

$(3, 0)$ $a^3 c^3 \Delta p' = 0,$ (3.156d)

$(2, 3)$ $\mu^2 c^2(3\mu^2 \Delta p' + 4\mu\gamma - 2c_0) = 0,$ (3.156e)

$(2, 2)$ $a^2 c^2[\mu(9\mu^2 \Delta p' + 9\mu\gamma - 4c_0) + 3(2\mu^2 - a^2 - c^2)/(2b^2)] = 0,$ (3.156f)

$(2, 1)$ $a^2 c^2(9\mu^2 \Delta p' + 6\mu\gamma - 2c_0) = 0,$ (3.156g)

$(2, 0)$ $a^3 c^2(3\mu\Delta p' + \gamma) = 0,$ (3.156h)

$(1, 3)$ $\mu^3 c(3\mu^2 \Delta p' + 5\mu\gamma - 4c_0) = 0,$ (3.156i)

$(1,2)$ $\mu^2 ac(9\mu^2 \Delta p' + 12\mu\gamma - 8c_0) = 0,$ $\qquad\qquad\qquad$ (3.156j)

$(1,1)$ $a^2 c[\mu(9\mu^2 \Delta p' + 9\mu\gamma - 4c_0) - 3(2\mu^2 - a^2 - c^2)/(2b^2)] = 0,$

$\qquad\qquad\qquad\qquad\qquad\qquad\qquad\qquad\qquad\qquad$ (3.156k)

$(1,0)$ $\mu a^3 c(3\mu\Delta p' + 2\gamma) = 0,$ $\qquad\qquad\qquad\qquad\qquad$ (3.156l)

$(0,3)$ $\mu^4(\mu^2 \Delta p' + 2\mu\gamma - 2c_0) = 0,$ $\qquad\qquad\qquad$ (3.156m)

$(0,2)$ $\mu^3 a(3\mu^2 \Delta p' + 5\mu\gamma - 4c_0) = 0,$ $\qquad\qquad$ (3.156n)

$(0,1)$ $\mu^2 a^2(3\mu^2 \Delta p' + 4\mu\gamma - 2c_0) = 0,$ $\qquad\qquad$ (3.156o)

$(0,0)$ $a^3[\mu^2(\mu\Delta p' + \gamma) + (2\mu^2 - a^2 - c^2)/(2b^2)] = 0.$ $\qquad$ (3.156p)

The condition (3.156d) shows that either

$$c = 0, \quad \Delta p' = 0 \quad \text{or} \quad c = 0, \quad \Delta p' \neq 0 \quad \text{or} \quad c \neq 0, \quad \Delta p' = 0. \quad (3.157)$$

The first case $(c = 0, \Delta p' = 0)$ refers to anchor ring vesicles and at the same time all conditions from (3.156a) up to (3.156l) are satisfied automatically. Besides, at the same time, conditions (3.156m) up to (3.156p) lead to

$$\gamma = c_0 = \lambda = 0 \quad \text{and} \quad \mu/a = 1/\sqrt{2}. \quad (3.158)$$

The second case $(c = 0, \Delta p' \neq 0)$ also refers to the anchor ring solution, for which conditions (3.156m) up to (3.156p) lead to

$$c_0 = \gamma\mu/2, \quad \text{i.e.,} \quad c_0 = 2\sqrt{2}/a - (8/a^2 - 2\lambda/k)^{1/2},$$
$$\Delta p' = -4c_0/a^2, \quad \mu/a = 1/\sqrt{2}. \quad (3.159)$$

The last case $(c \neq 0, \Delta p' = 0)$ refers to the non-axisymmetric Dupin cyclide solution. Here, Eqs. (3.156b, c, e, g, h, i, j, l, m, n and p) of (3.156) lead to

$$\gamma = 0, \quad c_0 = 0 \quad \text{and} \quad \lambda = 0. \quad (3.160)$$

It follows from Eqs. (3.156a, f, k and q) of (3.156) that

$$\mu^2 = (a^2 + c^2)/2. \quad (3.161)$$

The conditions $(c \neq 0, \Delta p' = 0)$, and Eqs. (3.160) and (3.161) are the necessary and sufficient conditions for the Dupin cyclide to be the vesicle solution. The condition Eq. (3.161) restricts the shape of the generating conic sections of the Dupin cyclide and makes the Dupin cyclide vesicle hard to appear.

3.9   Shape Equation for Axisymmetric Vesicles

To describe an axisymmetric shape, it is convenient to use the distance ρ of a point on the surface from the axis of rotational symmetry (the z-axis) as the independent parameter and to describe the angle ψ between the tangent line of the contour at that point and the ρ-axis as a function of ρ, $\psi = \psi(\rho)$, as shown in Fig. 3.16. It follows that:

$$x = \rho \cos \theta, \quad y = \rho \sin \theta, \quad \mathrm{d}z = \tan \psi \mathrm{d}\rho,$$

$$\mathrm{d}s^2 = \mathrm{d}\rho^2 + \rho^2 \mathrm{d}\theta^2 + \mathrm{d}z^2 = \sec^2 \psi \mathrm{d}\rho^2 + \rho^2 \mathrm{d}\theta^2, \quad g_{\rho\rho} = \sec^2 \psi, \quad g_{\theta\theta} = \rho^2,$$

$$n_\rho = -\sin \psi, \quad n_\theta = 0, \quad n_z = \cos \psi,$$

$$n_x = -\sin \psi \cos \theta, \quad n_y = -\sin \psi \sin \theta,$$

$$L_{\rho\rho} = \sec \psi \frac{\mathrm{d}\psi}{\mathrm{d}\rho}, \quad L_{\rho\theta} = 0, \quad L_{\theta\theta} = \rho \sin \psi,$$

$$2H = \cos \psi \frac{\mathrm{d}\psi}{\mathrm{d}\rho} + \frac{\sin \psi}{\rho}, \quad K = \frac{\sin \psi \cos \psi}{\rho} \frac{\mathrm{d}\psi}{\mathrm{d}\rho},$$

$$\nabla^2 = \frac{1}{\sqrt{g}} \frac{\partial}{\partial i} \left(g^{ij} \sqrt{g} \frac{\partial}{\partial j} \right) = \frac{\cos \psi}{\rho} \left[\frac{\partial}{\partial \rho} \left(\rho \cos \psi \frac{\partial}{\partial \rho} \right) + \frac{\partial}{\partial \theta} \left(\frac{\sec \psi}{\rho} \frac{\partial}{\partial \theta} \right) \right],$$

$$(3.162)$$

where $0 \leq \theta \leq 2\pi$ is the rotational angle. The outward normal **n** is analytically and uniquely defined everywhere on the surface. This kind of surface is called smooth surface.

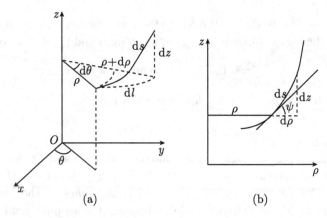

Fig. 3.16.   Coordinates for the description of axisymmetric shapes.

With Eq. (3.162), it is easy to show that the general shape Eq. (3.46) takes the form[27]:

$$\cos^3 \psi \frac{d^3\psi}{d\rho^3} - 4 \sin \psi \cos^2 \psi \frac{d\psi}{d\rho}\frac{d^2\psi}{d\rho^2} + \cos \psi \left(\sin^2 \psi - \frac{1}{2}\cos^2 \psi \right) \left(\frac{d\psi}{d\rho} \right)^3$$

$$+ \frac{2\cos^3 \psi}{\rho}\frac{d^2\psi}{d\rho^2} - \frac{7\sin \psi \cos^2 \psi}{2\rho} \left(\frac{d\psi}{d\rho} \right)^2$$

$$+ \left[\frac{\sin^2 \psi - 2\cos^2 \psi}{2\rho^2} - \left(\frac{\lambda}{k_c} + \frac{c_0^2}{2} \right) - \frac{2c_0 \sin \psi}{\rho} \right] \cos \psi \frac{d\psi}{d\rho}$$

$$+ \left[\frac{\sin \psi(1 + \cos^2 \psi)}{2\rho^3} - \left(\frac{\lambda}{k_c} + \frac{c_0^2}{2} \right) \frac{\sin \psi}{\rho} + \frac{\Delta p}{k_c} \right] = 0. \qquad (3.163)$$

Under no external force field, any variation of the physical or chemical condition such as the pressure difference or the chemical content will certainly cause variation of the surface of a vesicle with constant area and volume along the normal direction of its surface. This is the basic physical consideration in deriving the general shape Eq. (3.46). When the shape of the vesicle possesses axisymmetry, the shape equation reduces to Eq. (3.163), the so-called HO shape equation. However, there are other more mathematical than physical ways of deriving the shape equation for axisymmetric vesicles. That is to express the shape energy Eq. (3.38) in an action form and to find the conditions that the shape energy is at its extreme value.

Mathematically, this is simply a variational problem with or without subsidiary conditions.[49] There are different ways of choosing the independent variable. For simplicity, only two ways will be discussed. One way is to use the arc length s of the cross-sectional curve of the vesicle in the plane containing the axis of symmetry as the independent variable (the s-parameterization). Another way is to use the distance ρ of the surface from the axis of symmetry as the independent variable (the ρ-parameterization). The ρ-parameterization was initiated by Helfrich[29] and used by many workers in numerical calculations. From Fig. 3.16, it is clear that the area element dA and the volume element dV are given by

$$dA = 2\pi\rho(s)ds = 2\pi\rho d\rho/\cos\psi,$$

and

$$dV = \pi\rho^2(s)dz = \pi\rho^2(s)\sin\psi ds = \pi\rho^2 \tan\psi d\rho,$$

respectively. With Eq. (3.162), the shape energy Eq. (3.38) of an axisymmetric vesicle has the following forms:

In s-parameterization,

$$F_s = 2\pi k_c \int \left[\frac{\rho}{2} \left(\frac{d\psi}{d\rho} + \frac{\sin\psi}{\rho} + c_0 \right)^2 + \bar{\lambda}\rho + \frac{1}{2}\overline{\Delta p}\rho^2 \sin\psi \right] ds, \quad (3.164a)$$

In ρ-parameterization,

$$F_\rho = 2\pi k_c \int \left[\frac{\rho}{2\cos\psi} \left(\frac{d\psi}{d\rho} + \frac{\sin\psi}{\rho} + c_0 \right)^2 + \frac{\bar{\lambda}\rho}{\cos\psi} + \frac{1}{2}\overline{\Delta p}\rho^2 \tan\psi \right] d\rho,$$
$$(3.164b)$$

where $\bar{\lambda} = \lambda/k_c$ and $\overline{\Delta p} = \Delta p/k_c$. It should be pointed out that, in variational calculation, the solution depends not only on the energy integral itself but also on the constraints between the variables in the energy integral and the boundary conditions of the problem under consideration. The various constraints in the present case are

$$\dot{\rho} = \frac{d\rho}{ds} = \cos\psi, \quad \dot{z} = \frac{dz}{ds} = \sin\psi, \quad \frac{dz}{d\rho} = \tan\psi, \quad \left(\frac{d\rho}{ds} \right)^2 + \left(\frac{dz}{ds} \right)^2 = 1,$$
$$(3.165)$$

depending on the independent variable chosen.

In s-parameterization, the action function $\mathcal{L}$ is in the form

$$\mathcal{L}(\psi, \dot{\psi}, \rho, \dot{\rho}, \gamma, \eta, \epsilon, s)$$
$$= (1/2)\rho(\dot{\psi} + \sin\psi/\rho + c_0^2)^2 + \bar{\lambda}\rho + (1/2)\overline{\Delta p}\rho^2 \sin\psi$$
$$+ \gamma(s)(\dot{\rho} - \cos\psi) + \eta(s)(\dot{z} - \sin\psi) + \epsilon(s)(\dot{\rho}^2 + \dot{z}^2 - 1), \quad (3.166)$$

where $\gamma(s), \eta(s)$ and $\epsilon(s)$ are Lagrange multipliers for the three constraints. The three constraints are added in order to take the variation of $\mathcal{L}$ with respect to $\psi(s), \rho(s)$ and $z(s)$ independently. The variation of $\int_{s_0}^{s_1} \mathcal{L}ds$ is given by[50]

$$\delta \int_{s_0}^{s_1} \mathcal{L}ds$$
$$= \int_{s_0}^{s_1} \left\{ \left[\frac{\partial \mathcal{L}}{\partial \psi} - \frac{d}{ds} \left(\frac{\partial \mathcal{L}}{\partial \dot{\psi}} \right) \right] \delta\psi + \left[\frac{\partial \mathcal{L}}{\partial \rho} - \frac{d}{ds} \left(\frac{\partial \mathcal{L}}{\partial \dot{\rho}} \right) \right] \delta\rho - \frac{d}{ds} \left(\frac{\partial \mathcal{L}}{\partial \dot{z}} \right) \delta z \right.$$
$$\left. + \frac{\partial \mathcal{L}}{\partial \gamma} \delta\gamma + \frac{\partial \mathcal{L}}{\partial \eta} \delta\eta + \frac{\partial \mathcal{L}}{\partial \epsilon} \delta\epsilon \right\} ds - (\mathcal{H}_1 \Delta s_1 - \mathcal{H}_0 \Delta s_0)$$

$$+ \left[\left(\frac{\partial \mathcal{L}}{\partial \dot{\psi}} \right)_1 \Delta \psi_1 - \left(\frac{\partial \mathcal{L}}{\partial \dot{\psi}} \right)_0 \Delta \psi_0 \right] + \left[\left(\frac{\partial \mathcal{L}}{\partial \dot{\rho}} \right)_1 \Delta \rho_1 - \left(\frac{\partial \mathcal{L}}{\partial \dot{\rho}} \right)_0 \Delta \rho_0 \right]$$

$$+ \left[\left(\frac{\partial \mathcal{L}}{\partial \dot{z}} \right)_1 \Delta z_1 - \left(\frac{\partial \mathcal{L}}{\partial \dot{z}} \right)_0 \Delta z_0 \right], \tag{3.167}$$

where the subscripts 1 and 0 indicate values evaluated at the end points s_1 and s_0, respectively, and the Hamiltonian function $\mathcal{H}$ is denoted by

$$\mathcal{H} = -\mathcal{L} + \dot{\psi} \frac{\partial \mathcal{L}}{\partial \dot{\psi}} + \dot{\rho} \frac{\partial \mathcal{L}}{\partial \dot{\rho}} + \dot{z} \frac{\partial \mathcal{L}}{\partial \dot{z}}. \tag{3.168}$$

The last four terms in Eq. (3.167) come from the variations of the two end points s_1 and s_0 of the boundary. Since the variations $\delta\psi$, $\delta\rho$, δz, $\delta\gamma$, $\delta\eta$ and $\delta\epsilon$ are independent of each other, it requires that each coefficient of these terms, together with the four terms relating to the boundary in Eq. (3.167), should vanish separately. In this way, it leads to six Euler–Lagrange equations and four equations for the boundary.

In the case of Δs_1 and Δs_0 being free and independent of each other, the condition $\mathcal{H}_1 \Delta s_1 - \mathcal{H}_0 \Delta s_0 = 0$ requires that $\mathcal{H}_1 = \mathcal{H}_0 = 0$. Since both $\mathcal{L}$ and $\mathcal{H}$ do not contain s explicitly, $\mathcal{H}$ is conserved at both end points and in the bulk. This leads to another condition

$$\mathcal{H} = 0. \tag{3.169}$$

With Eq. (3.166) and some simplification, the six Euler–Lagrange equations and the four boundary conditions together with Eq. (3.169) lead to

$$\gamma \sin\psi - \eta \cos\psi = \rho \ddot{\psi} + \dot{\psi} \cos\psi - \sin\psi \cos\psi / \rho - (1/2) \overline{\Delta p} \rho^2 \cos\psi, \tag{3.170a}$$

$$\dot{\gamma} = (1/2)[(\dot{\psi} + c_0)^2 - \sin^2\psi / \rho^2] + \bar{\lambda} + \overline{\Delta p} \rho \sin\psi - 2\dot{\epsilon} \cos\psi + 2\epsilon \dot{\psi} \sin\psi, \tag{3.170b}$$

$$\dot{\eta} = -2\epsilon \dot{\psi} \cos\psi - 2\dot{\epsilon} \sin\psi, \tag{3.170c}$$

$$\dot{\rho} = \cos\psi, \tag{3.170d}$$

$$\dot{z} = \sin\psi, \tag{3.170e}$$

$$\dot{\rho}^2 + \dot{z}^2 = 1, \tag{3.170f}$$

$$\rho_1(\dot{\psi} + \sin\psi / \rho + c_0)_1 = \rho_0(\dot{\psi} + \sin\psi / \rho + c_0)_0 = 0, \tag{3.170g}$$

$$\gamma_1 + 2\dot{\rho}_1\epsilon_1 = \gamma_0 + 2\dot{\rho}_0\epsilon_0 = 0, \tag{3.170h}$$

$$\eta_1 + 2\dot{z}_1\epsilon_1 = \eta_0 + 2\dot{z}_0\epsilon_0 = 0, \tag{3.170i}$$

$$\gamma\cos\psi + \eta\sin\psi = -(1/2)\rho\dot{\psi}^2 + \sin^2\psi/(2\rho) + c_0\sin\psi + (1/2)c_0^2\rho$$
$$+ \bar{\lambda}\rho + (1/2)\overline{\Delta p}\rho^2\sin\psi - 2\epsilon, \tag{3.170j}$$

respectively. Equations (3.170a) and (3.170j) can be solved, which give

$$\gamma = \rho\sin\psi\ddot{\psi} + \sin\psi\cos\psi\dot{\psi} - (1/2)\rho\cos\psi\dot{\psi}^2 - \sin^2\psi\cos\psi/(2\rho)$$
$$+ (1/2)c_0^2\rho\cos\psi + c_0\sin\psi\cos\psi + \bar{\lambda}\rho\cos\psi - 2\epsilon\cos\psi, \tag{3.171}$$

$$\eta = -\rho\cos\psi\ddot{\psi} - (1/2)\rho\dot{\psi}^2\sin\psi - \cos^2\psi\dot{\psi} + \sin\psi(1 + \cos^2\psi)/(2\rho)$$
$$+ (1/2)c_0^2\rho\sin\psi + c_0\sin^2\psi + \bar{\lambda}\rho\sin\psi + (1/2)\overline{\Delta p}\rho^2 - 2\epsilon\sin\psi. \tag{3.172}$$

Substitution of Eq. (3.171) into Eq. (3.170b) and Eq. (3.172) into Eq. (3.170c), respectively, gives the following two equations:

$$\rho d^3\psi/ds^3 + (1/2)\rho\dot{\psi}^3 + 2\cos\psi\ddot{\psi} - (3/2)\sin\psi\dot{\psi}^2$$
$$- [(2\cos^2\psi - \sin^2\psi)/(2\rho) + 2c_0\sin\psi + (1/2)c_0^2\rho + \bar{\lambda}\rho]\dot{\psi}$$
$$+ [\sin\psi(1 + \cos^2\psi)/(2\rho)^2 - (1/2)c_0^2\sin\psi - \bar{\lambda}\sin\psi - \overline{\Delta p}\rho]$$
$$= -4\epsilon\dot{\psi} \tag{3.173a}$$

and

$$\rho d^3\psi/ds^3 + (1/2)\rho\dot{\psi}^3 + 2\cos\psi\ddot{\psi} - (3/2)\sin\psi\dot{\psi}^2$$
$$- [(2\cos^2\psi - \sin^2\psi)/(2\rho) + 2c_0\sin\psi + (1/2)c_0^2\rho + \bar{\lambda}\rho]\dot{\psi}$$
$$+ [\sin\psi(1 + \cos^2\psi)/(2\rho)^2 - (1/2)c_0^2\sin\psi - \bar{\lambda}\sin\psi - \overline{\Delta p}\rho] = 0, \tag{3.173b}$$

The two equations in (3.173) are the same equation if ψ is independent of s or if

$$\epsilon = 0. \tag{3.174}$$

In the first case, the Euler–Lagrange equation for the ψ-field becomes meaningless. In other words, for the cylinder $\psi = $ const, no Euler–Lagrange shape equation can be obtained. However, in this case, Eq. (3.173) leads directly to a constraint condition similar to Eq. (3.90). On the other hand, under condition (3.174), transformation of s into ρ makes Eq. (3.173) exactly Eq. (3.163) times ρ. Julicher and Seifert[28] have shown that without the

ϵ term in Eq. (3.166), it gives directly Eq. (3.163). However, it must be remembered that this is true only for genus zero surfaces where the two end points are free and independent of each other. It is interesting to see that if the η term in Eq. (3.166) is neglected, i.e., let $\eta = 0$, the same set of Eq. (3.173) exists, only in this case $\epsilon \neq 0$ and $\dot{\psi} = 0$. This is no wonder because only two of the constraints $\dot{\rho} = \cos\psi$, $\dot{z} = \sin\psi$ and $\dot{\rho}^2 + \dot{z}^2 = 1$ are independent. One notices that in case of $\epsilon = 0$, Eq. (3.170c) gives $\dot{\eta} = 0$, in other words, $\eta = \eta_0$ (a constant) is the first integral of Eq. (3.173b). The transformation of s into ρ now changes Eq. (3.172) into

$$\cos^2\psi \mathrm{d}^2\psi/\mathrm{d}\rho^2$$

$$- (1/2)\sin\psi\cos\psi(\mathrm{d}\psi/\mathrm{d}\rho)^2 + (\cos^2\psi/\rho)\mathrm{d}\psi/\mathrm{d}\rho - \sin 2\psi/(2\rho)^2$$

$$- [\sin\psi/(2\cos\psi)](\sin\psi/\rho + c_0)^2 - \bar{\lambda}\sin\psi/\cos\psi - \overline{\Delta p}\rho/(2\cos\psi)$$

$$= \eta_0/(\rho\cos\psi). \tag{3.175}$$

In case of $\eta_0 = 0$, Eq. (3.175) is another shape equation (called DH shape equation) which was first derived by Deuling and Helfrich.[29, 30] This fact has been pointed out by Zheng and Liu.[51]

It should be emphasized that Eq. (3.173b) comes from the existence of Eq. (3.169), i.e., the variations Δs_1 and Δs_0 are free and independent of each other. For vesicles such as of toroidal topology where s_1 coincides with s_0 and $\Delta s_0 = \Delta s_1$, Eq. (3.169) does not exist. In such cases, there exists only the condition $\mathcal{H}_1 = \mathcal{H}_0$ which does not have enough equations to solve for γ, η and ϵ, and no shape equation can be obtained.

In s-parameterization, the energy integral Eq. (3.164a) does not involve the functional z explicitly. Mathematically, it is perfectly acceptable to take the action function in the form

$$\mathcal{L}(\psi, \dot{\psi}, \rho, \dot{\rho}, s) = (1/2)\rho(\dot{\psi} + \sin\psi/\rho + c_0)^2 + \bar{\lambda}\rho$$

$$+ (1/2)\overline{\Delta p}\rho^2 \sin\psi + \gamma(s)(\dot{\rho} - \cos\psi), \tag{3.176}$$

without the extra η term and the ϵ term as given in Eq. (3.166). The three Euler–Lagrange equations for the ψ-field, the ρ-field and the γ-field are sufficient to give an equation

$$\mathrm{d}^3\psi/\mathrm{d}s^3 - (\cos\psi/\sin\psi)\dot{\psi}\ddot{\psi} + (2\cos\psi/\rho)\ddot{\psi} - [(2 + \sin^2\psi)/(2\rho\sin\psi)]\dot{\psi}^2$$

$$+ (\sin^2\psi/\rho^2 - c_0\sin\psi/\rho + \overline{\Delta p}\rho/(2\sin\psi))\dot{\psi} + [\sin\psi(2 - \sin^2\psi)/(2\rho^3)$$

$$- (2\bar{\lambda} + c_0^2)\sin\psi/(2\rho) - \overline{\Delta p}] = 0. \tag{3.177a}$$

By transformation of s into ρ, it becomes

$$\cos^3 \psi (d^3\psi/ds^3) - (3\sin\psi + 1/\sin\psi)\cos^2\psi(d\psi/d\rho)(d^2\psi/d\rho^2)$$

$$+ \sin^2\psi\cos\psi(d\psi/d\rho)^3 - [(2 + 5\sin^2\psi)\cos^2\psi/(2\rho\sin\psi)](d\psi/d\rho)^2$$

$$+ (2\cos^3\psi/\rho)(d^2\psi/d\rho^2) - (c_0\sin\psi/\rho - \sin^2\psi/\rho^2$$

$$- \overline{\Delta p}\rho/(2\sin\psi))\cos\psi(d\psi/d\rho)$$

$$+ [\sin\psi(2 - \sin^2\psi)/(2\rho^3) - (c_0^2 + 2\bar\lambda)\sin\psi/\rho - \overline{\Delta p}] = 0. \qquad (3.177b)$$

This equation was derived by Seifert, Berndl and Lipowsky,[36] which is called the SBL equation and is obviously different from the HO shape equation (3.163). Nevertheless, the DH shape equation is simply a first integral of the HO shape equation with vanishing constant of integration.

In ρ-parameterization, the action function becomes

$$\mathcal{L} = \frac{\rho}{2\cos\psi(\rho)}\left(\cos\psi\frac{d\psi}{d\rho} + \frac{\sin\psi}{\rho} + c_0\right)^2 + \frac{\bar\lambda\rho}{\cos\psi}$$

$$+ \frac{\overline{\Delta p}\rho^2\sin\psi}{2\cos\psi} + \eta(\rho)\left(\frac{dz}{d\rho} + \frac{\sin\psi}{\cos\psi}\right), \qquad (3.178)$$

where the last term takes care of the constraint on the boundary. The vanishing of the coefficients of the variations on the boundary in Eq. (3.167) gives nothing concerning the bulk, they may be left there as they are. The Euler–Lagrange equation for the ψ-field now gives exactly Eq. (3.175). Similar to s-parameterization, the energy integral Eq. (3.164b) does not involve z explicitly. Mathematically, it is also acceptable to take the action integral simply as

$$\mathcal{L} = \frac{\rho}{2\cos\psi(\rho)}\left(\cos\psi\frac{d\psi}{d\rho} + \frac{\sin\psi}{\rho} + c_0\right)^2 + \frac{\bar\lambda\rho}{\cos\psi} + \frac{\overline{\Delta p}\rho^2\sin\psi}{2\cos\psi}. \qquad (3.179)$$

The Euler–Lagrange equation for the ψ-field in this case directly gives the DH shape equation, i.e., a first integral of the HO shape equation with vanishing constant of integration.

Hu and Ou-Yang[27] have shown that, for spherical vesicle $\rho = R\sin\psi$, the HO equation, DH equation and SBL equation give the same constraint condition

$$\Delta p R^3 + 2\lambda R^2 + k_c c_0^2 R^2 - 2k_c c_0 R = 0.$$

In the case of cylindrical vesicles, for which $\rho = R$ and $\psi = \pi/2$, both HO equation and SBL equation give the constraint condition

$$\Delta p R^3 + \lambda R^2 + \frac{1}{2}k_c(c_0^2 R^2 - 1) = 0,$$

while the DH equation gives

$$\Delta p R^3 + 2\lambda R^2 + k_c(c_0 R + 1)^2 = 0,$$

which is also a balance equation for the cylinder (note that c_0 may be positive or negative).

In cases of Clifford torus, Fig. 3.13 shows that $\rho = R + r\sin\psi$, where $0 \le \psi \le 2\pi$. With the HO equation, the resulting constraints are

$$(R/r)^2 = 2, \quad \Delta p = -2k_c c_0/c^2, \quad \bar{\lambda} = -2k_c(c_0/c + c_0^2/4),$$

where $c^2 = r^2$. With the DH equation, it can be shown easily that the results are

$$(R/r)^2 = 2, \quad \overline{\Delta p} = k_c/r^3, \quad \lambda = -9k_c/(8r^2), \quad c_0 = 1/(2r).$$

With the SBL equation it is easy to show that the constraints are

$$(R/r)^2 = 2, \quad \Delta p = k_c/r^3, \quad \lambda = -k_c[1/(2r^2) + c_0/r + c_0^2/2].$$

It is clear that for $c_0 = 1/(2r)$, the above three equations are same.

All the special vesicle cases discussed above show that although the DH equation and the SBL equation are shape equations, unlike the HO equation, they give only restricted shapes of individual vesicles. Euler–Lagrange shape equations give only shape energies at these extreme values, not necessarily at their minimum value. The HO equation is more versatile and admits all the possible solutions. If we denote

$$H(\rho) = \rho\cos\psi\{\cos^2\psi(\mathrm{d}^2\psi/\mathrm{d}\rho^2) - \sin 2\psi(\mathrm{d}\psi/\mathrm{d}\rho)^2/4 + (\cos^2\psi/\rho)\mathrm{d}\psi/\mathrm{d}\rho$$
$$- \sin 2\psi/(2\rho^2) - [\sin\psi/(2\cos\psi)](\sin\psi/\rho + c_0)^2 - \bar{\lambda}\sin\psi/\cos\psi$$
$$- \overline{\Delta p}\rho/(2\cos\psi)\},$$

then the HO equation, DH equation and SBL shape equation are simply

$$\frac{\mathrm{d}H(\rho)}{\mathrm{d}\rho} = 0, \quad H(\rho) = \eta_0, \tag{3.180}$$

$$\frac{H(\rho)}{\rho} = 0, \tag{3.181}$$

and

$$\frac{dH(\rho)}{d\rho} - \frac{\cos\psi}{\sin\psi}\frac{d\psi}{d\rho}H(\rho) = 0, \quad H(\rho) = \text{const.} \times \sin\psi, \quad (3.182)$$

respectively. Podgornik, Svetina and Zeks using bilayer-coupled model, with area parameterization, obtained a shape equation in the form[52]:

$$\frac{dH(\rho)}{d\rho} - \frac{H(\rho)}{\rho} = 0, \quad H(\rho) = \text{const.} \times \rho. \quad (3.183)$$

The action function used by them is just $1/2$ of the function given by Eq. (3.178). The fact that $H(\rho)/\rho = 0$ is the common solution to all the shape equations is in consistency with parameterization invariance of variational calculus.[53]

The diversity of the shapes of vesicles are fascinating. For a long time no successful theoretical treatment on this problem (except numerical calculations) has been found. The appearance of the general shape Eq. (3.46) offers a starting point for an analytical study of this problem. However, the general shape equation (3.46) is a fourth-order nonlinear differential equation. So far, no detailed study on such a differential equation has ever been done. For the time being, it is only possible to try some known geometrical figures to see whether they are solutions of the shape equation. Geometrical figures with known analytical expressions are rare. It is necessary now to pay more attention in studying the solutions to the differential equation itself. The shape equation obtained comes from the Helfrich theory of lipid bilayer, which is developed based on the similarity between lipid bilayer and smectic liquid crystal. Lipid bilayers are closely related to biomembranes. The success of the shape equation in describing the shapes of some simpler vesicles indicates a step forward in our understanding of the formation of biomembranes. It means that our understanding on this part of biophysics has progressed from simple observations to possible theoretical predictions, a significant step forward. A new branch of biophysics is now in sight.

3.10 Circular Biconcave Discoid

Mature red blood cells of mammals and human beings have no nuclei. The balance of forces acting on the red blood cell membrane regulates the shape of the cell. This fact makes the shape of red blood cell an ideal subject for the study of biomembranes. Observations show that mature red blood cells of human beings are always in the shape of biconcave discoid.

In fact, it was the problem "why red blood cells are always in biconcave discoidal shape" that motivated Professor Helfrich to get interested in the study of biomembranes. However, as yet, there is no general mathematical representation for biconcave discoidal surfaces. Only recently, Naito et al.[54] suggested an expression which may give an answer.

The contour suggested by Naito et al. is described by

$$\psi = \arcsin[\rho(a\ln\rho + b)] = \arcsin[a\rho\ln(\rho/\rho_B)], \qquad (3.184)$$

where a and b are constants determined by the shape equation and the sign of the vesicle and $\rho_B = \exp(-b/a)$. Direct differentiation gives that

$$\frac{d\psi}{d\rho} = \frac{1}{\cos\psi}\left(\frac{\sin\psi}{\rho} + a\right),$$

$$\frac{d^2\psi}{d\rho^2} = \frac{\sin\psi}{\cos^3\psi}\left(\frac{\sin\psi}{\rho} + a\right)^2 + \frac{a}{\rho\cos\psi},$$

$$\frac{d^3\psi}{d\rho^3} = \left(\frac{3\sin^2\psi}{\cos^5\psi} + \frac{1}{\cos^3\psi}\right)\left(\frac{\sin\psi}{\rho} + a\right)^3$$

$$+ \frac{3a\sin\psi}{\rho\cos^3\psi}\left(\frac{\sin\psi}{\rho} + a\right) - \frac{a}{\rho^2\cos\psi}. \qquad (3.185)$$

Substitution of Eq. (3.185) into Eq. (3.163) gives

$$2(a + c_0)\left(\frac{\sin\psi}{\rho}\right)^2 + \left[(c_0 + a)^2 + \frac{2\lambda}{k_c}\right]\frac{\sin\psi}{\rho}$$

$$- \left[\frac{\Delta p}{k_c} - \frac{1}{2}a(c_0^2 - a^2) - \frac{a\lambda}{k_c}\right] = 0. \qquad (3.186)$$

Since Eq. (3.186) must be satisfied for any value of ψ, it requires that

$$a + c_0 = 0,$$

$$(c_0 + a)^2 + 2\lambda/k_c = 0, \qquad (3.187)$$

$$\Delta p/k_c - (1/2)a(c_0^2 - a^2) - a\lambda/k_c = 0.$$

Equation (3.187) shows that Eq. (3.184) is a rigorous solution of the general shape equation under the conditions

$$a = -c_0 \quad \text{and} \quad \Delta p = \lambda = 0. \qquad (3.188)$$

With dimensionless variable $x = \rho/\rho_B$, Eq. (3.184) may be written as

$$\sin\psi = \beta x \ln x, \quad \beta = -c_0\rho_B. \tag{3.189}$$

The function $z(x)$ is given by

$$z = \int \frac{dz}{d\rho}d\rho = \int \frac{dz}{dx}dx = \int \tan\psi(x)dx. \tag{3.190}$$

Unfortunately, no simple analytical expression of $z(x)$ can be obtained in the present case. For the convenience of discussion, consider the case of

$$\beta < 0 \quad \text{and} \quad |\beta| < e \tag{3.191}$$

first. It is easy to see that $\sin\psi = 0$ at $x = 0$ and $x = 1$ and that $\sin\psi$ has its maximum value at $x = e^{-1}$ with $\sin\psi_M = |\beta|e^{-1}$. The constraint $\sin\psi \leq 1$ limits the possible values of x in the range $0 \leq x \leq x_c$, where x_c satisfies the condition

$$x_c \ln x_c = |\beta|^{-1}. \tag{3.192}$$

Numerical calculation shows that under the condition (3.191) there exists only one value of $x > 1$ that satisfies Eq. (3.192). Corresponding to Eq. (3.189), there are two functions of $\cos\psi$:

$$\cos\psi = \pm(1 - x^2\ln^2 x)^{1/2}. \tag{3.193}$$

With the upper sign, it follows that $\tan\psi = 0$ at $x = 0$. As x increases, $\tan\psi$ increases at first and then decreases to zero at $x = 1$ and then becomes infinite at x_c. The profile of the function $z(x)$ is shown in Fig. 3.17. It describes the upper half of a circular biconcave discoidal surface. Likewise, the lower sign of Eq. (3.193) describes the lower half of the discoid. In fact, for

$$\beta > 0 \quad \text{and} \quad \beta < e, \tag{3.194}$$

the situation is exactly the same, except for an interchange of the plus and the minus sign of Eq. (3.193). In the special case of $a = c_0 = 0$, Eq. (3.184) reduces to $\rho = R\sin\psi$ with $b = 1/R$, the equation of a sphere with radius R.

The best-fitted contour line of red blood cell by Eq. (3.184) is also shown as a dotted line in Fig. 3.18, which is in good agreement with the experimental shape of red blood cell in the whole range from ρ_0 to ρ_c with slight deviation. In principle, the deviation does not have a serious meaning because Evans and Fung's experimental expression indeed represents

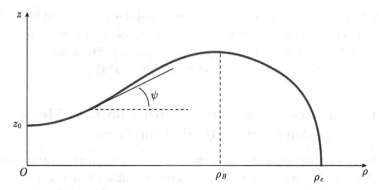

Fig. 3.17. One quadrant of the cross-section of the contour described by Eq. (3.184). There is rotational symmetry around the z-axis and reflection symmetry about the ρ-axis.

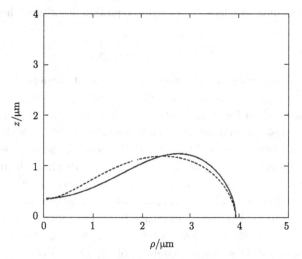

Fig. 3.18. Cross-section of cell shapes. Only one quadrant is shown. There is rotational symmetry around the z axis and reflection symmetry at the ρ axis. Solid line represents the experimentally obtained expression for RBCs by Evans and Fung.[55] Dashed line represents Eq. (3.184).

an averaged shape over 50 red blood cells. Furthermore, the parameters $\rho_B = 0.62\rho_c$ and $c_0 = -0.52$ μm for red blood cells can be fitted from the experiments.

It is interesting to note that instead of revolving around the z-axis, rotation of the contour given by Eq. (3.184) about the ρ-axis will lead to a dumb-bell shaped surface. However, no dumb-bell shaped red blood cell has ever been observed. Since no analytical expression $z(\rho)$ can be found, in the present case, it is not possible to examine the condition of the existence of dumb-bell shaped surfaces generated by Eq. (3.184).

3.11　Surfaces of Revolution with Constant Mean Curvature and Extended Surfaces

In 1841, Delaunay showed that surfaces of revolution with constant mean curvature in Euclidean space are spheres, circular cylinders, catenoids, unduloids and nodoids.[56] These surfaces are simply called the Delaunay surfaces. The general procedure of generating a Delaunay surface is to roll a conic section along a line in its plane and the rotation of the trace of one of the foci of the conic section around the given line (e.g. the z-axis) generates the Delaunay surface. The general equation describing the Delaunay surface is[57]

$$\sin \psi(\rho) = a\rho + d/\rho, \qquad (3.195)$$

where the parameters a and d determine the type of surfaces: (1) $a = 0$, the catenoid, the surface generated by the focus of a parabola, the catenary, i.e., the shape of an inextensible free-hanging chain with fixed two ends; (2) $0 < ad < 1/4$, the unduloid, the surface generated by one of the foci of an ellipse, a wave-like curve; and (3) $ad < 0$, the nodoid, the surface generated by one of the foci of a hyperbola as shown in Fig. 3.19 where the nodules are directed toward the axis of revolution. Since ρ is always positive and ψ can be counted either clockwise or counterclockwise, the parameter a may be considered positive. The mean curvature $H = (1/2)(\cos \psi d\psi/d\rho + \sin \psi/\rho)$ of the surface described by Eq. (3.195) is simply equal to a, which also implies that $a > 0$. One simple extension of Eq. (3.186) is[58]

$$\sin \psi(\rho) = a\rho + b + d/\rho, \qquad (3.196)$$

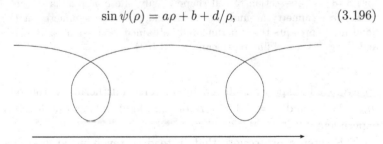

Fig. 3.19.　The nodoid with the bottom line as the axis of revolution.

with an additional constant b besides a and d in describing the Delaunay surfaces. Equation (3.196) describes surfaces of revolution not necessarily with constant mean curvature. The special case of $b = 0$ reduces Eq. (3.196) to Delaunay surfaces.

It is interesting to see under what conditions the surface described by Eq. (3.196) will be the solution of the general shape Eq. (3.163). Simple differentiation gives that

$$
\begin{aligned}
\mathrm{d}\psi/\mathrm{d}\rho &= (a - d/\rho^2)\sec\psi, \\
\mathrm{d}^2\psi/\mathrm{d}\rho^2 &= (a - d/\rho^2)^2\tan\psi\sec^2\psi + 2(d/\rho^3)\sec\psi, \\
\mathrm{d}^3\psi/\mathrm{d}\rho^3 &= (a - d/\rho^2)^3(\sec^5\psi + 2\tan^2\psi\sec^3\psi) \\
&\quad + (6d/\rho^3)(a - d/\rho^2)\tan\psi\sec^2\psi - 6(d/\rho^4)\sec\psi.
\end{aligned}
\tag{3.197}
$$

Substitution of Eq. (3.197) into Eq. (3.163) gives that

$$
\begin{aligned}
&\frac{1}{2\rho^3}\sin^3\psi - \frac{1}{2}\left(\frac{d}{\rho^4} + \frac{3a}{\rho^2}\right)\sin^2\psi \\
&- \left[\frac{d^2}{2\rho^5} + (ad + 2c_0d + 1)\frac{1}{\rho^3} - \left(\frac{3}{2}a^2 + 2ac_0 + \frac{\lambda}{k_c} + \frac{c_0^2}{2}\right)\frac{1}{\rho}\right]\sin\psi \\
&+ \frac{d^3}{2\rho^6} + d\left(1 - \frac{3}{2}ad\right)\frac{1}{\rho^4} + \left[\frac{3}{2}a^2d + a - d\left(\frac{\lambda}{k_c} + \frac{c_0^2}{2}\right)\right]\frac{1}{\rho^2} - \frac{1}{2}a^3 \\
&+ a\left(\frac{\lambda}{k_c} + \frac{c_0^2}{2}\right) - \frac{\Delta p}{k_c} = 0.
\end{aligned}
\tag{3.198}
$$

With Eq. (3.196), Eq. (3.198) becomes

$$
\begin{aligned}
&\left[-\frac{\Delta p}{k_c} + 2a^2c_0 + 2a\left(\frac{\lambda}{k_c} + \frac{c_0^2}{2}\right)\right] + b\left[2ac_0 + \left(\frac{\lambda}{k_c} + \frac{c_0^2}{2}\right)\right]\frac{1}{\rho} \\
&+ b\left(\frac{1}{2}b^2 - 2ad - 2c_0d - 1\right)\frac{1}{\rho^3} + d(b^2 - 4ad - 2c_0d)\frac{1}{\rho^4} = 0.
\end{aligned}
\tag{3.199}
$$

Since Eq. (3.199) should be satisfied by different values of ρ, therefore Eq. (3.196) is a solution of the general shape provided by Eq. (3.163):

$$
\begin{aligned}
&- \Delta p/k_c + 2a^2c_0 + 2a(\lambda/k_c + c_0^2/2) = 0, \\
&b[2ac_0 + (\lambda/k_c + c_0^2/2)] = 0, \\
&b(b^2/2 - 2ad - 2c_0d - 1) = 0, \\
&d(b^2 - 4ad - 2c_0d) = 0.
\end{aligned}
\tag{3.200}
$$

Equation (3.200) shows that there are four different cases:

(1) $b = d = 0$, $\sin\psi = a\rho$. With $\rho^2 = x^2 + y^2$ and $\tan\psi = dz/d\rho = \pm\sin\psi(1-\pm\sin^2\psi)^{-1/2}$, integration of $dz/d\rho$ leads to the solution of spherical surfaces with radius $R = |1/a|$. Here, the first equation of (3.200) is simply the condition Eq. (3.59) for spherical surfaces being possible vesicles.

(2) $b = 0$, $d \neq 0$, $\sin\psi = a\rho + d/\rho$. They are the Delaunay surfaces with constant mean curvature $H = a$. The requirements for Delaunay surfaces being vesicles are $c_0 = -2a$ and $\Delta p - 2a\lambda = 0$.

(3) $b \neq 0$, $d = 0$, $\sin\psi = a\rho + b$. In this case, the integration of $dz/d\rho$ gives the torus equation

$$[x^2 + y^2 + z^2 + (b^2 - 1)/a^2]^2 = 4(b/a)^2(x^2 + y^2), \qquad (3.201)$$

and Eq. (3.200) gives the constraints $b^2 = 2$, $\lambda/k_c = -(2ac_0 + c_0^2/2)$, $\Delta p = -2k_c a^2 c_0$. Comparison of Eq. (3.201) with Eqs. (3.136) and (3.137) shows that $|1/a|$ is simply the radius of the smaller generating circle of the Clifford torus.

(4) $bd \neq 0$, $\sin\psi = a\rho + b + d/\rho$. In this case, Eq. (3.200) leads to the following constraints:

$$\Delta p/k_c = -2a^2 c_0, \quad \lambda/k_c + c_0^2/2 = -2ac_0, \quad c_0 d = -1, \quad b^2 = -(2 + 4a/c_0). \tag{3.202}$$

In both cases (3) and (4), the surfaces do not have constant mean curvatures. Case (4) is interesting and needs to be studied in more detail.

In the case of $bd \neq 0$, to have real value of b, it is necessary that a and c_0 must have different signs and that $-2a/c_0 \geq 1$. Equation (3.202) shows that there are two possible values of b, $b_\pm = \pm(-4a/c_0 - 2)^{1/2}$, with corresponding surfaces described by

$$\sin\psi = a\rho + b_\pm - 1/(c_0\rho), \qquad (3.203a)$$

$$\rho = [1/(2a)][(\sin\psi - b_\pm) \pm (\sin^2\psi - 2b_\pm \sin\psi - 2)^{1/2}], \quad (3.203b)$$

where ρ is restricted to be real and positive. With $|\sin\psi| \leq 1$, Eq. (3.203b) indicates that b_+ should be discarded. Deuling and Helfrich have calculated Delaunay unduloids numerically to interpret the myelin shape of red blood cells.[59] Their calculation does not include the surfaces described by Eq. (3.203). It seems that Eq. (3.203) may be the answer to the pearling cylinder observed by Bar-ziv and Moses.[60]

Let $\rho = \rho_m$ be the point where $\mathrm{d}\sin\psi(\rho)/\mathrm{d}\rho = 0$, then $\rho_m = 1/\sqrt{-ac_0}$. The b_- branch surface may then be represented in terms of a dimensionless variable x in the form of

$$\sin\psi(x) = \alpha(x + 1/x) - (4\alpha^2 - 2)^{1/2},$$

$$x = \rho/\rho_m = (-ac_0)^{1/2}\rho, \quad \alpha = (-a/c_0)^{1/2}, \qquad (3.204)$$

$$\sin\psi_m = 2\alpha - (4\alpha^2 - 2)^{1/2}.$$

The constraint $|\sin\psi| \le 1$ requires that $\alpha = \sqrt{-a/c_0} \ge 3/4$. The possible values of x are confined in the range

$$x_1 = (1/2\alpha)\{1 + (4\alpha^2 - 2)^{1/2} - [2(4\alpha^2 - 2)^{1/2} - 1]^{1/2}\}$$

$$\le x \le (1/2\alpha)\{1 + (4\alpha^2 - 2)^{1/2} + [2(4\alpha^2 - 2)^{1/2} - 1]^{1/2}\} = x_2,$$
$$(3.205)$$

where $\sin\psi(x_1) = \sin\psi(x_2) = 1$, $\cos\psi(x_1) = \cos\psi(x_2) = 0$. Corresponding to each value of $\sin\psi(x)$, there are two values of $\tan\psi(x)$, one positive and the other negative with the same magnitude. At the two end points x_1 and x_2, the tangents to the curve $z(x)$ are both perpendicular to the ρ-axis (i.e., x-axis). Equation (3.204) shows that in the range $x_1 \le x \le x_2$, there is nowhere that the $z(\rho)$ function has a horizontal tangent. These two facts indicate that $z(\rho)$ is undulatory and that the surfaces represented by Eq. (3.204) are unduloid like.

It is interesting to look into the special case of $\alpha = \alpha_0 = 3/4$. In this case, Eq. (3.204) becomes

$$x^2 - 2(1 + 2\sin\psi)x/3 + 1 = 0.$$

The condition of the existence of a real root of x for this equation is that $(1 + 2\sin\psi)^2 \ge 9$. The only possibility to satisfy this condition is that $\sin\psi = 1$ and $x = 1$. In other words, the unduloid-like surface of revolution degenerates into a circular cylinder of radius $\rho_m = (ac_0)^{-1/2}$ with $a/c_0 = 9/16$. Here the constraint Eq. (3.91) for circular cylinder is also satisfied automatically. Figure 3.20 shows the half-period of the contour of the surfaces under different values of α.

Equation (3.204) does not include nodoids with $H = a = -c_0/2$. This does not mean that nodoid-like solutions with non-constant mean curvature are not solutions of the general shape Eq. (3.163). Surfaces described by Eq. (3.184) may be considered as another kind of extension of Delaunay surfaces.[58] It has been shown that under the conditions $a = -c_0$, $\Delta p = \lambda = 0$ and $|\beta| = |c_0|\rho_B < e$, Eq. (3.184) describes biconcave discoidal

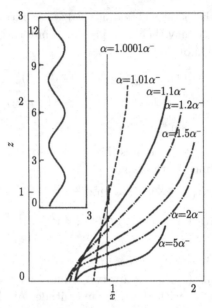

Fig. 3.20.   Half-period of unduloid-like shapes with different values of α in unit of $\alpha^- = 3/4$. The inset illustrates three periods of the shape with $\alpha = 1.1\alpha^-$.

surfaces. It is now interesting to look into the same equation under the condition

$$|\beta| > \mathrm{e}. \tag{3.206}$$

The significant difference between the present case and the case of $|\beta| < \mathrm{e}$ is that there exists three positions x_1, x_2 and x_3 at which $x \ln x = 1/|\beta|$ instead of only one position x_c. The three positions are located in the ranges

$$0 < x_1 < \mathrm{e}^{-1}, \quad \mathrm{e}^{-1} < x_2 < 1, \quad 1 < x_3, \tag{3.207}$$

respectively. Besides, on account of Eq. (3.193), at each point x, $x_1 < x < x_3$, generally there are two possible values of $\tan \psi$ with the same magnitude but opposite in sign. In particular,

$$\tan \psi(0) = 0, \quad \tan \psi(x_1) \to \pm\infty, \quad \tan \psi(x_2) \to \pm\infty,$$
$$\tan \psi(1) = 0, \quad \tan \psi(x_3) \to \pm\infty. \tag{3.208}$$

These properties of $\tan \psi(x)$ suggest that the contour $z(x)$ in the range $x_2 \le x \le x_3$ has nodules. Numerical calculation verifies these nodoid-like shapes as shown in Fig. 3.21.

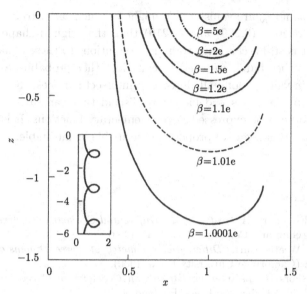

Fig. 3.21. Half-period of nodoid-like shapes with different values of β (in units of e). The inset illustrates three periods of the shape with $\beta = 1.0001$e.

The difference between the present nodoid-like shapes and the nodoids of Delaunay surfaces is that the nodules of the present case are out-stretched instead of inscribed onto the inner surface as shown in Fig. 3.19. From Fig. 3.21, it appears that the nodoid-like shape degenerates to a torus with a generating ring not of a right circle. As $\beta \to \infty$, the torus has an infinitesimal generating circle, and this surface is not a circular cylinder but a tape as shown in Sec. 3.5.

3.12 Challenge

Can we further find the other analytic solutions to the shape Eq. (3.163) or (3.175) which represent the closed vesicles without self-contact? Under certain conditions, Eq. (3.175) can be extremely simplified. If we chose a new variable

$$\xi = \frac{\sin\psi}{\rho} + \frac{d(\sin\psi)}{d\rho} - c_0, \tag{3.209}$$

Eq. (3.175) can be expressed as a very concise form[61]:

$$\frac{d}{d\rho}\left(\frac{\cos\psi}{\xi}\right) + \frac{\tan\psi}{2} = 0, \tag{3.210}$$

when λ, Δp and η_0 are vanishing. It might be much easier to find solutions to the above Eqs. (3.209) and (3.210) than the original shape equation. However, it is still a challenge to find the solutions to these equations.

On the other hand, we need to consider other probabilities if all our efforts are in vain. Among all closed non-intersect surfaces, there are probably only sphere, torus and biconcave discoid that can satisfy the shape equation and can be expressed as the elementary functions. It is also valuable to make this negative proposition verifiable or falsifiable.

References

[1] L. P. Eisenhart, *A Treatise of the Differential Geometry of Curves and Surfaces* (Ginn and Company, New York 1909).

[2] C. E. Weatherburn, *Differential Geometry of Three Dimensions*, Vols. I and II (Cambridge University Press, 1930).

[3] A. R. Forsyth, *Lectures on Differential Geometry of Curves and Surfaces* (Cambridge University Press, England, 1912).

[4] W. Flugge, *Tensor Analysis and Continuum Mechanics* (Springer-Verlag, New York, 1972).

[5] B. Spain, *Tensor Analysis* (Intersience Publishers, INC. New York 1953).

[6] W. Helfrich, *Z. Naturuforsch. C* **28** (1973) 693.

[7] H. J. Deuling and W. Helfrich, *Biophys. J.* **16** (1976) 861.

[8] Ou-Yang Zhong-Can and W. Helfrich, *Phys. Rev. Lett.* **59** (1989) 2486; *Phys. Rev. Lett.* **60** (1990) 1209.

[9] Ou-Yang Zhong-Can and W. Helfrich, *Phys. Rev. A* **39** (1989) 5280.

[10] L. I. Schiff, *Quantum Mechanics*, 3rd edn. (McGraw-Hill, New York 1968).

[11] H. J. Deuling and W. Helfrich, *J. Phys.* (Paris) **37** (1976) 1335.

[12] M. Bessis, *Living Blood Cells and Their Ultrastructure* (Springer-Verlag, Berlin 1973).

[13] M. A. Peterson, *J. Math. Phys.* **26** (1985) 711; *J. Appl. Phys.* **57** (1985) 1739; *Mol. Cryst. Liq. Cryst.* **127** (1985) 159; **127** (1985) 257.

[14] S. T. Milner and S. A. Safran, *Phys. Rev. A* **36** (1987) 437.

[15] H. J. Deuling and W. Helfrich, *Blood Cells* **3** (1977) 713.

[16] D. W. Fawcett, *The Cell* (W. B. Saunders Corp., Philadelphia, 1981).

[17] K. Blinzinger, *J. Cell. Biol.* **25** (1965) 293.

[18] U. Seifert, *Phys. Rev. A* **43** (1991) 6803.

[19] S. G. Zhang and Z. C. Ou-Yang, *Phys. Rev. E* **53** (1996) 4206.

[20] S. G. Zhang, *Ada Physica Sinica* (Oversea ed.) **6** (1997) 641.

[21] V. Vassilev, P. Djondjorov, and I. Mladenov, *J. Phys. A Math. Theor.* **41** (2008) 435201.

[22] F. Losch and Jahnke-Emde-Losch, *Tables of Higher Functions*, 6th edn. (McGraw Hill, New York, 1960).

[23] M. Abramowitz and I. A. Stegen, *Handbook of Mathematical Functions* (Dover, New York, 1970).

[24] B. O. Pierce, *A Short Table of Integrals*, 3rd edn. (Grinn and Co., New York 1929).

[25] P. F. Byrd and M. D. Friedman, *Handbook of Elliptic Integrals for Engineer and Physicists* (Springer-Verlag, New York, 1971).

[26] W. Harbich and W. Helfrich, *Chem. Phys. Lipids* **36** (1984) 39.

[27] J. G. Hu and Z. C. Ou-Yang, *Phys. Rev. E* **47** (1993) 461.

[28] F. Jülicher and U. Seifert, *Phys. Rev. E* **49** (1994) 4728.

[29] W. Helfrich, *Z. Naturforsch. C* **28** (1973) 693.

[30] H. J. Deuling and W. Helfrich, *J. Phys., Paris*, **37** (1976) 1335.

[31] M. A. Peterson, *J. Math. Phys.* **26** (1985) 711; *J. Appl. Phys.* **57** (1985) 1739; *Mol. Cryst. Liq. Cryst.* **127**(1985) 159; **127** (1985) 257.

[32] J. Jenkins, *J. Math. Biophys.* **4** (1977) 149.

[33] S. Svetina and B. Zeks, *Euro. Biophys. J.* **17** (1989) 101.

[34] L. Miao, B. Fourcade, M. Rao, M. Wortis, and R. Zia, *Phys. Rev. A* **43** (1991) 6843.

[35] U. Seifert, *Phys. Rev. Lett.* **66** (1991) 2404.

[36] U. Seifert, K. Berndl, and R. Lipowsky, *Phys. Rev. A* **44** (1991) 1182.

[37] K. Berndl, J. Kas, R. Lipowsky, E. Sackmann, and U. Seifert, *Europhy. Lett.* **13** (1990) 659.

[38] R. Lipowsky, *Nature* **349** (1991) 475.

[39] L. Miao, U. Seifert, M. Wortis, and H. G. Dobereiner, *Phys. Rev. E* **49** (1994) 5389.

[40] Z. C. Ou-Yang, *Phys. Rev. E* **41** (1990) 4517.

[41] P. M. Morse and H. Feshbach, *Methods of Theoretical Physics I* (McGraw-Hill, New York, 1953).

[42] E. W. Hobson, *The Theory of Spherical and Elliptical Harmonics* (Cambridge University Press, England, 1955).

[43] M. Mutz and D. Bensimon, *Phys. Rev. A* **43** (1991) 4525.

[44] B. Fourcade, M. Mutz, and D. Bensimon, *Phy. Rev. Lett.* **68** (1992) 2551.

[45] X. Michalet and D. Bensimon, *J. Phys. II* (Paris) 5 (1995) 263.

[46] Z. C. Ou-Yang, *Phys. Rev. E* **47** (1993) 747.

[47] A. R. Forsyth, *Lectures on Differential Geometry of Curves and Surface* (Cambridge University Press, England, 1912).

[48] Xie Yu-Zhang, *Ada Physica Sinica (Oversea ed.)* 2 (1993) 881; 3 (1994) 240.

[49] R. Courant and D. Hilbert, *Methods of Mathematical Physics* (In Terscience, New York, 1953).

[50] H. Jeffreys, and B. S. Jeffreys, *Methods of Mathematical Physic*, Chap. 10 (Cambridge University Press, England, 1950).

[51] Wei-mo Zheng and Jixing Liu, *Phys. Rev. E* **48** (1993) 2856.

[52] R. Podgornik, S. Svetina, and B. Zeks, *Phys. Rev. E* **51** (1995) 544.

[53] A. R. Forsyth, *Calculus of Variations* (Dover, New York, 1960).

[54] H. Naito, M. Okuda, and Z. C. Ou-Yang, *Phy. Rev. E* **48** (1993) 2304; **54** (1996) 2816

[55] E. A. Evans and Y. C. Fung, *Microvasc. Res.* **4** (1972) 335.

[56] C. Delaunay, *J. Math. Pure et Appl. Ser.* 1 (1841) 309.

[57] J. Bells, *Math. Intelligen* **9** (1987) 53.

[58] H. Naito, M. Okuda, and Z. C. Ou-Yang, *Phys. Rev. Lett.* **74** (1995) 4345.

[59] H. Deuling and W. Helfrich, *Blood Cells* **3** (1977) 713.

[60] R. Bar-Ziv and E. Moses, *Phys. Rev. Lett.* **73** (1994) 1392.

[61] Z. C. Tu, *Chin. Phys. B* **22** (2013) 028701.

4

Governing Equations for Open Lipid Membranes and Their Solutions

In this chapter, we will introduce another geometric method to deal with variational problems on a surface. This approach is called the moving frame method, which can simplify the calculus of variation.[1] The governing equations for an open lipid membrane with a free edge are derived with this method.[2] The possibility to find analytic solutions is also discussed.[3, 4]

4.1 Mathematical Preliminary

4.1.1 Surface Theory Based on Moving Frame Method

A membrane can be regarded as a 2D smooth orientable surface embedded in $\mathbb{E}^3$. The properties of the surface such as mean curvature and Gaussian curvature determine the shape of membrane. As shown in Fig. 4.1, each point on surface $\mathcal{M}$ can be represented by a position vector $\mathbf{r}$. At that point, we construct three unit orthonormal vectors $\mathbf{e}_1$, $\mathbf{e}_2$, and $\mathbf{e}_3$ with $\mathbf{e}_3$ being the normal vector of surface $\mathcal{M}$ at point $\mathbf{r}$. The set of right-handed orthonormal triple-vectors $\{\mathbf{e}_1, \mathbf{e}_2, \mathbf{e}_3\}$ is called a frame at point $\mathbf{r}$. Different points on the surface have different vectors $\mathbf{r}$, $\mathbf{e}_1$, $\mathbf{e}_2$, and $\mathbf{e}_3$, thus the set $\{\mathbf{r}; \mathbf{e}_1, \mathbf{e}_2, \mathbf{e}_3\}$ is called a moving frame.

Let us imagine a mass point that moves from position $\mathbf{r}$ to its neighbor position $\mathbf{r}'$ on the surface. The length of the path is denoted by Δs. Then we can define the differentiation of the frame as

$$d\mathbf{r} = \lim_{\Delta s \to 0} (\mathbf{r}' - \mathbf{r}) = \omega_1 \mathbf{e}_1 + \omega_2 \mathbf{e}_2, \qquad (4.1)$$

and

$$d\mathbf{e}_i = \lim_{\Delta s \to 0} (\mathbf{e}_i' - \mathbf{e}_i) = \omega_{ij} \mathbf{e}_j, \quad (i = 1, 2, 3), \qquad (4.2)$$

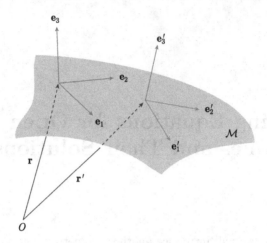

Fig. 4.1.   The moving frame on a surface.

where ω_1, ω_2, and ω_{ij} $(i, j = 1, 2, 3)$ are 1-forms, and "d" is the exterior differential operator.[5] The area element can be expressed as[5]

$$dA \equiv \omega_1 \wedge \omega_2. \tag{4.3}$$

The 1-form ω_{ij} is anti-symmetric with respect to i and j, that is, $\omega_{ij} = -\omega_{ji}$. Here and in the following contents without special statements, the repeated subscripts represent summation from 1 to 3. Additionally, the structure equations of the surface can be expressed as[5]

$$d\omega_1 = \omega_{12} \wedge \omega_2,$$
$$d\omega_2 = \omega_{21} \wedge \omega_1, \tag{4.4}$$
$$d\omega_{ij} = \omega_{ik} \wedge \omega_{kj} \quad (i, j = 1, 2, 3),$$

and

$$\begin{pmatrix} \omega_{13} \\ \omega_{23} \end{pmatrix} = \begin{pmatrix} a & b \\ b & c \end{pmatrix} \begin{pmatrix} \omega_1 \\ \omega_2 \end{pmatrix}, \tag{4.5}$$

where "$\wedge$" represents the wedge production between two differential forms. The repeated sub-index represents the Einstein summation convention. The matrix $\begin{pmatrix} a & b \\ b & c \end{pmatrix}$ is the representation matrix of the curvature tensor. Its trace and determinant are two invariants under the coordinate rotation around $\mathbf{e}_3$ which are denoted by

$$2H = a + c \quad \text{and} \quad K = ac - b^2. \tag{4.6}$$

H and K are called the mean curvature and Gaussian curvature, respectively. They determine the shape of the surface and can be expressed as $2H = -(1/R_1 + 1/R_2)$ and $K = 1/(R_1 R_2)$ with two principal curvature radii R_1 and R_2 at each point on the surface. In particular, we can obtain

$$K \mathrm{d}A = -\mathrm{d}\omega_{12} \tag{4.7}$$

from Eqs. (4.4)–(4.6).

Now, consider a curve on surface $\mathcal{M}$. Its tangent vector is denoted by $\mathbf{t}$. Let ϕ be the angle between $\mathbf{t}$ and $\mathbf{e}_1$ at the same point. Then the geodesic curvature k_g, the geodesic torsion τ_g, and the normal curvature k_n along the direction of $\mathbf{t}$ can be expressed as[1]

$$k_g = (\mathrm{d}\phi + \omega_{12})/\mathrm{d}s,$$
$$\tau_g = b\cos 2\phi + (c - a)\cos\phi\sin\phi, \tag{4.8}$$
$$k_n = a\cos^2\phi + 2b\cos\phi\sin\phi + c\sin^2\phi,$$

where $\mathrm{d}s$ is the arc length element along $\mathbf{t}$. If $\mathbf{t}$ aligns with $\mathbf{e}_1$, then $\phi = 0$, $k_g = \omega_{12}/\mathrm{d}s$, $\tau_g = b$, and $k_n = a$. In the principal frame, the geodesic torsion and normal curvature can be expressed as

$$k_n = -\frac{\cos^2\phi}{R_1} - \frac{\sin^2\phi}{R_2}, \quad \tau_g = (1/R_1 - 1/R_2)\cos\phi\sin\phi. \tag{4.9}$$

4.1.2   Hodge Star, Stokes' Theorem and Some Important Geometric Relations

The Hodge star ($*$) operator satisfies $*\omega_1 = \omega_2$ and $*\omega_2 = -\omega_1$.[6] The generalized Hodge star ($\tilde{*}$) operator satisfies $\tilde{*}\omega_{13} = \omega_{23}$ and $\tilde{*}\omega_{23} = -\omega_{13}$.[1] The generalized differential ($\tilde{\mathrm{d}}$) operator satisfies $\tilde{\mathrm{d}}f = f_1\omega_{13} + f_2\omega_{23}$ if $\mathrm{d}f = f_1\omega_1 + f_2\omega_2$.[1]

Stokes' theorem is a crucial theorem in differential geometry, which reads[5]

$$\oint_{\partial \mathfrak{D}} \omega = \int_{\mathfrak{D}} \mathrm{d}\omega, \tag{4.10}$$

where $\mathfrak{D}$ is a domain with boundary $\partial \mathfrak{D}$. ω is a differential form on $\partial \mathfrak{D}$. In particular, $\int_{\mathfrak{D}} \mathrm{d}\omega = 0$ for a closed domain $\mathfrak{D}$ free of singular points.

From Stokes' theorem, we can derive several identities listed as follows[1,7]:

(1) For smooth functions f and h on 2D subdomain $\mathfrak{D} \subseteq \mathcal{M}$,

$$\int_{\mathfrak{D}} (f\mathrm{d}*\mathrm{d}h - h\mathrm{d}*\mathrm{d}f) = \oint_{\partial\mathfrak{D}} (f*\mathrm{d}h - h*\mathrm{d}f),$$

$$\int_{\mathfrak{D}} (f\mathrm{d}*\tilde{\mathrm{d}}h - h\mathrm{d}*\tilde{\mathrm{d}}f) = \oint_{\partial\mathfrak{D}} (f*\tilde{\mathrm{d}}h - h*\tilde{\mathrm{d}}f), \qquad (4.11)$$

$$\int_{\mathfrak{D}} (f\mathrm{d}\tilde{*}\tilde{\mathrm{d}}h - h\mathrm{d}\tilde{*}\tilde{\mathrm{d}}f) = \oint_{\partial\mathfrak{D}} (f\tilde{*}\tilde{\mathrm{d}}h - h\tilde{*}\tilde{\mathrm{d}}f).$$

(2) If $\mathbf{u}$ and $\mathbf{v}$ are two vector fields defined on 2D subdomain $\mathfrak{D} \subseteq \mathcal{M}$, then

$$\int_{\mathfrak{D}} (\mathbf{u} \cdot \mathrm{d}*\mathrm{d}\mathbf{v} - \mathbf{v} \cdot \mathrm{d}*\mathrm{d}\mathbf{u}) = \oint_{\partial\mathfrak{D}} (\mathbf{u} \cdot *\mathrm{d}\mathbf{v} - \mathbf{v} \cdot *\mathrm{d}\mathbf{u}),$$

$$\int_{\mathfrak{D}} (\mathbf{u} \cdot \mathrm{d}*\tilde{\mathrm{d}}\mathbf{v} - \mathbf{v} \cdot \mathrm{d}*\tilde{\mathrm{d}}\mathbf{u}) = \oint_{\partial\mathfrak{D}} (\mathbf{u} \cdot *\tilde{\mathrm{d}}\mathbf{v} - \mathbf{v} \cdot *\tilde{\mathrm{d}}\mathbf{u}), \qquad (4.12)$$

$$\int_{\mathfrak{D}} (\mathbf{u} \cdot \mathrm{d}\tilde{*}\tilde{\mathrm{d}}\mathbf{v} - \mathbf{v} \cdot \mathrm{d}\tilde{*}\tilde{\mathrm{d}}\mathbf{u}) = \oint_{\partial\mathfrak{D}} (\mathbf{u} \cdot \tilde{*}\tilde{\mathrm{d}}\mathbf{v} - \mathbf{v} \cdot \tilde{*}\tilde{\mathrm{d}}\mathbf{u}),$$

where "$\cdot$" represents the inner product of vectors. For simplicity, Eqs. (4.11) and (4.12) are still called Stokes' theorem.

Additionally, we can also define the 2D gradient, curl, divergent, the Laplace operators, etc. on the surface in terms of the differential operators and the Hodge stars. They are summarized as follows[7]:

$$(\nabla \times \mathbf{u})\,\mathrm{d}A = \mathrm{d}(\mathbf{u} \cdot \mathrm{d}\mathbf{r}), \quad (\nabla \cdot \mathbf{u})\,\mathrm{d}A = \mathrm{d}(*\mathbf{u} \cdot \mathrm{d}\mathbf{r}),$$

$$(\tilde{\nabla} \cdot \mathbf{u})\,\mathrm{d}A = \mathrm{d}(\tilde{*}\mathbf{u} \cdot \tilde{\mathrm{d}}\mathbf{r}), \quad (\bar{\nabla} \cdot \mathbf{u})\,\mathrm{d}A = \mathrm{d}(*\mathbf{u} \cdot \tilde{\mathrm{d}}\mathbf{r}),$$

$$\nabla f \cdot \mathrm{d}\mathbf{r} = \mathrm{d}f, \quad \bar{\nabla} f \cdot \mathrm{d}\mathbf{r} = \tilde{\mathrm{d}}f, \quad \tilde{\nabla} f = 2H\nabla f - \bar{\nabla} f,$$

$$(\nabla^2 f)\,\mathrm{d}A = \mathrm{d}*\mathrm{d}f, \quad (\nabla \cdot \bar{\nabla} f)\,\mathrm{d}A = \mathrm{d}*\tilde{\mathrm{d}}f, \quad (\nabla \cdot \tilde{\nabla} f)\,\mathrm{d}A = \mathrm{d}\tilde{*}\tilde{\mathrm{d}}f,$$

$$(\mathbf{u} \cdot \nabla f)\,\mathrm{d}A = \mathbf{u} \cdot \mathrm{d}\mathbf{r} \wedge *\mathrm{d}f, \qquad (4.13)$$

$$(\mathbf{u} \cdot \bar{\nabla} f)\,\mathrm{d}A = \mathbf{u} \cdot \mathrm{d}\mathbf{r} \wedge *\tilde{\mathrm{d}}f,$$

$$(\mathbf{u} \cdot \tilde{\nabla} f)\,\mathrm{d}A = \mathbf{u} \cdot \mathrm{d}\mathbf{r} \wedge \tilde{*}\tilde{\mathrm{d}}f,$$

$$(\nabla^2 \mathbf{u})\,\mathrm{d}A = \mathrm{d}*\mathrm{d}\mathbf{u}, \quad (\nabla \mathbf{u}) \cdot \mathrm{d}\mathbf{r} = \mathrm{d}\mathbf{u},$$

$$(\nabla \mathbf{u}: \nabla \mathbf{v})\mathrm{d}A = \mathrm{d}\mathbf{u} \wedge \mathrm{d}\mathbf{v},$$

where "$\wedge$" represents calculating dot production and wedge production simultaneously.

4.1.3   Variational Theory Based on the Moving Frame Method

Any infinitesimal deformation of a surface can be achieved by a displacement vector

$$\delta \mathbf{r} \equiv \mathbf{\Omega} = \Omega_i \mathbf{e}_i \qquad (4.14)$$

at each point on the surface, where δ can be understood as a variational operator. The frame is also changed because of the deformation of the surface, which is denoted as

$$\delta \mathbf{e}_i = \Omega_{ij} \mathbf{e}_j \quad (i = 1, 2, 3), \qquad (4.15)$$

where $\Omega_{ij} = -\Omega_{ji}$ $(i, j = 1, 2, 3)$ corresponds to the rotation of the frame due to the deformation of the surface. From $\delta \mathrm{d}\mathbf{r} = \mathrm{d}\delta\mathbf{r}$, $\delta \mathrm{d}\mathbf{e}_j = \mathrm{d}\delta\mathbf{e}_j$, and Eqs. (4.1)–(4.5), we can derive[1,7]

$$\delta\omega_1 = \mathrm{d}\mathbf{\Omega} \cdot \mathbf{e}_1 - \omega_2\Omega_{21} = \mathrm{d}\Omega_1 + \Omega_2\omega_{21} + \Omega_3\omega_{31} - \omega_2\Omega_{21}, \qquad (4.16\mathrm{a})$$

$$\delta\omega_2 = \mathrm{d}\mathbf{\Omega} \cdot \mathbf{e}_2 - \omega_1\Omega_{12} = \mathrm{d}\Omega_2 + \Omega_1\omega_{12} + \Omega_3\omega_{32} - \omega_1\Omega_{12}, \qquad (4.16\mathrm{b})$$

$$\delta\omega_{ij} = \mathrm{d}\Omega_{ij} + \Omega_{il}\omega_{lj} - \omega_{il}\Omega_{lj}, \qquad (4.16\mathrm{c})$$

$$\Omega_{13}\omega_1 + \Omega_{23}\omega_2 = \mathrm{d}\mathbf{\Omega} \cdot \mathbf{e}_3 = \mathrm{d}\Omega_3 + \Omega_1\omega_{13} + \Omega_2\omega_{23}. \qquad (4.16\mathrm{d})$$

These equations are the essential equations of the variational method based on the moving frame. They are proved one by one as follows.

On the one hand, with the consideration of Eqs. (4.2), (4.4) and (4.5), we can derive

$$\begin{aligned} \mathrm{d}\delta\mathbf{r} = \mathrm{d}\mathbf{\Omega} &= \mathrm{d}(\Omega_1\mathbf{e}_1 + \Omega_2\mathbf{e}_2 + \Omega_3\mathbf{e}_3) \\ &= \mathrm{d}\Omega_1\mathbf{e}_1 + \Omega_1\mathrm{d}\mathbf{e}_1 + \mathrm{d}\Omega_2\mathbf{e}_2 + \Omega_2\mathrm{d}\mathbf{e}_2 + \mathrm{d}\Omega_3\mathbf{e}_3 + \Omega_3\mathrm{d}\mathbf{e}_3 \\ &= \mathrm{d}\Omega_1\mathbf{e}_1 + \Omega_1(\omega_{12}\mathbf{e}_2 + \omega_{13}\mathbf{e}_3) + \mathrm{d}\Omega_2\mathbf{e}_2 + \Omega_2(\omega_{21}\mathbf{e}_1 + \omega_{23}\mathbf{e}_3) \\ &\quad + \mathrm{d}\Omega_3\mathbf{e}_3 + \Omega_3(\omega_{31}\mathbf{e}_1 + \omega_{32}\mathbf{e}_2) \\ &= (\mathrm{d}\Omega_1 + \Omega_2\omega_{21} + \Omega_3\omega_{31})\mathbf{e}_1 + (\Omega_1\omega_{12} + \mathrm{d}\Omega_2 + \Omega_3\omega_{32})\mathbf{e}_2 \\ &\quad + (\Omega_1\omega_{13} + \Omega_2\omega_{23} + \mathrm{d}\Omega_3)\mathbf{e}_3. \end{aligned}$$

On the other hand, with the consideration of Eqs. (4.1), (4.4) and (4.5), we can derive

$$\delta d\mathbf{r} = \delta(\omega_1\mathbf{e}_1 + \omega_2\mathbf{e}_2) = \delta\omega_1\mathbf{e}_1 + \omega_1\delta\mathbf{e}_1 + \delta\omega_2\mathbf{e}_2 + \omega_2\delta\mathbf{e}_2$$

$$= \delta\omega_1\mathbf{e}_1 + \omega_1(\Omega_{12}\mathbf{e}_2 + \Omega_{13}\mathbf{e}_3) + \delta\omega_2\mathbf{e}_2 + \omega_2(\Omega_{21}\mathbf{e}_1 + \Omega_{23}\mathbf{e}_3)$$

$$= (\delta\omega_1 + \omega_2\Omega_{21})\mathbf{e}_1 + (\delta\omega_2 + \omega_1\Omega_{12})\mathbf{e}_2 + (\omega_1\Omega_{13} + \omega_2\Omega_{23})\mathbf{e}_3.$$

Considering $\delta d\mathbf{r} = d\delta\mathbf{r}$, we can derive Eqs. (4.16a), (4.16b) and (4.16d) from the above two equations.

In addition, from Eqs. (4.2), (4.4) and (4.15), we have

$$d\delta\mathbf{e}_i = d(\Omega_{ij}\mathbf{e}_j) = d\Omega_{ij}\mathbf{e}_j + \Omega_{ij}d\mathbf{e}_j = d\Omega_{ij}\mathbf{e}_j + \Omega_{ij}\omega_{jk}\mathbf{e}_k$$

$$= d\Omega_{ij}\mathbf{e}_j + \Omega_{il}\omega_{lj}\mathbf{e}_j = (d\Omega_{ij} + \Omega_{il}\omega_{lj})\mathbf{e}_j,$$

and

$$\delta d\mathbf{e}_i = \delta(\omega_{ij}\mathbf{e}_j) = \delta\omega_{ij}\mathbf{e}_j + \omega_{ij}\delta\mathbf{e}_j = \delta\omega_{ij}\mathbf{e}_j + \omega_{ij}\Omega_{jk}\mathbf{e}_k$$

$$= \delta\omega_{ij}\mathbf{e}_j + \omega_{il}\Omega_{lj}\mathbf{e}_j = (\delta\omega_{ij} + \omega_{il}\Omega_{lj})\mathbf{e}_j.$$

Considering $\delta d\mathbf{e}_i = d\delta\mathbf{e}_i$, we can derive Eqs. (4.16c) from the above two equations.

With essential equation (4.16), we can derive

$$\delta dA = (\nabla \cdot \mathbf{\Omega} - 2H\Omega_3)dA, \tag{4.17a}$$

$$\delta(2H) = [\nabla^2 + (4H^2 - 2K)]\Omega_3 + \nabla(2H) \cdot \mathbf{\Omega}, \tag{4.17b}$$

$$\delta K = \nabla \cdot \tilde{\nabla}\Omega_3 + 2KH\Omega_3 + \nabla K \cdot \mathbf{\Omega}. \tag{4.17c}$$

Let us prove the above Eqs. (4.17a)–(4.17c) one by one. First, from Eqs. (4.1) and (4.16), we have $\mathbf{\Omega} \cdot d\mathbf{r} = \Omega_1\omega_1 + \Omega_2\omega_2$, and further

$$*\mathbf{\Omega} \cdot d\mathbf{r} = \Omega_1\omega_2 - \Omega_2\omega_1. \tag{4.18}$$

Then Eq. (4.13) follows that

$$(\nabla \cdot \mathbf{\Omega})dA = d(*\mathbf{\Omega} \cdot d\mathbf{r}) = d(\Omega_1\omega_2 - \Omega_2\omega_1)$$

$$= d\Omega_1 \wedge \omega_2 + \Omega_1 d\omega_2 - d\Omega_2 \wedge \omega_1 - \Omega_2 d\omega_1. \tag{4.19}$$

In addition, Eq. (4.5) leads to

$$\omega_{13} \wedge \omega_2 + \omega_1 \wedge \omega_{23} = (a\omega_1 + b\omega_2) \wedge \omega_2 + \omega_1 \wedge (b\omega_1 + c\omega_2)$$

$$= (a + c)\omega_1 \wedge \omega_2 = 2Hd A. \tag{4.20}$$

Thus, using Eqs. (4.3) and (4.16), we derive

$$\delta dA = \delta(\omega_1 \wedge \omega_2) = \delta\omega_1 \wedge \omega_2 + \omega_1 \wedge \delta\omega_2$$

$$= (d\Omega_1 - \omega_{12}\Omega_2 - \omega_{13}\Omega_3 - \omega_2\Omega_{21}) \wedge \omega_2$$

$$+ \omega_1 \wedge (d\Omega_2 - \omega_{21}\Omega_1 - \omega_{23}\Omega_3 - \omega_1\Omega_{12})$$

$$= d\Omega_1 \wedge \omega_2 + \Omega_1 d\omega_2 - d\Omega_2 \wedge \omega_1 - \Omega_2 d\omega_1 - \Omega_3 (\omega_{13} \wedge \omega_2 + \omega_1 \wedge \omega_{23})$$

$$= d(*\mathbf{\Omega} \cdot d\mathbf{r}) - 2H\Omega_3 dA = (\nabla \cdot \mathbf{\Omega} - 2H\Omega_3)dA. \qquad (4.21)$$

Thus, we derive Eq. (4.17a).

Next, using Eqs. (4.16) and (4.20), we can obtain

$$\delta(2H dA) = \delta(\omega_{13} \wedge \omega_2 + \omega_1 \wedge \omega_{23})$$

$$= \delta\omega_{13} \wedge \omega_2 + \omega_{13} \wedge \delta\omega_2 + \delta\omega_1 \wedge \omega_{23} + \omega_1 \wedge \delta\omega_{23}$$

$$= (d\Omega_{13} + \Omega_{12}\omega_{23} - \omega_{12}\Omega_{23}) \wedge \omega_2$$

$$+ \omega_{13} \wedge (d\Omega_2 + \Omega_1\omega_{12} + \Omega_3\omega_{32} - \omega_1\Omega_{12})$$

$$+ (d\Omega_1 + \Omega_2\omega_{21} + \Omega_3\omega_{31} - \omega_2\Omega_{21}) \wedge \omega_{23}$$

$$+ \omega_1 \wedge (d\Omega_{23} + \Omega_{21}\omega_{13} - \omega_{21}\Omega_{13})$$

$$= d\Omega_{13} \wedge \omega_2 + \Omega_{13}d\omega_2 - \Omega_{23}d\omega_1 - d\Omega_{23} \wedge \omega_1 + 2\Omega_3 d\omega_{12}$$

$$- d\Omega_2 \wedge \omega_{13} - \Omega_2 d\omega_{13} + \Omega_1 d\omega_{23} + d\Omega_1 \wedge \omega_{23}$$

$$= d(\Omega_{13}\omega_2) - d(\Omega_{23}\omega_1) - 2K\Omega_3 dA - d(\Omega_2\omega_{13}) + d(\Omega_1\omega_{23})$$

$$= d(\Omega_{13}\omega_2 - \Omega_{23}\omega_1 + \Omega_1\omega_{23} - \Omega_2\omega_{13}) - 2K\Omega_3 dA$$

$$= d[*(\Omega_{13}\omega_1 + \Omega_{23}\omega_2) + \Omega_1\omega_{23} - \Omega_2\omega_{13}] - 2K\Omega_3 dA$$

$$= d[*(\Omega_1\omega_{13} + \Omega_2\omega_{23} + d\Omega_3) + \Omega_1\omega_{23} - \Omega_2\omega_{13}] - 2K\Omega_3 dA$$

$$= d[*d\Omega_3 + \Omega_1 * \omega_{13} + \Omega_2 * \omega_{23} + \Omega_1\omega_{23} - \Omega_2\omega_{13}] - 2K\Omega_3 dA$$

$$= d * d\Omega_3 + d[\Omega_1(a\omega_2 - b\omega_1) + \Omega_2(b\omega_2 - c\omega_1)$$

$$+ \Omega_1(b\omega_1 + c\omega_2) - \Omega_2(a\omega_1 + b\omega_2)] - 2K\Omega_3 dA$$

$$= \nabla^2\Omega_3 dA - 2K\Omega_3 dA + d[2H(\Omega_1\omega_2 - \Omega_2\omega_1)]$$

$$= (\nabla^2\Omega_3 - 2K\Omega_3)dA + d(2H * \mathbf{\Omega} \cdot d\mathbf{r}).$$

By considering the above equation and Eq. (4.21), we have

$$\delta(2H)dA = \delta(2HdA) - 2H\delta(dA)$$

$$= (\nabla^2\Omega_3 - 2K\Omega_3)dA + d(2H * \boldsymbol{\Omega} \cdot \mathbf{dr})$$

$$- 2H[d(*\boldsymbol{\Omega} \cdot \mathbf{dr}) - 2H\Omega_3 dA]$$

$$= [\nabla^2\Omega_3 + (4H^2 - 2K)\Omega_3]dA + d(2H * \boldsymbol{\Omega} \cdot \mathbf{dr})$$

$$- 2Hd(*\boldsymbol{\Omega} \cdot \mathbf{dr})$$

$$= [\nabla^2\Omega_3 + (4H^2 - 2K)\Omega_3]dA + d(2H) \wedge *\boldsymbol{\Omega} \cdot \mathbf{dr}$$

$$= [\nabla^2\Omega_3 + (4H^2 - 2K)\Omega_3]dA + [\nabla(2H) \cdot \boldsymbol{\Omega}]dA,$$

which reduces to Eq. (4.17b).

Finally, we will derive Eq. (4.17c). With the consideration of $\delta d(\cdot) = d\delta(\cdot)$, $dd(\cdot) = 0$ and Eq. (4.16), we have

$$\delta d\omega_{12} = d\delta\omega_{12} = d[d\Omega_{12} + \Omega_{13}\omega_{32} - \omega_{13}\Omega_{32}] = -d(\Omega_{13}\omega_{23} - \Omega_{23}\omega_{13}),$$
$$(4.22)$$

and

$$d\Omega_3 = \omega_1\Omega_{13} + \omega_2\Omega_{23} - (\Omega_1\omega_{13} + \Omega_2\omega_{23})$$

$$= (\Omega_{13} - a\Omega_1 - b\Omega_2)\omega_1 + (\Omega_{23} - b\Omega_1 - c\Omega_2)\omega_2,$$

which leads to

$$\tilde{d}\Omega_3 = (\Omega_{13} - a\Omega_1 - b\Omega_2)\omega_{13} + (\Omega_{23} - b\Omega_1 - c\Omega_2)\omega_{23},$$

and then

$$\tilde{*}\tilde{d}\Omega_3 = (\Omega_{13} - a\Omega_1 - b\Omega_2)\omega_{23} - (\Omega_{23} - b\Omega_1 - c\Omega_2)\omega_{13}$$

$$= \Omega_{13}\omega_{23} - \Omega_{23}\omega_{13} + (b^2 - ac)(\Omega_1\omega_2 - \Omega_2\omega_1)$$

$$= \Omega_{13}\omega_{23} - \Omega_{23}\omega_{13} - K * \boldsymbol{\Omega} \cdot \mathbf{dr}. \qquad (4.23)$$

With the consideration of Eqs. (4.7), (4.22) and (4.23), we can derive

$$\delta(KdA) = -\delta d\omega_{12} = d(\Omega_{13}\omega_{23} - \Omega_{23}\omega_{13}) = d\tilde{*}\tilde{d}\Omega_3 + d(K * \boldsymbol{\Omega} \cdot \mathbf{dr}).$$

Thus, from the above equation and Eq. (4.21), we derive

$$\delta KdA = \delta(KdA) - K\delta dA$$

$$= d\tilde{*}\tilde{d}\Omega_3 + d(K * \boldsymbol{\Omega} \cdot \mathbf{dr}) - K[d(*\boldsymbol{\Omega} \cdot \mathbf{dr}) - 2H\Omega_3 dA]$$

$$= (\nabla \cdot \tilde{\nabla}\Omega_3 + 2HK\Omega_3)dA + dK \wedge *\boldsymbol{\Omega} \cdot \mathbf{dr}$$

$$= (\nabla \cdot \tilde{\nabla}\Omega_3 + 2HK\Omega_3 + \nabla K \cdot \boldsymbol{\Omega})dA,$$

which is equivalent to Eq. (4.17c).

Using the above equations (4.4)–(4.8), (4.11)–(4.13), (4.16) and (4.17), we can deal with almost all variational problems on surfaces. As an example, we will derive the Euler–Lagrange equation and boundary conditions corresponding to the minimal of $F = \int_{\mathcal{M}} G(2H, K)\mathrm{d}A + \gamma \oint_C \mathrm{d}s$ on an open smooth surface ($\mathcal{M}$) with a boundary curve (C).

Since G is a function of $2H$ and K, we have

$$\delta(G\mathrm{d}A) = \delta G \mathrm{d}A + G\delta \mathrm{d}A = \left[\frac{\partial G}{\partial(2H)}\delta(2H) + \frac{\partial G}{\partial K}\delta K\right]\mathrm{d}A + G\delta \mathrm{d}A$$

$$= \frac{\partial G}{\partial(2H)}[\nabla^2 \Omega_3 + (4H^2 - 2K)\Omega_3]\mathrm{d}A + \frac{\partial G}{\partial(2H)}\mathrm{d}(2H) \wedge *\mathbf{\Omega} \cdot \mathrm{d}\mathbf{r}$$

$$+ \frac{\partial G}{\partial K}(\nabla \cdot \tilde{\nabla}\Omega_3 + 2HK\Omega_3)\mathrm{d}A$$

$$+ \frac{\partial G}{\partial K}\mathrm{d}K \wedge *\mathbf{\Omega} \cdot \mathrm{d}\mathbf{r} + G[\mathrm{d}(*\mathbf{\Omega} \cdot \mathrm{d}\mathbf{r}) - 2H\Omega_3 \mathrm{d}A]$$

$$= \frac{\partial G}{\partial(2H)}\mathrm{d}*\mathrm{d}\Omega_3 + (4H^2 - 2K)\frac{\partial G}{\partial(2H)}\Omega_3 \mathrm{d}A$$

$$+ \frac{\partial G}{\partial K}\mathrm{d}\tilde{*}\mathrm{d}\Omega_3 + 2HK\frac{\partial G}{\partial K}\Omega_3 \mathrm{d}A - 2HG\Omega_3 \mathrm{d}A$$

$$+ \frac{\partial G}{\partial(2H)}\mathrm{d}(2H) \wedge *\mathbf{\Omega} \cdot \mathrm{d}\mathbf{r} + \frac{\partial G}{\partial K}\mathrm{d}K \wedge *\mathbf{\Omega} \cdot \mathrm{d}\mathbf{r} + G\mathrm{d}(*\mathbf{\Omega} \cdot \mathrm{d}\mathbf{r})$$

$$= \frac{\partial G}{\partial(2H)}\mathrm{d}*\mathrm{d}\Omega_3 + \frac{\partial G}{\partial K}\mathrm{d}\tilde{*}\mathrm{d}\Omega_3$$

$$+ \left[(4H^2 - 2K)\frac{\partial G}{\partial(2H)} + 2HK\frac{\partial G}{\partial K} - 2HG\right]\Omega_3 \mathrm{d}A$$

$$+ \mathrm{d}(G*\mathbf{\Omega} \cdot \mathrm{d}\mathbf{r}).$$

Considering the Green Identity (4.11), we derive

$$\delta \int_{\mathcal{M}} G\mathrm{d}A = \int_{\mathcal{M}} \delta(G\mathrm{d}A)$$

$$= \int_{\mathcal{M}} \Omega_3 \mathrm{d}*\mathrm{d}\frac{\partial G}{\partial(2H)} + \oint_C \left[\frac{\partial G}{\partial(2H)}*\mathrm{d}\Omega_3 - \Omega_3 *\mathrm{d}\frac{\partial G}{\partial(2H)}\right]$$

$$+ \int_{\mathcal{M}} \Omega_3 \mathrm{d}\tilde{*}\mathrm{d}\frac{\partial G}{\partial K} + \oint_C \left(\frac{\partial G}{\partial K}\tilde{*}\mathrm{d}\Omega_3 - \Omega_3 \tilde{*}\mathrm{d}\frac{\partial G}{\partial K}\right)$$

$$+ \int_{\mathcal{M}} \left[\left(4H^2 - 2K\right) \frac{\partial G}{\partial(2H)} + 2HK\frac{\partial G}{\partial K} - 2HG \right] \Omega_3 \mathrm{d}A$$

$$+ \oint_C G * \boldsymbol{\Omega} \cdot \mathrm{d}\mathbf{r} \tag{4.24}$$

from the above equation.

On the other hand, in the boundary curve, we have $\mathrm{d}s = \omega_1, \omega_2 = 0$, $\omega_{12} = k_g \mathrm{d}s, a = k_n$. Thus, $\delta \mathrm{d}s = \delta\omega_1 = \mathrm{d}\Omega_1 + \Omega_2\omega_{21} + \Omega_3\omega_{31} - \omega_2\Omega_{21} = \mathrm{d}\Omega_1 - \Omega_2 k_g \mathrm{d}s - k_n\Omega_3 \mathrm{d}s$, and then

$$\delta \oint_C \mathrm{d}s = \oint_C \delta \mathrm{d}s = \oint_C (\mathrm{d}\Omega_1 - \Omega_2 k_g \mathrm{d}s - k_n\Omega_3 \mathrm{d}s)$$

$$= \oint_C (-\Omega_2 k_g - k_n\Omega_3)\mathrm{d}s. \tag{4.25}$$

With the consideration of the above equation and Eq. (4.24), we obtain

$$\delta F = \int_{\mathcal{M}} \left[\nabla^2 \frac{\partial G}{\partial(2H)} + \nabla \cdot \tilde{\nabla} \left(\frac{\partial G}{\partial K} \right) \right.$$

$$+ \left(4H^2 - 2K\right) \frac{\partial G}{\partial(2H)} + 2HK\frac{\partial G}{\partial K} - 2HG \Big] \Omega_3 \mathrm{d}A$$

$$+ \oint_C \left[\frac{\partial G}{\partial(2H)} * \mathrm{d}\Omega_3 - \Omega_3 * \mathrm{d}\frac{\partial G}{\partial(2H)} + \frac{\partial G}{\partial K}\tilde{*}\mathrm{d}\Omega_3 - \Omega_3\tilde{*}\mathrm{d}\frac{\partial G}{\partial K} \right]$$

$$+ \oint_C G * \boldsymbol{\Omega} \cdot \mathrm{d}\mathbf{r} - \gamma \oint_C (\Omega_2 k_g + k_n\Omega_3)\,\mathrm{d}s \tag{4.26}$$

Let us consider the first variational mode $\Omega_1 \neq 0, \Omega_2 = \Omega_3 = 0$. In this case,

$$\delta F = \oint_C G * \boldsymbol{\Omega} \cdot \mathrm{d}\mathbf{r} = \oint_C G(\Omega_1\omega_2 - \Omega_2\omega_1) \equiv 0, \tag{4.27}$$

from which we will achieve nothing.

Next, considering the second variational mode $\Omega_2 \neq 0, \Omega_1 = \Omega_3 = 0$, we have

$$\delta F = \oint_C G * \boldsymbol{\Omega} \cdot \mathrm{d}\mathbf{r} - \gamma \oint_C (\Omega_2 k_g)\,\mathrm{d}s = \oint_C G(\Omega_1\omega_2 - \Omega_2\omega_1) - \gamma \oint_C \Omega_2 k_g \mathrm{d}s$$

$$= - \oint_C (G + \gamma k_g)\,\Omega_2 \mathrm{d}s = 0,$$

from which we derive

$$[G + \gamma k_g]_C = 0. \tag{4.28}$$

Finally, we consider the third variational mode $\Omega_3 \neq 0, \Omega_1 = \Omega_2 = 0$. In this case, we have $d\Omega_3 = \Omega_{13}\omega_1 + \Omega_{23}\omega_2$, $*d\Omega_3 = \Omega_{13}\omega_2 - \Omega_{23}\omega_1$ and $\tilde{*}\tilde{d}\Omega_3 = \Omega_{13}\omega_{23} - \Omega_{23}\omega_{13}$. We further derive

$$\oint_C \frac{\partial G}{\partial(2H)} * d\Omega_3 = -\oint_C \frac{\partial G}{\partial(2H)}\Omega_{23}ds,$$

$$\oint_C \Omega_3 * d\frac{\partial G}{\partial(2H)} = \oint_C \Omega_3 \nabla\left[\frac{\partial G}{\partial(2H)}\right] \cdot *dr = -\oint_C \Omega_3 \mathbf{b} \cdot \nabla\left[\frac{\partial G}{\partial(2H)}\right] ds,$$

$$\oint_C \frac{\partial G}{\partial K}\tilde{*}\tilde{d}\Omega_3 = \oint_C \frac{\partial G}{\partial K}[\Omega_{13}\omega_{23} - \Omega_{23}\omega_{13}] = \oint_C \frac{\partial G}{\partial K}(\Omega_{13}bds - \Omega_{23}ads)$$

$$= \oint_C b\frac{\partial G}{\partial K}\Omega_{13}ds - \oint_C \frac{\partial G}{\partial K}\Omega_{23}ads$$

$$= \oint_C b\frac{\partial G}{\partial K}d\Omega_{3C} - \oint_C \frac{\partial G}{\partial K}\Omega_{23}ads$$

$$= -\oint_C \Omega_{3C}d\left(b\frac{\partial G}{\partial K}\right) - \oint_C a\frac{\partial G}{\partial K}\Omega_{23}ds$$

$$= -\oint_C \Omega_{3C}\frac{d}{ds}\left(\tau_g\frac{\partial G}{\partial K}\right)ds - \oint_C k_n\frac{\partial G}{\partial K}\Omega_{23}ds,$$

$$\oint_C \Omega_3\tilde{*}\tilde{d}\frac{\partial G}{\partial K} = -\oint_C \Omega_3\mathbf{b} \cdot \tilde{\nabla}\left(\frac{\partial G}{\partial K}\right)ds.$$

Thus,

$$\delta F = \int_{\mathcal{M}}\left[\nabla^2\frac{\partial G}{\partial(2H)} + \nabla \cdot \tilde{\nabla}\left(\frac{\partial G}{\partial K}\right) + (4H^2 - 2K)\frac{\partial G}{\partial(2H)}\right.$$

$$\left. + 2HK\frac{\partial G}{\partial K} - 2HG\right]\Omega_3 dA$$

$$+ \oint_C \Omega_3\left[\mathbf{b} \cdot \left(\nabla\frac{\partial G}{\partial(2H)} + \tilde{\nabla}\frac{\partial G}{\partial K}\right) - \frac{d}{ds}\left(\tau_g\frac{\partial G}{\partial K}\right) - \gamma k_n\right]ds$$

$$- \oint_C \left(\frac{\partial G}{\partial(2H)} + k_n\frac{\partial G}{\partial K}\right)\Omega_{23}ds,$$

from which we have

$$\nabla^2\frac{\partial G}{\partial(2H)} + \nabla \cdot \tilde{\nabla}\left(\frac{\partial G}{\partial K}\right) + (4H^2 - 2K)\frac{\partial G}{\partial(2H)} + 2HK\frac{\partial G}{\partial K} - 2HG = 0,$$

$$(4.29)$$

and

$$\left[\mathbf{b} \cdot \left(\nabla \frac{\partial G}{\partial (2H)} + \tilde{\nabla} \frac{\partial G}{\partial K} \right) - \frac{d}{ds} \left(\tau_g \frac{\partial G}{\partial K} \right) - \gamma k_n \right]_C = 0, \quad (4.30)$$

$$\left[\frac{\partial G}{\partial (2H)} + k_n \frac{\partial G}{\partial K} \right]_C = 0. \quad (4.31)$$

4.2 Governing Equations for Open Lipid Membranes

The opening-up process of liposomal membranes by talin[8] was observed, which gives rise to the study of equilibrium equation and boundary conditions of lipid membranes with free exposed edges. This problem was theoretically investigated by Capovilla et al.[9] and Tu et al.[2] respectively using different methods. In this section, we will present these theoretical results and subsequent advancements.

As shown in Fig. 4.2, a lipid membrane with a free edge can be expressed as an open smooth surface $(\mathcal{M})$ with a boundary curve (C) in geometry. Because the free exposed edge is energetically unfavorable, we assign the line tension (energy cost per length) to be $\gamma > 0$. Then the free energy functional that we need to minimize can be expressed as

$$F = \int_{\mathcal{M}} \left[\frac{k_c}{2} (2H + c_0)^2 + \bar{k} K \right] dA + \lambda A + \gamma L, \quad (4.32)$$

where L is the total length of the free edge.

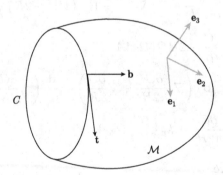

Fig. 4.2.   An open smooth surface $(\mathcal{M})$ with a boundary curve (C). $\{\mathbf{e}_1, \mathbf{e}_2, \mathbf{e}_3\}$ forms the frame at some point on the surface. $\{\mathbf{t}, \mathbf{b}, \mathbf{e}_3\}$ also forms the right-handed frame for the point in C such that $\mathbf{t}$ is the tangent of C and $\mathbf{b}$ points to the side where the surface is located.

4.2.1   Governing Equations

Taking $G = \frac{k_c}{2}(2H + c_0)^2 + \bar{k}K + \lambda$, we can derive the shape equation

$$(2H + c_0)(2H^2 - c_0 H - 2K) - 2\tilde{\lambda}H + \nabla^2(2H) = 0, \qquad (4.33)$$

and three boundary conditions

$$\left[(2H + c_0) + \tilde{k}\kappa_n\right]\Big|_C = 0, \qquad (4.34)$$

$$\left[-2\mathbf{b} \cdot \nabla H + \tilde{\gamma}\kappa_n + \tilde{k}\dot{\tau}_g\right]\Big|_C = 0, \qquad (4.35)$$

$$\left[(1/2)(2H + c_0)^2 + \tilde{k}K + \tilde{\lambda} + \tilde{\gamma}\kappa_g\right]\Big|_C = 0, \qquad (4.36)$$

from Eqs. (4.28)–(4.31). Here, $\tilde{\lambda} \equiv \lambda/k_c$, $\tilde{k} \equiv \bar{k}/k_c$, and $\tilde{\gamma} \equiv \gamma/k_c$ are the reduced surface tension, reduced bending modulus, and reduced line tension, respectively. κ_n, κ_g, and τ_g are the normal curvature, geodesic curvature, and geodesic torsion of the boundary curve, respectively. The "dot" represents the derivative with respect to the arc length of the edge. Equation (4.33) expresses the normal force balance of the membrane. Equations (4.34)–(4.36) represent the force and moment balances at each point in curve C. Thus, in general, the above four equations are independent of each other and available for an open membrane with several edges.

Now, we consider axisymmetric membranes. When a planar curve AC shown in Fig. 4.3 revolves around z-axis, an axisymmetric surface is generated. Let ψ represent the angle between the tangent line and the horizontal plane. Each point in the surface can be expressed as vector

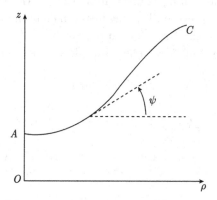

Fig. 4.3.   Outline of an axisymmetric open surface. Each open surface can be generated by a planar curve AC rotating around z-axis. ψ is the angle between the tangent line and the horizontal plane.

$\mathbf{r} = \{\rho\cos\phi, \rho\sin\phi, z(\rho)\}$, where ρ and ϕ are the radius and azimuth angle which the point corresponds to. Introduce a notation σ such that $\sigma = 1$ if $\mathbf{t}$ is parallel to $\partial\mathbf{r}/\partial\phi$, and $\sigma = -1$ if $\mathbf{t}$ is antiparallel to $\partial\mathbf{r}/\partial\phi$ in the boundary curve generated by point C. Equations (4.33)–(4.36) are transformed into

$$(h - c_0)\left(\frac{h^2}{2} + \frac{c_0 h}{2} - 2K\right) - \tilde{\lambda}h + \frac{\cos\psi}{\rho}(\rho\cos\psi h')' = 0, \quad (4.37)$$

$$\left[h - c_0 + \tilde{k}\sin\psi/\rho\right]_C = 0, \quad (4.38)$$

$$[-\sigma\cos\psi h' + \tilde{\gamma}\sin\psi/\rho]_C = 0, \quad (4.39)$$

$$\left[\frac{1}{2}(h - c_0)^2 + \tilde{k}K + \tilde{\lambda} - \sigma\tilde{\gamma}\frac{\cos\psi}{\rho}\right]_C = 0, \quad (4.40)$$

with $h \equiv \sin\psi/\rho + (\sin\psi)'$ and $K \equiv \sin\psi(\sin\psi)'/\rho$. The "prime" represents the derivative with respect to ρ.

4.2.2 Compatibility Conditions

The free energy (4.32) can be written in another form:

$$F = \int [(k_c/2)(2H)^2 + \bar{k}K]dA$$
$$+ 2k_c c_0 \int H dA + (\lambda + k_c c_0^2/2)A + \gamma L. \quad (4.41)$$

Let us consider the scaling transformation $\mathbf{r} \to \Lambda\mathbf{r}$, where the vector $\mathbf{r}$ represents the position of each point in the membrane and Λ is a scaling parameter.[10] Under this transformation, we have $A \to \Lambda^2 A$, $L \to \Lambda L$, $H \to \Lambda^{-1}H$, and $K \to \Lambda^{-2}K$. Thus, Eq. (4.41) is transformed into

$$F(\Lambda) = \int [(k_c/2)(2H)^2 + \bar{k}K]dA$$
$$+ 2k_c c_0 \Lambda \int H dA + (\lambda + k_c c_0^2/2)\Lambda^2 A + \gamma\Lambda L. \quad (4.42)$$

The equilibrium configuration should satisfy $\partial F/\partial\Lambda = 0$ when $\Lambda = 1$. Thus, we obtain

$$2c_0 \int H dA + (2\tilde{\lambda} + c_0^2)A + \tilde{\gamma}L = 0. \quad (4.43)$$

This equation is an additional constraint for open membranes.

Shape equation (4.37) is integrable, which reduces to a second-order differential equation

$$\cos\psi h' + (h - c_0)\sin\psi\psi' - \tilde{\lambda}\tan\psi + \frac{\eta_0}{\rho\cos\psi} - \frac{\tan\psi}{2}(h - c_0)^2 = 0, \quad (4.44)$$

with an integral constant η_0.[3] The configuration of an axisymmetric open lipid membrane should satisfy shape Eq. (4.44) and boundary conditions (4.38)–(4.40). In particular, the points in the boundary curve should satisfy not only the boundary conditions but also shape Eq. (4.44) because they are also located in the surface, that is, Eqs. (4.38)–(4.40) and (4.44) should be compatible with each other in the edge. Substituting Eqs. (4.38)–(4.40) into (4.44), we derive the compatibility condition[3] to be

$$\eta_0 = 0. \quad (4.45)$$

It is a necessary (not sufficient) condition for the existence of axisymmetric open membranes. Under this condition, the shape equation is reduced to

$$\cos\psi h' + (h - c_0)\sin\psi\psi' - \tilde{\lambda}\tan\psi - \frac{\tan\psi}{2}(h - c_0)^2 = 0, \quad (4.46)$$

while three boundary conditions are reduced to two equations, i.e., Eqs. (4.38) and (4.40).

4.3 Analytic Solutions

Now, our task is to find analytic solutions that satisfy both the shape equation and boundary conditions. An obvious but trivial one is a circular disk with radius R. In this case, Eqs. (4.33)–(4.36) degenerate to

$$\tilde{\lambda}R + \tilde{\gamma} = 0. \quad (4.47)$$

Can we find nontrivial analytic solutions? We have known some analytic solutions that satisfy the shape Eq. (4.33), which include surfaces with constant mean curvature, biconcave discoid, torus and invert catenoid. Can we find a closed curve on these surfaces satisfying the boundary conditions (4.34)–(4.36)? We will prove the following theorem of non-existence: For finite line tension, there does not exist an open membrane as part of surfaces with constant (non-vanishing) mean curvature, biconcave discoid (valid for axisymmetric case), or Willmore surfaces (torus, invert catenoid[11]). Several typical impossible open membranes with free edges are shown in Fig. 4.4.

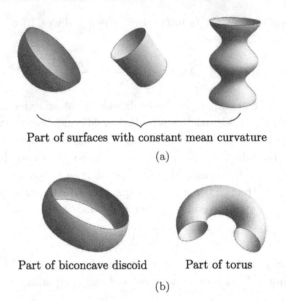

Part of surfaces with constant mean curvature

(a)

Part of biconcave discoid   Part of torus

(b)

Fig. 4.4.   Schematic diagrams of several impossible open membranes with free edges: (a) parts of sphere, cylinder and unduloid, and (b) parts of biconcave discoid and torus.

The original version of this theorem was proposed in Refs. [3, 4]. Here, we refine the original proof of this theorem and correct some flaws simultaneously.

First, it is easy to prove there is no open membrane which is a part of a spherical vesicle or cylindrical surface. For a sphere with radius R, we can calculate $H = -1/R$, $\kappa_n = -1/R$ and $\tau_g = 0$ in terms of Eq. (4.8) because $a = c = -1/R$ and $b = 0$ for a sphere. Boundary condition (4.35) cannot be abided. Thus, an open membrane cannot be a part of a spherical vesicle. For any line element on the surface of a cylinder with radius R, we can calculate $\kappa_n = -\cos^2\theta/R$ from Eq. (4.9), where θ is the angle between the line element and the circumferential direction. Additionally, $H = -1/R$ is a constant. If $\tilde{k} = 0$, then boundary condition (4.35) results in $\kappa_n = 0$, that is, $\theta = \pi/2$. The line in this direction is not a closed curve and so cannot be an edge of a membrane. If $\tilde{k} \neq 0$, then boundary condition (4.34) results in $\kappa_n = (c_0 - 1/R)/\tilde{k}$, which implies θ should be a constant. The unique closed curve is a circle, i.e. $\theta = 0$ and $\kappa_n = -1/R$. But $\tau_g = 0$ if $\theta = 0$, which contradicts boundary condition (4.35). Thus, an open membrane cannot be a part of a spherical surface. We emphasize that the key obstacle happens in boundary condition (4.35) which implies that the out-of-plane forces cannot balance in the edge.

Second, we will prove there is no open membrane as a part of a curved surface with non-vanishing constant mean curvature. From the shape Eq. (4.33), we derive $H = -c_0/2 \neq 0$ and $\tilde{\lambda} = 0$ in this case, which contradict the compatibility condition (4.43) for $\gamma \neq 0$.

Third, we will prove there is no axisymmetric open membrane as a part of a biconcave discoidal surface generated by a planar curve expressed by $\sin \psi = c_0 \rho \ln(\rho/\rho_B)$. Substituting this equation into shape Eq. (4.44), we obtain $\tilde{\lambda} = 0$ and $\eta_0 = -2c_0 \neq 0$, which contradicts compatibility condition (4.45).

Finally, we consider the Willmore surface[11] which satisfies the special form of Eq. (4.33) with vanishing $\tilde{\lambda}$ and c_0. Thus, the compatibility condition (4.43) cannot be satisfied when $\tilde{\lambda} = 0$ and $c_0 = 0$ because $\tilde{\gamma}L > 0$. That is, there is no open membrane as a part of Willmore surface which includes the torus and invert catenoid.

So far, we have proven the theorem of non-existence, which implies that it is hopeless to find analytic solutions to the shape equation and boundary conditions of open lipid membranes. Thus, the numerical simulations[3, 12] are highly appreciated.

4.4   Quasi-exact Solutions

Here, the quasi-exact solution is defined as a surface with free edge(s) such that the points on that surface exactly satisfy the shape equation, and most of the points in the edge(s) abide by boundary conditions. In fact, the proof of the theorem of non-existence implies two possible solutions as shown in Fig. 4.5. One is a straight stripe along the axial direction of cylinder, another is a twist ribbon which is a part of a minimal surface $(H = 0)$.

Let us consider a long enough straight stripe along the axial direction of cylinder that satisfies shape Eq. (4.33), that is, $\tilde{\lambda} = (1 - c_0^2 R^2)/2R^2$. The long enough configuration ensures we omit the boundary of two ends. The

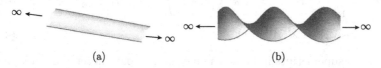

(a)   (b)

Fig. 4.5.   Schematics of two quasi-exact solutions: (a) straight stripe along axial direction of cylinder; (b) twist ribbon which is a part of a minimal surface.

lateral edges are straight lines which have $\kappa_n = \kappa_g = \tau_g = 0$. Thus, boundary conditions (4.34) and (4.35) are trivial. The third boundary condition results in $\tilde{\lambda} = (1 - c_0 R)^2/(2R^2)$. Thus, we arrive at $\tilde{\lambda} = 0$ and $R = 1/c_0$, that is, a long enough straight stripe along the axial direction of cylinder with $R = 1/c_0$ is a quasi-exact solution.

Next, a twisted ribbon with pitch T and width $2u_0$ can be expressed as vector form $\{u\cos\varphi, u\sin\varphi, \alpha\varphi\}$ with $|u| \leq u_0$, $|\varphi| < \infty$ and $|\alpha| = T/(2\pi)$. From simple calculations, we have

$$H = 0, \quad K = -\alpha^2/(u^2 + \alpha^2)^2 \tag{4.48}$$

for points on the surface, and

$$\kappa_n = 0, \quad K = -\alpha^2/(u_0^2 + \alpha^2)^2,$$
$$\kappa_g = u_0/(u_0^2 + \alpha^2), \tag{4.49}$$
$$\tau_g = \alpha/(u_0^2 + \alpha^2)$$

for points on the edges.

It is easy to see that Eq. (4.48) can satisfy shape Eq. (4.33) when $c_0 = 0$. Then Eq. (4.49) naturally validates boundary conditions (4.34) and (4.35). The last boundary condition (4.36) leads to $\tilde{\lambda} = [\tilde{k}\alpha^2 - \tilde{\gamma}u_0(u_0^2+\alpha^2)]/(u_0^2+\alpha^2)^2$, which can be satisfied by proper parameters $\tilde{\lambda}$, $\tilde{k}$, $\tilde{\gamma}$, u_0 and α, that is, the twist ribbon is indeed a quasi-exact solution.

4.5   Challenge

4.5.1   Neck Condition of Two-Phase Vesicles in the Budding State

The governing equations of open lipid membranes can be extended to a lipid vesicle with two phases separated by a boundary curve C as shown in Fig. 4.6.

The free energy of the two-phase vesicle can be expressed as

$$F_T = \int_{\mathrm{I}} \left[\frac{k_c^{\mathrm{I}}}{2}(2H + c_0^{\mathrm{I}})^2 + \bar{k}^{\mathrm{I}}K\right]\mathrm{d}A + \int_{\mathrm{II}} \left[\frac{k_c^{\mathrm{II}}}{2}(2H + c_0^{\mathrm{II}})^2 + \bar{k}^{\mathrm{II}}K\right]\mathrm{d}A$$
$$+ \lambda^{\mathrm{I}}A^{\mathrm{I}} + \lambda^{\mathrm{II}}A^{\mathrm{II}} + pV + \gamma L, \tag{4.50}$$

where the superscripts indicate the mechanical parameters for each phase, for example, c_0^{I} and c_0^{II} are, respectively, the spontaneous curvatures for phases I and II.

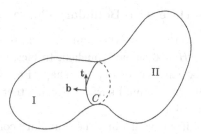

Fig. 4.6.   A vesicle with two phases (I and II) separated by curve C. $\mathbf{t}$ and $\mathbf{b}$ are located in the tangent plane of the surface. The former is the tangent vector of C while the latter is perpendicular to $\mathbf{t}$ and points to the side of phase I.

Usually, we can derive the matching conditions that the curve C should satisfy the variation of the above free energy. But if we notice that the physical meanings of Eqs. (4.34)–(4.36) are the force or moment balances in the boundary, we can directly write down the matching conditions as follows[1]:

$$\left[k_c^{\mathrm{I}}(2H + c_0^{\mathrm{I}}) - k_c^{\mathrm{II}}(2H + c_0^{\mathrm{II}}) + (\bar{k}^{\mathrm{I}} - \bar{k}^{\mathrm{II}})\kappa_n \right]_C = 0, \tag{4.51}$$

$$\left[\gamma \kappa_n + (\bar{k}^{\mathrm{I}} - \bar{k}^{\mathrm{II}})\dot{\tau}_g - 2(k_c^{\mathrm{I}} - k_c^{\mathrm{II}})\partial H / \partial \mathbf{b} \right]_C = 0, \tag{4.52}$$

$$\left[\frac{k_c^{\mathrm{I}}}{2}(2H + c_0^{\mathrm{I}})^2 - \frac{k_c^{\mathrm{II}}}{2}(2H + c_0^{\mathrm{II}})^2 + (\bar{k}^{\mathrm{I}} - \bar{k}^{\mathrm{II}})K + (\lambda^{\mathrm{I}} - \lambda^{\mathrm{II}}) + \gamma \kappa_g \right]_C = 0. \tag{4.53}$$

We note that Das et al. also obtained the equivalent form of above matching conditions in the axisymmetric case.[13]

Jülicher and Lipowsky investigated the budding of axisymmetric vesicles and found a limit shape which is the state of two vesicles connected by a small neck. They also derived the neck condition,[14, 15]

$$k_c^{\mathrm{I}} M^{\mathrm{I}} + k_c^{\mathrm{II}} M^{\mathrm{II}} = \frac{1}{2}\left(k_c^{\mathrm{I}} c_0^{\mathrm{I}} + k_c^{\mathrm{II}} c_0^{\mathrm{II}} + \gamma \right), \tag{4.54}$$

without considering the Gaussian bending terms. Here, M^{I} and M^{II} correspond to $-H^{\mathrm{I}}$ and $-H^{\mathrm{II}}$ for the points near the neck in domains I and II, respectively. They also conjectured that this neck condition holds for the asymmetric case and claimed the lack of a general proof to this conjecture.[16] It is not straightforward to derive the neck condition from the general matching conditions (4.51)–(4.53), which is a challenge to be overcome in the forthcoming years.[17]

4.5.2   Minimal Surface with Boundary Curve

If we carefully analyze the above theorem and its proof, we will find that the minimal surface $(H = 0)$ is not touched. In fact, when $c_0 = 0$, $H = 0$ with non-vanishing $\tilde{\lambda}$ can also satisfy the shape Eq. (4.33). Additionally, provided that $\tilde{\lambda} < 0$, the minimal surface is consistent with compatibility conditions (4.45) and (4.43).

On the one hand, if $\tilde{k}$ is vanishing, the boundary condition (4.34) holds naturally. Then boundary condition (4.35) suggests $\kappa_n = 0$. Further, boundary condition (4.36) requires $\kappa_g = -\tilde{\lambda}/\tilde{\gamma} = \text{constant}$.

On the other hand, if $\tilde{k} \neq 0$, the boundary condition (4.34) gives $\kappa_n = 0$. Then boundary condition (4.35) suggests $\tau_g = \text{constant}$. Since classical differential geometry tells us $K = -\tau_g^2$ when $\kappa_n = 0$, boundary condition (4.36) still requires $\kappa_g = \text{constant}$.

In short, the big challenge is whether we can find a smooth closed curve with vanishing normal curvature and constant geodesic curvature on some minimal surface except the planar circular disk. We conjecture the planar circular disk might be the unique minimal surface with a smooth boundary curve which has vanishing normal curvature and constant geodesic curvature. Recently, we have noted that, following the work on flat points of minimal surfaces by Koch and Fischer,[18] Giomi and Mahadevan argued that there does not exist a simple domain bounded by a smooth asymptotic curve in a minimal surface with nonvanishing Gauss curvature.[19]

References

[1] Z. C. Tu and Z. C. Ou-Yang, *J. Phys. A: Math. Gen.* **37** (2004) 11407.
[2] Z. C. Tu and Z. C. Ou-Yang, *Phys. Rev. E* **68** (2003) 061915.
[3] Z. C. Tu, *J. Chem. Phys.* **132** (2010) 084111.
[4] Z. C. Tu, *J. Geom. Symmetry Phys.* **24** (2011) 45.
[5] S. S. Chern and W. H. Chern, *Lecture on Differential Geometry* (Beijing University Press, Beijing, 1983).
[6] C. V. Westenholz, *Differential Forms in Mathematical Physics* (North-Holland, Amsterdam, 1981).
[7] Z. C. Tu and Z. C. Ou-Yang, *J. Comput. Theor. Nanosci.* **5** (2008) 422.
[8] A. Saitoh, K. Takiguchi, Y. Tanaka, and H. Hotani, *Proc. Natl. Acad. Sci.* **95** (1998) 1026.
[9] R. Capovilla, J. Guven, and J. A. Santiago, *Phys. Rev. E* **66** (2002) 021607.
[10] R. Capovilla and J. Guven, *J. Phys. A: Math. Gen.* **35** (2002) 6233.
[11] T. Willmore, *An Introduction to Differential Geometry* (Oxford Univ. Press, Oxford, 1982)

[12] Q. Du, C. Liu, and X. Wang, *J. Comput. Phys.* **212** (2006) 757.
[13] S. L. Das, J. T. Jenkins, and T. Baumgart, *EPL* **86** (2009) 48003.
[14] F. Jülicher and R. Lipowsky, *Phys. Rev. E* **53** (1996) 2670.
[15] U. Seifert, *Adv. Phys.* **46** (1997) 13.
[16] T. Baumgart, S. Das, M. Deserno, Q. Du, R. Lipowsky, Z. Ou-Yang, Z. Tu, Y. Yin, and P. Zhang, *Report on the KITPC Program on "Membrane Biophysics | Theory and Experiment"* (Kavli Institute for Theoretical Physics China, Beijing, 2012).
[17] This problem has been solved. Refer to P. Yang, Q. Du, and Z. C. Tu, *Phys. Rev. E* **95** (2017) 042403.
[18] E. Koch and W. Fischer, *Acta Cryst. A* **46** (1990) 33.
[19] L. Giomi and L. Mahadevan, *Proc. R. Soc. A* **468** (2012) 1851.

5

Theory of Tilted Chiral Lipid Bilayers

The expression of the shape energy given by Eq. (3.38) holds only for the L_α phase of membrane in which the more or less flexible hydrocarbon chains are directed in the normal direction of the bilayer (Fig. 5.1(b)). It holds neither for the L_β phase of membrane in which the lipid molecules are tilted nor for the L_β^* phase with tilted chiral molecules. Experiments demonstrate that the chirality of the component amphiphiles in the bilayer plays a crucial role in the formation of the helical structures.[1-7] Two types of helical structures have been reported. One of them looks like a ribbon wound around a cylinder with a spiral gap,[5] where the gradient angle of the spiral is nearly 45°. Sometimes, the gap closes up and the ribbon transforms into a prolate tube. Another type is a twisted ribbon which seems to be a crossover from vesicular dispersion to the first type of helical structures.[1,2] Technically, these structures can be used to make electro-optical elements, microelectronic elements, reagent delivery vesicles and microsurgical materials.[8]

Helfrich treated the wound-ribbon helices by assuming a competition of the spontaneous torsion of the edges with the bending of membranes.[9] The tube formation was also explained by de Gennes in terms of a buckling of flat solid ribbon due to ferroelectric polarization charges on their edges.[10] Helfrich and Prost later on improved the original Helfrich theory by employing a new linear term linked to molecular chirality in the bending of membrane with C_2 or D_2 symmetry.[11] Different from them, Ou-Yang and Liu[12,13] in analogue to cholesteric liquid crystals, with the assumption of strong chirality, developed a theory of helical structures. In addition, Komura and Ou-Yang,[14] with the full expression of Frank free energy of cholesterics, explained the new experimental observation of low- and high-pitch helical structures. Recently, Tu and Seifert[15] developed a concise theory for chiral lipid membranes which could interpret most of the experimental results on chiral lipid membranes. In this chapter, these theories are presented.

5.1 Theory of Tilted Chiral Lipid Bilayers with Strong Chirality

5.1.1 Geometric Description and Free Energy

The lipid bilayer may be treated as a 2D surface $\mathbf{Y}(u, v)$ with normal $\mathbf{n}$ and uniform thickness $t \cos \theta_0$ where θ_0 is the tilt angle of the molecules in the membrane as illustrated in Figs. 5.1(a) and 5.1(b). It has been shown that the two-parameter function $\mathbf{Y}(u, v)$ itself is enough to determine a uniform L_α phase on account of $\theta_0 = 0$. However, in L_β phase (tilted lipid bilayer (TLB)) and L_β^* phase (tilted chiral lipid bilayer, TCLB) besides the surface $\mathbf{Y}(u, v)$, it is also necessary to know the orientational direction $\mathbf{d}(u, v)$ of the molecule, where $\mathbf{d}$ depends on both the tilt angle θ and the azimuthal

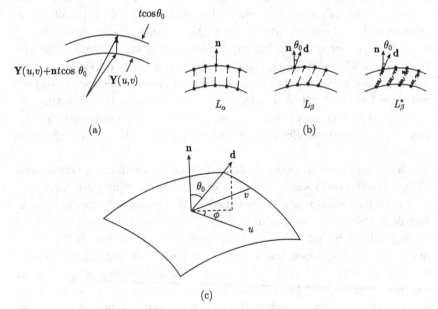

Fig. 5.1. Configuration of lipid bilayer: (a) geometry of bilayer, (b) molecule directions for the bilayer in different phases, (c) geometry configuration on the curved surface.

angle ϕ as shown in Figs. 5.1(b) and 5.1(c). For uniform L_β and L_β^* phases, $\theta = \theta_0$ is a constant. On a curved surface, the use of a scalar function $\phi(u, v)$ for the azimuthal angle is inconvenient because it depends on the curve-coordinate frame of reference. Instead, one may use a coordinate-free description of the unit orientational vector $\mathbf{d}$:

$$\mathbf{d} = d_u(u, v)\mathbf{Y}_{,u} + d_v(u, v)\mathbf{Y}_{,v} + \mathbf{n}\cos\theta_0 = d_l\mathbf{Y}_{,l} + \mathbf{n}\cos\theta_0, \qquad (5.1)$$

where d_u and d_v are two scalar functions of u and v independent of $\mathbf{n}$, the unit normal vector of the surface $\mathbf{Y}(u, v)$. In comparison with the Frank theory of liquid crystal, $\mathbf{d}$ just takes the place of the director in the case of liquid crystals. With Eq. (3.1), the definition of g_{ij}, and Eq. (3.5), the definition of $\mathbf{n}$, the self-dot product of Eq. (5.1) leads to the relation

$$g_{ij}d_id_j - \sin^2\theta_0 = 0 \quad (i, j = u, v). \qquad (5.2)$$

Experimental observations demonstrate that in TCLB, the chirality of the molecules plays a crucial role in the formation of its helical structure. For simplicity, let us first consider only the chiral effect and neglect all other terms, i.e., the consideration of a TCLB with strong chirality effect. In the Frank theory, the chirality of a cholesteric liquid crystal is characterized by the term

$$g_{\mathrm{ch}} = -k_2\mathbf{d}\cdot(\nabla \times \mathbf{d}) \qquad (5.3)$$

in the free energy density expression, where $k_2 > 0$ corresponds to the right-handed helix and $k_2 < 0$ corresponds to the left-handed helix. Since the calculation concerns the two surfaces $\mathbf{Y}(u, v)$ and $\mathbf{Y}(u, v) + \mathbf{n}(u, v)t\cos\theta_0$, as shown in Chapter 2, the gradient operator in Eq. (5.3) should be represented in the form

$$\nabla = \nabla' + \mathbf{n}\frac{\partial}{\partial n}, \qquad (5.4)$$

where ∇' is the 2D gradient operator on the surface $\mathbf{Y}(u, v)$ defined by

$$\nabla' = g^{ij}\mathbf{Y}_{,i}\frac{\partial}{\partial j} \quad (i, j = u, v). \qquad (5.5)$$

(For the convenience of the reader, simple derivation of Eqs. (5.4) and (5.5) together with some related results will be given in Appendix B). Thus, we have

$$g_{\mathrm{ch}} = -k_2\mathbf{d}\cdot(\nabla' \times \mathbf{d} + \mathbf{n} \times \mathbf{d}_{,n})$$
$$= -k_2\mathbf{d}\cdot(g^{ij}\mathbf{Y}_{,i} \times \mathbf{d}_{,j} + \mathbf{n} \times \mathbf{d}_{,n}). \qquad (5.6)$$

With $\mathbf{d}$ given by Eq. (5.1) there are two terms containing $\mathbf{n}_{,n}$ and $\mathbf{Y}_{,ln}$, respectively, which need to be considered first. Since $\mathbf{n}$ is a unit vector with constant magnitude, the variation of $\mathbf{n}$ along its own direction, naturally, is null, i.e.,

$$\mathbf{n}_{,n} = 0. \tag{5.7}$$

On the other hand, if we give $\mathbf{Y}$ a small variation $l\mathbf{n}$ at a point P on the surface $\mathbf{Y}(u, v)$, then the corresponding variation of $\mathbf{Y}$ is simply l. By definition $\partial \mathbf{Y}/\partial n = \lim_{l \to 0}(l\mathbf{n}/l)$, and we have

$$\mathbf{Y}_{,ln} = \mathbf{n}_{,l} = \mathbf{Y}_{,nl} = -L_{lm}g^{mk}\mathbf{Y}_{,k}$$

according to Eq. (3.13). Therefore, by Eq. (3.2), we have

$$\begin{aligned}
d_l\mathbf{n} \times \mathbf{Y}_{,ln} &= d_l L_{lm}g^{mk}(\mathbf{Y}_{,k} \times \mathbf{n}) \\
&= g^{-1/2}d_l L_{lm}g^{mk}(g_{kv}\mathbf{Y}_{,u} - g_{ku}\mathbf{Y}_{,v}) \\
&= g^{-1/2}d_l e^{3mn}L_{ln}\mathbf{Y}_{,m},
\end{aligned} \tag{5.8}$$

where e^{3mn} is the permutation symbol defined by Eq. (3.6).

The term $\nabla' \times \mathbf{d}$ is now given by

$$\begin{aligned}
\nabla' \times \mathbf{d} &= g^{ij}\mathbf{Y}_{,i} \times (\partial/\partial j)(d_l\mathbf{Y}_{,l} + \mathbf{n}\cos\theta_0) \\
&= g^{ij}d_{l,j}\mathbf{Y}_{,i} \times \mathbf{Y}_{,l} + g^{ij}d_l\mathbf{Y}_{,i} \times \mathbf{Y}_{,lj} + \cos\theta_0 g^{ij}\mathbf{Y}_{,i} \times \mathbf{n}_{,j} \\
&= g^{1/2}(g^{uj}d_{v,j} - g^{vj}d_{u,j})\mathbf{n} + g^{1/2}(g^{uj}\Gamma^v_{lj} - g^{vj}\Gamma^u_{lj})d_l\mathbf{n} \\
&\quad + g^{1/2}g^{ij}d_l L_{lj}(g_{iv}\mathbf{Y}_{,u} - g_{iu}\mathbf{Y}_{,v}) \\
&\quad - \cos\theta_0 g^{1/2}L_{jl}(g^{uj}g^{lv} - g^{vj}g^{lu})\mathbf{n},
\end{aligned}$$

under the application of Eqs. (3.4), (3.5) and (3.15). By simple expansion, it is easy to show that

$$g^{uj}d_{v,j} - g^{vj}d_{u,j} = (1/g)e^{3mn}g_{kn}d_{k,m},$$

$$g^{uj}\Gamma^v_{lj} - g^{vj}\Gamma^u_{lj} = (1/g)e^{3mu}g_{kn}\Gamma^k_{ml},$$

$$g^{ij}L_{jl}(g_{iv}\mathbf{Y}_{,u} - g_{iu}\mathbf{Y}_{,v}) = e^{3mn}L_{ln}\mathbf{Y}_{,m},$$

$$L_{jl}(g^{uj}g^{lv} - g^{vj}g^{lu}) = 0,$$

where $j, k, l, m, n = u, v$. Thus, we have

$$\nabla' \times \mathbf{d} = g^{-1/2}e^{3mn}[g_{kn}(d_{k,m} + \Gamma^k_{ml}d_l)\mathbf{n} + L_{ln}d_l\mathbf{Y}_{,m}]. \tag{5.9}$$

On the other hand, with Eqs. (5.7) and (5.8), we have

$$\mathbf{n} \times \mathbf{d}_{,n} = \mathbf{n} \times (d_l \mathbf{Y}_{,l} + \mathbf{n} \cos \theta_0)_{,n}$$

$$= g^{-1/2} d_l e^{3mn} L_{ln} \mathbf{Y}_m. \tag{5.10}$$

Combining Eqs. (5.9) and (5.10), we find that

$$\nabla \times \mathbf{d} = g^{-1/2} e^{3mn} [g_{kn}(d_{k,m} + \Gamma_{ml}^k d_l)\mathbf{n} + 2L_{ln} d_l \mathbf{Y}_{,m}]. \tag{5.11}$$

Equation (5.3) now becomes

$$g_{\text{ch}} = -k_2 g^{-1/2} e^{3mn} [g_{kn}(d_{k,m} + \Gamma_{ml}^k d_l) \cos \theta_0 + 2L_{ln} g_{mk} d_l d_k]. \tag{5.12}$$

The total free energy of the TCLB bulk is given by

$$F = \int g_{\text{ch}} \mathrm{d}V.$$

Since the thickness $t \cos \theta_0$ of the TCLB is about twice the length $t/2$ of the amphiphilic molecule and is negligibly small in comparison with the linear size of the membrane, the volume element may be approximated by $t \cos \theta_0 \mathrm{d}A$, where $\mathrm{d}A$ is the area element of the surface $\mathbf{Y}(u, v)$. In this way, we have

$$F = t \cos \theta_0 \int g_{\text{ch}} \mathrm{d}A = \int g^{1/2} g_{\text{TCLB}} \mathrm{d}u \mathrm{d}v, \tag{5.13}$$

where

$$g_{\text{TCLB}} = -k_2 t \cos \theta_0 g^{-1/2} e^{3mn} [g_{kn}(d_{k,m} + \Gamma_{ml}^k d_l) \cos \theta_0 + 2g_{km} L_{ln} d_k d_l]$$

$$+ \lambda(u, v)(g_{mn} d_m d_n - \sin^2 \theta_0) \quad (k, l, m, n = u, v). \tag{5.14}$$

The λ term comes from the constraint Eq. (5.2). The unknown Lagrange multiplier $\lambda(u, v)$ may be considered as a tensile stress acting on the surface induced by the coupling of the tilt of the molecule and the curvature of the surface.

5.1.2 Tilt-Equilibrium and Surface-Equilibrium Equations in Case of Strong Chiral Effect

In the equilibrium state, the free energy of the system is at its minimum. In other words, the condition

$$\delta F = \delta \int g^{-1/2} g_{\text{TCLB}} \mathrm{d}u \mathrm{d}v = 0$$

determines the equilibrium state of the system. The corresponding Euler–Lagrange equations are

$$\left(\frac{\partial}{\partial d_j} - \frac{\partial}{\partial m}\frac{\partial}{\partial d_{j,m}}\right)g^{1/2}g_{\text{TCLB}} = 0 \quad (j, m = u, v). \quad (5.15)$$

With g_{TCLB} given by Eq. (5.14), the Euler–Lagrange equation (5.15) gives the condition for the equilibrium state as

$$2g^{1/2}\lambda g_{ij}d_i - k_2 t e^{3mn}[(g_{kn}\Gamma^k_{mj} - g_{nj,m})\cos^2\theta_0$$

$$+2(g_{km}L_{jn} + g_{jm}L_{kn})d_k\cos\theta_0] = 0 \quad (i, j, k, m, n = u, v). \quad (5.16)$$

In principle, for given surface $\mathbf{Y}(u, v)$, Eq. (5.16) offers two linear equations of d_u, d_v and λ. They may be solved to give d_u and d_v as functions of λ. The substitution of these solutions into Eq. (5.2) determines λ as a function of g_{ij}, L_{ij} and Γ^k_{ij} which are known for a given surface $\mathbf{Y}(u, v)$. Therefore, Eq. (5.16) may be called the tilt-equilibrium equation.

On the other hand, in order to determine the equilibrium surface, one has to find a surface-equilibrium equation. To find the surface-equilibrium equation, it is necessary to find the variation of the free energy resulting from a variation of the surface. Let $\mathbf{Y}(u, v)$ be the equilibrium shape of the surface and

$$\mathbf{Y}'(u, v) = \mathbf{Y}(u, v) + \psi(u, v)\mathbf{n}, \quad (5.17)$$

where ψ is a sufficiently small smooth function of u and v, be a slightly distorted surface. The variation of the free energy caused by ψ is given by

$$\delta F = \delta \int g_{\text{TCLB}}dA = \delta \int g^{1/2}g_{\text{TCLB}}dudv. \quad (5.18)$$

It has already been shown in Chapter 3 that up to the first order of ψ and its derivatives with respect to u and to v, we have the following variations:

$$\delta g_{ij} = -2L_{ij}\psi,$$

$$\delta g = -4gH\psi,$$

$$\delta g^{1/2} = -2g^{1/2}H\psi, \quad (5.19)$$

$$\delta L_{ij} = \psi_{,lj} - \Gamma^k_{ij}\psi_{,k} - (2HL_{ij} - Kg_{ij})\psi,$$

$$\delta(dA) = -2H\psi dA.$$

From Eq. (5.19), it is easy to show that:

$$\delta g_{ij,k} = (\delta g_{ij})_{,k} = -2(L_{ij}\psi)_{,k},$$

$$\delta g^{-1/2} = -(1/2)g^{-3/2}\delta g = 2g^{-1/2}H\psi. \quad (5.20)$$

With these relations and $g_{in,j} + g_{jn,i} - g_{ij,n} = 2g_{kn}\Gamma_{ij}^k$, one finds that

$$\delta(g_{kn}\Gamma_{ml}^k) = (1/2)\delta(g_{mn,l} + g_{ln,m} - g_{ml,n})$$
$$= -[(L_{mn}\psi)_{,l} + (L_{ln}\psi)_{,m} - (L_{lm}\psi)_{,n}]. \qquad (5.21)$$

The variation of F can be written as

$$\delta F = \int (\delta g_{\text{TCLB}})\mathrm{d}A + \int g_{\text{TCLB}}\delta(\mathrm{d}A)$$
$$= \int g^{1/2}(\delta g_{\text{TCLB}})\mathrm{d}u\mathrm{d}v - 2\int H g_{\text{TCLB}}\psi\mathrm{d}A,$$

where

$$g^{1/2}(\delta g_{\text{TCLB}}) = (\delta g^{-1/2})g_{\text{TCLB}}$$
$$- k_2 t \cos\theta_0 e^{3mn}\{[\delta(g_{kn}(d_{k,m} + \Gamma_{ml}^k d_l))]\cos\theta_0$$
$$+ 2(\delta L_{ln})g_{km}d_k d_l + 2L_{ln}(\delta g_{km})d_k d_l]\}$$
$$+ \lambda(u,v)g^{1/2}(\delta g_{ij})d_i d_j.$$

Thus, we have that

$$\delta F = \int [-k_2 t \cos\theta_0 e^{3mn}\{\delta[g_{kn}(d_{k,m} + \Gamma_{ml}^k d_l)]\cos\theta_0 + 2(\delta L_{ln})g_{km}d_k d_l$$
$$+ 2L_{ln}(\delta g_{km})d_k d_l\} + \lambda(\delta g_{ij})g d_i d_j]\mathrm{d}u\mathrm{d}v. \qquad (5.22)$$

One notices that with $\delta g_{kn}d_{k,m}$ being of second order of ψ,

$$e^{3mn}\delta[g_{kn}(d_{k,m} + \Gamma_{ml}^k d_l)] = e^{3mn}[(L_{mn}\psi)_{,l} + (L_{ln}\psi)_{,m} - (L_{lm}\psi)_{,n}] = 0,$$
$$e^{3mn}L_{ln}(\delta g_{km})d_k d_l = -2e^{3mn}L_{ln}L_{km}\psi d_k d_l = 0,$$
$$e^{3mn}(\delta L_{ln})g_{km}d_k d_l = e^{3mn}[\psi_{,ln} - \Gamma_{ln}^p\psi_{,p} - (2HL_{ln} - Kg_{ln})\psi]g_{km}d_k d_l$$
$$= e^{3mn}(\psi_{,ln} - \Gamma_{ln}^p\psi_{,p} - 2HL_{ln}\psi)g_{km}d_k d_l,$$
$$\lambda(\delta g_{ij})d_i d_j = -2\lambda L_{ij}d_i d_j\psi. \qquad (5.23)$$

Therefore, it gives

$$\delta F = \int \{-k_2 t \cos\theta_0 e^{3mn}[2(\psi_{,ln} - \Gamma_{ln}^p\psi_{,p} - 2HL_{ln}\psi)g_{km}d_k d_l]$$
$$- 2\lambda g^{1/2}L_{mn}d_m d_n\psi\}\mathrm{d}u\mathrm{d}v.$$

With consideration of Eq. (3.41) and the vanishing variation on the boundary, it follows that

$$\delta F = \oint 2 \left[k_2 tg^{-1/2} \cos\theta_0 e^{3mn} \left(2HL_{ln} - \frac{\partial}{\partial p}\Gamma^p_{ln} - \frac{\partial}{\partial l}\frac{\partial}{\partial n} \right) \right.$$

$$\left. \times g_{km} d_k d_l - \lambda L_{mn} d_m d_n \right] \psi dA.$$

This leads to the surface-equilibrium equation

$$k_2 tg^{-1/2} \cos\theta_0 e^{3mn} \left(2HL_{ln} - \frac{\partial}{\partial l}\frac{\partial}{\partial n} - \frac{\partial}{\partial p}\Gamma^p_{ln} \right)$$

$$\times g_{km} d_k d_l - \lambda L_{mn} d_m d_n = 0 \quad (k, l, m, n, p = u, v). \quad (5.24)$$

The set of Eqs. (5.2), (5.16) and (5.24) comprises basic equations to determine the nature of the problem. Some practical applications will be given in the following sections.

5.1.3   Wound Ribbon Helix

A helicoid is generated by a curve, plane or twisted, rotating about a fixed line and at the same time translated along the direction of the axis with a constant rate proportional to the velocity of rotation. With z-axis as the axis of rotation, ρ as the distance of a point on the helicoid from the z-axis and ϕ as the angle made by the plane through the point and the axis with the xz-plane in the positive direction, the general equation of the helicoid surface may be represented by[16, 17]

$$\mathbf{Y}(\rho, \phi) = (\rho\cos\phi, \rho\sin\phi, h(\rho) + b\phi), \quad (5.25)$$

where $z = h(\rho)$ is the generating curve in any position of its plane. The constant b characterizes the parity of the rotation, i.e., $b > 0$ represents a right-handed rotation and $b < 0$ a left-handed rotation. The pitch of the twist is given by $2\pi|b|$. When $b = 0$, $\mathbf{Y}(\rho, \phi)$ represents a surface of revolution. When $h(\rho)$ is a constant, the curves $\phi = $ const. are straight lines perpendicular to the ρ-axis and the surface is called a right helicoid or right concoid. The curves $\rho = R$ (a constant) are helices on the helicoid and circles on the surface of revolution. If $h(\rho)$ varies linearly with z in the range $(0, a)$, where $2\pi|b| > a > 0$, then the right helicoid becomes a wound ribbon helix with a spiraling gap of $2\pi|b| - a$ along the z-direction. It looks like a ribbon of width $a\cos\psi$ wound around a right circular cylinder of radius R, where ψ is the gradient angle of the helix with respect to the

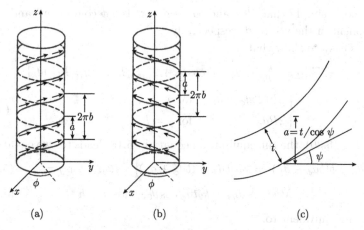

Fig. 5.2. Schematic illustrations of wound ribbon helix, (a) Right-handed helix. (b) Left-handed helix. The arrows represent the local tilt direction. (c) Section of the ribbon, t is the thickness of the ribbon, $a = t/\cos\psi$.

xy-plane. The wound ribbon helix surface is no longer a function of ρ and ϕ, but a function of ϕ and h. It may be represented by

$$\mathbf{Y}(\phi, h) = (R\cos\phi, R\sin\phi, h + b\phi). \tag{5.26}$$

It is easy to see that the gradient angle ψ of the helix satisfies the relation

$$\tan\psi = 2\pi b/(2\pi R) = b/R. \tag{5.27}$$

Figure 5.2 is a schematic illustration of the wound ribbon helix. The various fundamental quantities associated with the surface (5.26) are

$$\begin{aligned}
&\mathbf{Y}_{,\phi} = (-R\sin\phi, R\cos\phi, b), \quad \mathbf{Y}_{,h} = (0,0,1), \\
&g_{\phi\phi} = R^2 + b^2, \quad g_{\phi h} = g_{h\phi} = b, \quad g_{hh} = 1, \\
&g = R^2, \quad g_{ij,k} = 0, \\
&\mathbf{n} = (\cos\phi, \sin\phi, 0), \\
&L_{\phi\phi} = -R, \quad L_{\phi h} = L_{h\phi} = L_{hh} = L = 0, \\
&2H = -1/R, \quad K = 0, \quad \Gamma^k_{ij} = 0 \quad (i, j, k = \phi, h).
\end{aligned} \tag{5.28}$$

The constraint on $\mathbf{d}$, Eq. (5.2), now takes the form:

$$R^2 d_\phi^2 + (b d_\phi + d_h)^2 = \sin^2\theta_0. \tag{5.29}$$

This shows simply that Rd_ϕ and $bd_\phi + d_h$ are the ϕ component and the z component of the vector $\mathbf{d}$, respectively.

It is easy to show that

$$e^{3mn}[(g_{kn}\Gamma^k_{jm} - g_{jn,m})\cos\theta_0 + 2(g_{km}L_{jn} + g_{jm}L_{kn})d_k]$$

$$= \begin{cases} 2R(2bd_\phi + d_h) & \text{for } j = \phi, \\ 2Rd_\phi & \text{for } j = h. \end{cases} \tag{5.30}$$

With Eq. (5.30), the tilt-equilibrium equation (5.16) leads to two equations:

$$\lambda[R^2 d_\phi + b(bd_\phi + d_h)] = k_2 t[bd_\phi + (bd_\phi + d_h)]\cos\theta_0, \quad \text{for } j = \phi, \tag{5.31a}$$

$$\lambda(bd_\phi + d_h) = k_2 td_\phi \cos\theta_0, \quad \text{for } j = h. \tag{5.31b}$$

They are equivalent to:

$$R^2 d_\phi^2 = (bd_\phi + d_h)^2, \tag{5.32a}$$

$$\lambda^2 = k_2^2 t^2 \cos^2\theta_0/R^2. \tag{5.32b}$$

Equations (5.29) and (5.32a) give

$$R^2 d_\phi^2 = \sin^2\theta_0/2, \tag{5.33a}$$

$$(bd_\phi + d_h)^2 = \sin^2\theta_0/2. \tag{5.33b}$$

The four pairs of solutions for d_ϕ and d_h determined by Eq. (5.33) are

(1)

$$d_\phi = \sin\theta_0/(\sqrt{2}R), \quad bd_\phi + d_h = \sin\theta_0/\sqrt{2},$$
$$d_h = (R - b)\sin\theta_0/(\sqrt{2}R), \quad Rd_\phi(bd_\phi + d_h) = \sin^2\theta_0/2, \tag{5.34a}$$
$$\lambda = k_2 t\cos\theta_0/R;$$

(2)

$$d_\phi = -\sin\theta_0/(\sqrt{2}R), \quad bd_\phi + d_h = -\sin\theta_0/\sqrt{2},$$
$$d_h = -(R - b)\sin\theta_0/(\sqrt{2}R), \quad Rd_\phi(bd_\phi + d_h) = \sin^2\theta_0/2, \tag{5.34b}$$
$$\lambda = k_2 t\cos\theta_0/R;$$

(3)

$$d_\phi = \sin\theta_0/(\sqrt{2}R), \quad bd_\phi + d_h = -\sin\theta_0/\sqrt{2},$$
$$d_h = -(R + b)\sin\theta_0/(\sqrt{2}R), \quad Rd_\phi(bd_\phi + d_h) = -\sin^2\theta_0/2,$$
$$\lambda = -k_2 t\cos\theta_0/R;$$

$$\tag{5.34c}$$

(4)

$$d_\phi = -\sin\theta_0/(\sqrt{2}R), \quad bd_\phi + d_h = \sin\theta_0/\sqrt{2},$$
$$d_h = (R+b)\sin\theta_0/(\sqrt{2}R), \quad Rd_\phi(bd_\phi + d_h) = -\sin^2\theta_0/2, \quad (5.34d)$$
$$\lambda = -k_2 t \cos\theta_0/R.$$

The corresponding values of λ in Eqs. (5.34) are determined from Eq. (5.31b). The orientation vector $\mathbf{d}$ now takes the form

$$\mathbf{d} = d_\phi(-R\sin\phi, R\cos\phi, b) + d_h(0,0,1) + \cos\theta_0(\cos\phi, \sin\phi, 0). \quad (5.35)$$

In comparison with the liquid crystal theory, where the directors $\mathbf{n}$ and $-\mathbf{n}$ are equivalent, here the orientation vectors $\mathbf{d}$ and $-\mathbf{d}$ may also be considered equivalent. Both b and k_2 are positive for the right-handed twist and negative for the left-handed twist. These conditions imply that the four sets of solutions (1),(2), (3) and (4) given by Eq. (5.35) are equivalent and may be grouped into two groups: the first group, (1) and (2), with positive $d_\phi(bd_\phi + d_h)$, and the second group, (3) and (4), with negative $d_\phi(bd_\phi + d_h)$. Since R is always positive, with $Rd_\phi(bd_\phi + d_h) > 0$, a positive increment Rd_ϕ means a positive increase of the vertical component $bd_\phi + d_h$. This case corresponds to a right-handed helix. On the other hand, $Rd_\phi(bd_\phi + d_h) < 0$ refers to a left-handed helix.

It is interesting to note that at any point on the surface, the horizontal component Rd_ϕ, and the vertical component $bd_\phi + d_h$ of the vector $\mathbf{d}$ have the same magnitude but opposite signs for right-handed and left-handed helices, respectively. This fact implies that the gradient angle of $\mathbf{d}$ is equal to $45°(-135°)$ for $b > 0$ and $135°(-45°)$ for $b < 0$. This agrees with the gradient angle of the helix with $R = |b|$ as shown in Eq. (5.27). Therefore, the orientation vector $\mathbf{d}$ is parallel to the edge line of the helix with $R = |b|$. Since the gradient angle of such a helix is $\pm 45°$ or $\mp 135°$, it follows that the pitch of the helix is equal to $2\pi R$. Such a conclusion seems to give an explanation to the experimental observations[1,2] that the gradient angle of wound ribbon helices are practically $45°$. However, more experiments are needed to check whether there exists wound ribbon helices with gradient angle differing from $45°$ or not. The $45°$ gradient angle asserts that the orientation vector $\mathbf{d}$ is parallel to the edge of the helix. The fact that both d_ϕ and d_h given in Eq. (5.35) are independent of ϕ and h imposes on the surface-equilibrium equation (5.24) the form:

$$d_\phi[-k_2 t \cos\theta_0(bd_\phi + d_h) + \lambda R^2 d_\phi]/R = 0. \quad (5.36)$$

This equation is automatically satisfied by all the solutions given in Eq. (5.34).

The function g_{TCLB}, Eq. (5.14), now takes the form:

$$g_{\text{TCLB}} = -2k_2 t g^{-1/2} \cos\theta_0 R d_\phi (b d_\phi + d_h). \qquad (5.37)$$

The area of the wound ribbon helix is given by

$$A = \int_0^{2n\pi} \int_0^{na} R d\phi dh = 2n^2 \pi a R, \qquad (5.38)$$

where $a = \sqrt{2}t$. Consequently, the free energy F_W of wound ribbon helix becomes

$$F_W = -\int_0^{2n\pi} \int_0^{na} 2k_2 t \cos\theta_0 R d_\phi (b d_\phi + d_h) d\phi dh$$

$$= -2k_2 t A \cos\theta_0 d_\phi (b d_\phi + d_h). \qquad (5.39)$$

For right-handed wound ribbon helices ($b > 0$), Eqs. (5.34a) and (5.34b) show that

$$F_W^R = -k_2 t A \cos\theta_0 \sin^2\theta_0 / R. \qquad (5.40a)$$

Since θ_0 is defined as an acute angle, it means that right-handed wound ribbon helices with $k_2 > 0$ molecules are more stable than those with $k_2 < 0$. On the other hand, for left-handed wound ribbon helices ($b < 0$), Eqs. (5.34c) and (5.34d) lead to

$$F_W^L = k_2 t A \cos\theta_0 \sin^2\theta_0 / R. \qquad (5.40b)$$

This shows that for left-handed wound ribbon helices ($b < 0$) with $k_2 < 0$ molecules are more stable. As a whole, wound ribbon helices with b and k_2 of the same sign have lower free energy than those with opposite signs and are more favorable. This agrees with the case of cholesteric liquid crystals for which $k_2 > 0$ and $k_2 < 0$ corresponding to right-handed and left-handed patterns, respectively.[18] The calculation also indicates that the director is parallel to the edge of the helix.

It has been stated at the beginning of this section that in case of $b = 0$ the helicoid degenerates into a cylinder. The parity of the twist now depends on the sign of k_2 alone: right-handed for $k_2 > 0$ and left-handed for $k_2 < 0$. The four sets of solutions for d_ϕ and d_h given by Eq. (5.34) now reduce to[13]

$$d_\phi = \alpha_1 \sin\theta_0/(\sqrt{2}R), \quad d_h = \alpha_2 \sin\theta_0/\sqrt{2}, \quad \lambda = \alpha_1 k_2 t \cos\theta_0/(\alpha_2 R),$$
(5.41)

where $(\alpha_1, \alpha_2) = (\pm 1, \pm 1)$ for $k_2 > 0$ (right-handed) or $(\pm 1, \mp 1)$ for $k_2 < 0$ (left-handed). The orientation vector $\mathbf{d}$ now is given by

$$\mathbf{d} = \frac{\alpha_1 \sin\theta_0}{\sqrt{2}}(-\sin\phi, \cos\phi, 0) + \frac{\alpha_2 \sin\theta_0}{\sqrt{2}}(0, 0, 1) + \cos\theta_0(\cos\phi, \sin\phi, 0),$$
(5.42)

and the area A of the cylinder is equal to

$$A = 2\pi RL, \tag{5.43}$$

where L is the length of the cylinder. Finally, the free energy F_W^C of the cylinder is given by

$$F_W^C = -\alpha_1 \alpha_2 k_2 t A \cos\theta_0 \sin^2\theta_0/R. \tag{5.44}$$

Again, the right-handed helix is the favorable one for $k_2 > 0$ and it favors the left-handed helix for $k_2 < 0$.

Since for wound ribbon helix the area A is given by $A = 2n^2 \pi aR$, the free energy F_W, Eqs. (5.40), may be written as

$$F_W = \mp k_2 t 2n^2 \pi a \cos\theta_0 \sin^2\theta_0.$$

For wound ribbon helix with radius R, the maximum value of a is $2\pi|b|$. In that case, the gap of the helix vanishes and the free energy reaches its minimum. This explains the experimental fact of the self-transformation of the wound ribbon helix into a prolate tube. In case of $a = \sqrt{2}t > 2\pi|b|$, the surface becomes a multilayer prolate tube-like a soda straw. The general properties of these two cases are similar to what has already been described. Both cases have also been observed in experiments.[1-3]

5.1.4 Twisted Ribbon

In case of $h(\rho) = 0$, $|\rho| \leq R$ (a constant) and $0 \leq \phi \leq 2n\pi$ (n is the spiraling number), the helicoid surface represented by Eq. (5.25) becomes

$$\mathbf{Y}(\rho, \phi) = (\rho\cos\phi, \rho\sin\phi, b\phi), \quad |\rho| \leq R, \quad 0 \leq \phi \leq 2n\pi. \tag{5.45}$$

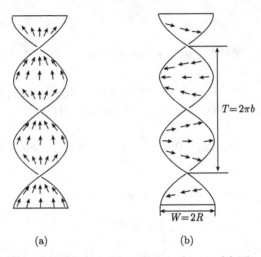

(a)   (b)

Fig. 5.3. Schematic illustrations of the twisted ribbons: (a) The tilt field is along the edge line, (b) the tilt field is perpendicular to the edge line. The arrows represent the local tilt direction.

It describes a twisted ribbon with z-axis as the axis of twist. The geometry of this type of helical structure is shown in Fig. 5.3. The related fundamental quantities are

$$\mathbf{Y}_{,\rho} = (\cos\phi, \sin\phi, 0), \quad \mathbf{Y}_{,\phi} = (-\rho\sin\phi, \rho\cos\phi, b),$$

$$g_{\rho\rho} = 1, \quad g_{\rho\phi} = g_{\phi\rho} = 0, \quad g_{\phi\phi} = \rho^2 + b^2 = g,$$

$$\mathbf{n} = (\rho^2 + b^2)^{-1/2}(b\sin\phi, -b\cos\phi, \rho),$$

$$L_{\rho\rho} = 0, \quad L_{\rho\phi} = L_{\phi\rho} = -b(\rho^2 + b^2)^{-1/2}, \tag{5.46}$$

$$L_{\phi\phi} = 0, \quad L = -b^2/(\rho^2 + b^2),$$

$$H = 0, \quad K = -b^2/(\rho^2 + b^2)^2,$$

$$\Gamma^\rho_{\phi\phi} = -\rho, \quad \Gamma^\phi_{\rho\phi} = \rho/(\rho^2 + b^2), \quad \Gamma^\rho_{\rho\rho} = \Gamma^\rho_{\rho\phi} = \Gamma^\rho_{\phi\rho} = \Gamma^\phi_{\rho\rho} = \Gamma^\phi_{\phi\phi} = 0.$$

With these known quantities, the equation of constraint on $\mathbf{d}$, Eq. (5.2), becomes

$$d_\rho^2 + (\rho^2 + b^2)d_\phi^2 - \sin^2\theta_0 = 0. \tag{5.47}$$

In the tilt-equilibrium equation (5.16), the following terms for $j = \rho$ and for $j = \phi$ are given, respectively, by

$$g_{ij}d_i = d_\rho,$$

$$e^{3mn}(g_{kn}\Gamma^k_{mj} - g_{nj,m}) = 0 \quad (j = \rho), \tag{5.48a}$$

$$e^{3mn}(g_{km}L_{jn} + g_{jm}L_{kn})d_k = -2b(\rho^2 + b^2)^{-1/2}d_\rho$$

and

$$g_{ij}d_i = (\rho^2 + b^2)d_\phi,$$

$$e^{3mn}(g_{kn}\Gamma^k_{mj} - g_{nj,m}) = 0 \quad (j = \phi), \tag{5.48b}$$

$$e^{3mn}(g_{km}L_{jn} + g_{jm}L_{kn})d_k = 2b(\rho^2 + b^2)^{1/2}d_\phi.$$

Consequently, Eq. (5.16) gives the following two relations:

$$d_\rho(\rho^2 + b^2)^{1/2}[\lambda + 2k_2tb\cos\theta_0/(\rho^2 + b_2)] = 0 \quad (j = \rho), \tag{5.49a}$$

$$d_\phi(\rho^2 + b^2)^{3/2}[\lambda - 2k_2tb\cos\theta_0/(\rho^2 + b_2)] = 0 \quad (j = \phi). \tag{5.49b}$$

The two sets of solutions of Eqs. (5.47) and (5.49) are
Case (1)

$$d_\rho = 0,$$

$$d_\phi = \pm(\rho^2 + b^2)^{-1/2}\sin\theta_0, \tag{5.50a}$$

$$\lambda = 2k_2tb\cos\theta_0/(\rho^2 + b^2);$$

Case (2)

$$d_\rho = \pm\sin\theta_0,$$

$$d_\phi = 0, \tag{5.50b}$$

$$\lambda = -2k_2tb\cos\theta_0/(\rho^2 + b^2),$$

respectively. The two functions d_ρ and d_ϕ represent the two components of the unit orientation vector $\mathbf{d}$ in the (ρ, ϕ) plane which is perpendicular to the normal of the surface. Together, they represent the average direction of the molecules at a point in the lipid bilayer. Lipid bilayers are formed by two layers of molecules with opposite molecular orientations. Obviously, in the (ρ, ϕ) plane $\mathbf{d}$ and $-\mathbf{d}$ are equivalent. This is equivalent to the fact that the directors $\mathbf{n}$ and $-\mathbf{n}$ are equivalent in the liquid crystal theory. In the case of twisted ribbons, Eq. (5.50) shows that there are two sets of solutions for d_ρ and d_ϕ, either $d_\rho = 0$ or $d_\phi = 0$. The equivalence of the components of $\mathbf{d}$ and $-\mathbf{d}$ in the (ρ, ϕ) plane implies that the plus and minus signs in Eq. (5.50) are irrelevant and can be neglected. Without the consideration of the $\mathbf{n}$ term, in case (1), $\mathbf{d}$ has a constant component in the z-direction, while in case (2) $\mathbf{d}$ has no component in the z-direction. Thus, case (1) corresponds to the case of $\mathbf{d}$ being along the edge of the helix, while case (2) corresponds to $\mathbf{d}$ being perpendicular to the edge of the helix. Furthermore, case (2) shows that in any vertical plane $\phi = \text{const.}$,

$\mathbf{d}$ is independent of b, only the component of $\mathbf{d}$ perpendicular to the surface depends on b. In the case of wound ribbon helix there does not exist the case of $\mathbf{d}$ being perpendicular to the edge of the helix. In view of this, case (2) may not be a realistic physical case requiring serious consideration.

With Eq. (5.46), by direct expansion into individual terms, it is easy to show that

$$e^{3mn}g_{kn}(d_{km} + \Gamma^k_{ml}d_l) = (d/d\rho)[(\rho^2 + b^2)d_\phi],$$
$$e^{3mn}L_{ln}g_{km}d_kd_l = -b(\rho^2 + b^2)^{-1/2}d^2_\rho + b(\rho^2 + b^2)^{1/2}d^2_\phi,$$
$$g_{ij}d_id_j = d^2_\rho + (\rho^2 + b^2)d^2_\phi, \tag{5.51}$$
$$e^{3mn}(\partial/\partial l)(\partial/\partial n)(g_{km}d_kd_l) = -(\partial^2/\partial\rho^2)[(\rho^2 + b^2)d_\rho d_\phi],$$
$$e^{3mn}(\partial/\partial p)(\Gamma^p_{ln}g_{km}d_kd_l) = -(\partial/\partial\rho)(\rho d_\rho d_\phi),$$
$$L_{mn}d_md_n = -2b(\rho^2 + b^2)^{-1/2}d_\rho d_\phi.$$

The surface-equilibrium equation for twisted ribbon now becomes

$$k_2t(\rho^2 + b^2)^{-1/2}\left[\frac{\partial^2}{\partial\rho^2}(\rho^2 + b^2) + \frac{\partial}{\partial\rho}\rho + 2b\right]d_\rho d_\phi = 0. \tag{5.52}$$

Obviously, this equation is automatically satisfied by the set of solutions d_ρ, d_ϕ and λ given by Eq. (5.50a) or by Eq. (5.50b).

The area of the twisted ribbon is given by

$$A = \int_{-R}^{R}\int_0^{2n\pi}(\rho^2 + b^2)^{-1/2}d\phi d\rho$$
$$= 2n\pi\{R(R^2 + b^2)^{1/2} + (b^2/2)\ln[(R^2 + b^2)^{1/2} + R]$$
$$- (b^2/2)\ln[(R^2 + b^2)^{1/2} - R]\}.$$

Obviously, for given $|b|$, the right-handed twisted ribbon and the left-handed twisted ribbon are symmetric with each other and have the same area. Therefore, area A may be written as

$$A = 2n\pi\{R(R^2 + b^2)^{1/2} + b^2\ln[(R^2 + b^2)^{1/2} + R] - b^2\ln|b|\}.$$

Let

$$x = R/|b| > 0, \tag{5.53}$$

then the area A of the twisted ribbon is given by

$$A = 2n\pi b^2\{x(1 + x^2)^{1/2} + \ln[(1 + x^2)^{1/2} + x]\}, \tag{5.54}$$

where x may be greater or less than one depending on whether R is greater or less than $|b|$.

With Eq. (5.51), the function g_{TCLB} takes the form

$$g_{\text{TCLB}} = -k_2 tg^{-1/2} \cos\theta_0 \{ d/d\rho[(\rho^2 + b^2)d_\phi] \cos\theta_0$$
$$- 2b(\rho^2 + b^2)^{-1/2}d_\rho^2 + 2b(\rho^2 + b^2)^{1/2}d_\phi^2 \}.$$

Corresponding to the two sets of solutions of d_ρ and d_ϕ given by Eq. (5.50), g_{TCLB} takes the following forms:

Case (1) $d_\rho = 0$, $d_\phi = \pm(\rho^2 + b^2)^{-1/2}\sin\theta_0$:

$$g_{\text{TCLB}}^{(1)} = -k_2 tg^{-1/2} \cos\theta_0 \{ d/d\rho[\pm(\rho^2 + b^2)^{1/2}] \sin\theta_0 \cos\theta_0$$
$$+ 2b(\rho^2 + b^2)^{-1/2}\sin^2\theta_0 \}; \tag{5.55a}$$

Case (2) $d_\rho = \pm\sin\theta_0$, $d_\phi = 0$:

$$g_{\text{TCLB}}^{(2)} = 2k_2 tbg^{-1/2} \cos\theta_0 \sin^2\theta_0 (\rho^2 + b^2)^{-1/2}. \tag{5.55b}$$

In the integration of g_{TCLB} to get the free energy of the twisted ribbon, it should be noted that b^2 in the term $(\rho^2 + b^2)^{-1/2}$ refers to $|b^2|$, while the b factor on $(\rho^2 + b^2)^{-1/2}$ determines the parity of the twist. Consequently, the free energies $F_T^{(1)}$ and $F_T^{(2)}$ of the twisted ribbon corresponding to case (1) and case (2), respectively, are given by

$$F_T^{(1)} = -F_T^{(2)} = -8n\pi k_2 tb \sin^2\theta_0 \cos\theta_0 \ln[(1 + x^2)^{1/2} + x]. \tag{5.56a}$$

Since $(1 + x^2)^{1/2} + x > 1$, it follows that when k_2 and b have the same sign, $F_T^{(1)}$ is less than $F_T^{(2)}$ and is the favorable state. On the other hand, when k_2 and b have different signs, $F_T^{(2)}$ will be the favorable state. According to the general point of view,[18] the sign of k_2 should be determined by the chirality of the molecules. The present analysis offers an interesting structure with helical senses of macrostructure different from the microscopic chiral senses of the component molecules.

To further compare with wound ribbon helices, the free energy F_T may be written as

$$F_T^{(1)} = -F_T^{(2)} = -\frac{4k_2 tA \cos\theta_0 \sin^2\theta_0 \ln[(1 + x^2)^{1/2} + x]}{b\{x(1 + x^2)^{1/2} + \ln[(1 + x^2)^{1/2} + x]\}}. \tag{5.56b}$$

According to Eq. (5.53), b may be expressed as $\pm R/x$, where the upper sign holds for $b > 0$ and the lower sign for $b < 0$. In this way, F_T may be written as

$$F_T^{(1)} = -F_T^{(2)} = \mp 2k_2 At \cos\theta_0 \sin^2\theta_0 f(x)/R, \tag{5.56c}$$

where the upper sign holds for $b > 0$, the lower sign for $b < 0$ and the function $f(x)$ is given by

$$f(x) = \frac{2x \ln[(1 + x^2)^{1/2} + x]}{x(1 + x^2)^{1/2} + \ln[(1 + x^2)^{1/2} + x]}. \tag{5.57}$$

Numerical analysis shows that $f(x)$ takes its maximum value at

$$x = 2.34, \quad \text{with } f(x) \leq f(2.34) = 0.984. \tag{5.58}$$

In comparison with Eq. (5.44), the free energy F_W of wound ribbon helix, it shows that with the same area and the same radius R, the free energy of the favorable state of wound ribbon helix is less than that of the twisted ribbon. This seems to give an explanation on the spontaneous transition from the twisted strip to wound ribbon observed in experiments.[1]

5.1.5   Spherical Vesicle

Some experimental studies[1,3] observed the formation of spherical vesicles of chiral fibrous bilayers before growth into a helical structure. Therefore, it is interesting to study the orientation field of chiral molecules on a spherical surface. In spherical coordinates (θ, ϕ), the spherical surface of radius R is represented by

$$\mathbf{Y}(\theta, \phi) = R(\sin\theta\cos\phi, \sin\theta\sin\phi, \cos\theta). \tag{5.59}$$

The related basic quantities are

$$\begin{aligned}
& g_{\theta\theta} = R^2, \quad g_{\theta\phi} = g_{\phi\theta} = 0, \quad g_{\phi\phi} = R^2\sin^2\theta, \quad g = R^4\sin^2\theta, \\
& \mathbf{n} = (\sin\theta\cos\phi, \sin\theta\sin\phi, \cos\theta), \\
& L_{\theta\theta} = -R, \quad L_{\theta\phi} = L_{\phi\theta} = 0, \quad L_{\phi\phi} = -R^2\sin^2\theta, \quad L = R^2\sin^2\theta, \\
& H = -1/R, \quad K = 1/R^2, \\
& \Gamma^\theta_{\theta\theta} = \Gamma^\theta_{\theta\phi} = \Gamma^\theta_{\phi\theta} = \Gamma^\phi_{\theta\theta} = \Gamma^\phi_{\phi\phi} = 0, \\
& \Gamma^\theta_{\phi\phi} = -\sin^2\theta\Gamma^\phi_{\theta\phi} = -\sin^2\theta\Gamma^\phi_{\theta\phi} = -\sin\theta\cos\theta.
\end{aligned} \tag{5.60}$$

The constraint on $\mathbf{d}$, Eq. (5.2), and the tilt-equilibrium equation (5.16) now become

$$R^2 d_\theta^2 + R^2\sin^2\theta\, \theta_\phi^2 - \sin^2\theta_0 = 0, \tag{5.61a}$$

$$\lambda d_\theta = 0, \tag{5.61b}$$

$$\lambda\sin^3\theta d_\phi = 0, \tag{5.61c}$$

respectively. Apparently, the two sets of solutions of Eq. (5.61) are

Case (1)

$$d_\theta = 0, \quad d_\phi = \pm \sin\theta_0/(R\sin\theta), \quad \lambda = 0; \qquad (5.62a)$$

Case (2)

$$d_\theta = \pm \sin\theta_0/R, \quad d_\phi = 0, \quad \lambda = 0. \qquad (5.62b)$$

The surface-equilibrium equation (5.24) in this case is in the form

$$\left(\frac{\partial^2}{\partial\theta^2}\sin^2\theta + \frac{\partial}{\partial\theta}\sin\theta\cos\theta\right) d_\theta d_\phi = 0.$$

It is obvious that both sets of Eqs. (5.62a) and (5.62b) satisfy this equation. Since $\mathbf{d}$ and $-\mathbf{d}$ are equivalent, the plus and minus signs in Eq. (5.62) are irrelevant and can be neglected. The azimuthal angle fields of the director corresponding to Eq. (5.62) are sketched in Figs. 5.4(a) and 5.4(b) as latitudinal lines and longitudinal lines on the spherical surface, respectively. In both cases, the orientation of the director is discontinuous at the north pole and the south pole. This is a natural consequence of the theorem in differential geometry that on a closed surface of topological sphere a line field has at least two singular points.[19] The two singular points correspond to defects of the TCLB vesicle. If it is assumed that these defects will cause leakage and fusion of vesicles, then the feature of the orientation field provides a reasonable explanation for the experimentally observed fact of transition from vesicular dispersion to helical superstructure formation of the TCLB.

In their experiment,[1] Nakashima et al. observed that with the decrease of temperature, there is a transition from vesicle to vesicle fusion and finally

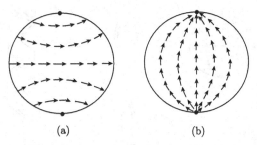

(a)   (b)

Fig. 5.4. Schematic illustration of spherical vesicles: (a) The tilt field corresponding to Eq. (5.62a), (b) the tilt field corresponding to Eq. (5.62b). The arrows represent the local tilt direction.

the formation of aggregated helical structures. It seems that this process may be considered as a sequence of a transition from the L_α phase of the bilayer material without tilt and defect of orientation to the L_β^* phase. At the L_β^* phase, the tilt of the molecule will cause orientation defects leading to the vesicle fusion and the formation of helical structure. In general, the transition from vesicle to helical structures can be seen from the energy point of view. The function g_{TCLB} in the present case is equal to

$$g_{\text{TCLB}} = -k_2 t g^{-1/2} \cos^2 \theta_0 [g_{\phi\phi}(d_{\phi,\theta} + \Gamma^\phi_{\theta\phi} d_\phi) - g_{\theta\theta}(d_{\theta,\phi} + \Gamma^\theta_{\phi\phi} d_\phi)].$$

Since d_θ and d_ϕ are not functions of ϕ, the free energy of spherical vesicles becomes

$$F_S = -k_2 t \cos^2 \theta_0 \int_{-\pi/2}^{\pi/2} \int_0^{2\pi} \frac{\mathrm{d}}{\mathrm{d}\theta}(\sin^2 \theta d_\phi) \mathrm{d}\theta \mathrm{d}\phi.$$

Therefore, for both cases (1) and (2), it gives that

$$F_S = 0. \tag{5.63}$$

In comparison with Eqs. (5.40) and (5.56) it shows that

$$F_S > F_T > F_W. \tag{5.64}$$

This confirms the observed transition from vesicular dispersion to twisted strip and to wound ribbon. Figure 5.5 illustrates the transition sequence of different shapes of the membrane. Conventionally, the local tilt directions are chosen to be along the edge line of the helical structure. The second kind of twisted strip illustrated in Fig. 5.3(b) is ruled out in this sequence.

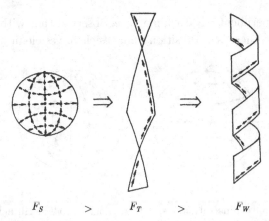

$$F_S \qquad > \qquad F_T \qquad > \qquad F_W$$

Fig. 5.5.   Schematic illustration of the transition sequence from the vesicle to the helical structure. The arrows represent the local tilt field direction.

5.2 General Theory for TCLB

5.2.1 General Formula of Chiral Free Energy

The free energy expression given by Eqs. (5.13) and (5.14) is convenient in discussing the given shapes of TCLB. In this section, another general formula of free energy of TCLB in terms of differential geometry terms will be discussed. Although some of the differential geometry terms have been introduced in Chapter 2, they will be introduced again in a somewhat more detailed way. A surface $\mathbf{Y}$ in space may be represented by the positions of the points on the surface in terms of the two parameters u and v, i.e., by $\mathbf{Y} = \mathbf{Y}(u, v)$. Any curve on the surface $\mathbf{Y}(u, v)$ may be expressed by the positions of the points $\mathbf{r}(s) = \mathbf{Y}(u(s), v(s)) = \mathbf{Y}(s)$ along the curve in terms of a parameter s which, for convenience, may be chosen as the arc length along the curve from some given point $\mathbf{r}_0 = \mathbf{r}(s_0)$ on the curve.

Consider two neighboring points P and P' on a smooth continuous curve $\mathbf{r}(s)$ (not a straight line), the secant line PP' approaches the tangent line of the curve at P as P' approaches P along the curve. The unit vector along the tangent line at P in the direction of increasing s (the arc length of the curve), $\mathbf{t}(s)$, is called the unit tangent vector of the curve and

$$\mathbf{t}(s) = \mathbf{r}_{,s}(s) = \mathbf{Y}_{,s}(s). \tag{2.1}$$

The normal plane at point P is that plane which is orthogonal to the tangent at P. For a space curve $\mathbf{r} = \mathbf{r}(s)$ possessing nonvanishing $\mathbf{r}_{,s}$ and $\mathbf{r}_{,ss}$, the limiting plane containing the tangent line at P on the curve and a neighboring point P' on the curve as P' approaches P as limit is called the osculating plane of the curve at the point P. The principal (main) normal of the curve at P is the line of intersection of the normal plane and the osculating plane at P. The unit principal normal $\mathbf{m}(s)$ is defined by

$$\mathbf{m}(s) = \mathbf{Y}_{,ss}(s)/|\mathbf{Y}_{,ss}(s)|. \tag{2.2}$$

This is obvious, since $\mathbf{Y}_{,s} \cdot \mathbf{Y}_{,s} = 1$ and consequently $\mathbf{Y}_{,ss}$ is perpendicular to $\mathbf{Y}_{,s}$. Thus, the plane determined by $\mathbf{t}(s)$ and $\mathbf{m}(s)$ is the osculating plane. The unit binormal vector $\mathbf{b}(s)$ is defined by

$$\mathbf{b}(s) = \mathbf{t}(s) \times \mathbf{m}(s) = \mathbf{Y}_{,s}(s) \times \mathbf{Y}_{,ss}(s)/|\mathbf{Y}_{,ss}(s)|. \tag{2.3}$$

The three unit vectors $\mathbf{t}(s)$, $\mathbf{m}(s)$ and $\mathbf{b}(s)$ are mutually perpendicular to each other forming a right-handed trio with

$$\mathbf{t}(s) \times \mathbf{m}(s) \cdot \mathbf{b}(s) = 1.$$

The arc rate at which the tangent changes direction as point P moves along the curve is called the curvature $c(s)$ of the curve. The magnitude of $c(s)$ is given by

$$c(s) = |\mathbf{t}_{,s}(s)| = \mathbf{Y}_{,ss}(s). \tag{2.4}$$

The inverse of $c(s)$ is called the radius of curvature R of the curve at P and

$$1/R(s) = c(s) = [\mathbf{Y}_{,ss}(s) \cdot \mathbf{Y}_{,ss}(s)]^{1/2}.$$

There is a unique circle tangent to the curve at P in the osculating plane with center on the principal normal and radius $R(s)$. This circle is known as the osculating circle or the circle of curvature. Consider a point P on the curve C (not a straight line): $u = u(s), v = v(s)$, which is on the surface $\mathbf{Y} = \mathbf{Y}(u, v)$. Let ϕ be the angle between the principal normal $\mathbf{m}$ to C at P and the normal $\mathbf{n}$ of the surface, then

$$\frac{\mathrm{d}\mathbf{t}}{\mathrm{d}s} \cdot \mathbf{n} = \frac{\cos\phi}{R}$$

defines a radius of curvature of the curve. In case of $\phi = 0$ or π, $R = R_n$ is called the radius of normal curvature $c_n = 1/R_n$. It is not difficult to show that the normal curvature c_n is given by

$$c_n = \frac{1}{R_n} = \pm \left[L_{uu} \left(\frac{\mathrm{d}u}{\mathrm{d}s} \right)^2 + 2L_{uv} \frac{\mathrm{d}u}{\mathrm{d}s} \frac{\mathrm{d}v}{\mathrm{d}s} + L_{vv} \left(\frac{\mathrm{d}v}{\mathrm{d}s} \right)^2 \right].$$

With Eqs. (2.2) and (2.4), the differentiation of Eq. (2.1) gives that

$$\mathbf{t}_{,s} = \frac{\mathrm{d}\mathbf{t}}{\mathrm{d}s} = c(s)\mathbf{m}(s). \tag{2.7}$$

Similar to the curvature of a curve, as the point P moves along the curve, the arc rate at which the osculating plane turns about the tangent, i.e., the arc rate of change of the binormal, is called the torsion $\tau(s)$ of the curve at P:

$$|\tau(s)| = \left| \frac{\mathrm{d}\mathbf{b}(s)}{\mathrm{d}s} \right|. \tag{2.6}$$

Since $\mathbf{b}(s)$ is perpendicular to the osculating plane determined by $\mathbf{t}(s)$, $\mathbf{m}(s)$, and $\mathbf{b}_{,s}(s)$ is perpendicular to $\mathbf{b}(s)$ on account of $\mathbf{b}(s) \cdot \mathbf{b}(s) = 1$, therefore $\mathbf{b}_{,s}$ is in the osculating plane. On the other hand, $\mathbf{b} \cdot \mathbf{t}_{,s} = 0$, it follows that $\mathbf{b}_{,s}$ is orthogonal to $\mathbf{t}$. Therefore $\mathbf{b}_{,s}$ must be parallel to $\mathbf{m}(s)$.

These lead to

$$\frac{d\mathbf{b}(s)}{ds} = \tau(s)\mathbf{m}(s). \tag{2.9}$$

Differentiation of the relation $\mathbf{m}(s) = \mathbf{b}(s) \times \mathbf{t}(s)$ gives $\mathbf{m}_{,s} = \mathbf{b}_{,s} \times \mathbf{t} + \mathbf{b} \times \mathbf{t}_{,s}$. Under Eqs. (2.7) and (2.9), it becomes

$$\mathbf{m}_{,s} = -c(s)\mathbf{t} - \tau(s)\mathbf{b}. \tag{2.8}$$

The set of Eqs. (2.7), (2.8) and (2.9) is called the Frenet–Serret formula. On a surface, a curve which is not a straight line, with principal normal at every point on the curve coinciding with the normal to the surface at that point is known as a geodesic on the surface.[20] Thus, on the geodesic, the principal normal $\mathbf{m}$ coincides with the normal $\mathbf{n}$ of the surface and

$$\mathbf{b} = \mathbf{t} \times \mathbf{n} = \mathbf{Y}_{,s} \times \mathbf{n}.$$

The Frenet–Serret formula (2.8) now reads

$$\frac{d\mathbf{m}(s)}{ds} = \frac{d\mathbf{n}(s)}{ds} = -c(s)\mathbf{t} - \tau(s)\mathbf{b}.$$

The dot product of this equation with b gives the geodesic torsion τ_g as

$$\tau_g = |\tau(s)| = \mathbf{b} \cdot \frac{d\mathbf{n}(s)}{ds} = \frac{d\mathbf{n}}{ds} \cdot (\mathbf{Y}_{,s} \times \mathbf{n}). \tag{5.65}$$

Now, we return back to the problem of the free energy of TCLB. The last several sections have shown that the free energy of TCLB depends only on the project field casted by the orientation vector $\mathbf{d}$ on the TCLB surface apart from some constant factor of θ_0, which is the orientation of the molecules in the macrostructure. In fact, the project field of $\mathbf{d}$ is nothing but $d_i\mathbf{Y}_{,i}(i = u, v)$. Now, consider the single infinite family of curves on the surface $\mathbf{Y}(u, v)$ described by a variable, s, $\mathbf{Y}(s) = \mathbf{Y}(u(s), v(s))$, and taking ds along the direction of $d_i\mathbf{Y}_{,i}$. In this way, du/ds and dv/ds will be proportional to d_u and d_v, respectively. Let the constant of proportionality be γ, then

$$\dot{\mathbf{Y}} = \mathbf{Y}_{,s} = \mathbf{Y}_{,u}\frac{du}{ds} + \mathbf{Y}_{,v}\frac{dv}{ds} = \gamma(d_u\mathbf{Y}_{,u} + d_v\mathbf{Y}_{,v}).$$

Since $\mathbf{Y}_{,s}$ is the unit tangent vector $\mathbf{t}$, it requires that

$$\mathbf{Y}_{,s} \cdot \mathbf{Y}_{,s} = \gamma^2 g_{ij}d_id_j = 1.$$

With Eq. (5.2), it shows that

$$\gamma = 1/\sin\theta_0$$

and

$$du/ds = d_u/\sin\theta_0, \quad dv/ds = d_v/\sin\theta_0,$$
$$\dot{\mathbf{Y}} = \mathbf{Y}_{,s} = d_l\mathbf{Y}_{,l}/\sin\theta_0 \quad (l = u, v). \tag{5.66a}$$

Similarly, for the unit vector $\mathbf{n}$ of the surface, there is

$$\dot{\mathbf{n}} = \mathbf{n}_{,s} = d_i\mathbf{n}_{,i}/\sin\theta_0 = -L_{ij}g^{jk}\mathbf{Y}_{,k}d_i/\sin\theta_0 \quad (i, j, k = u, v), \tag{5.66b}$$

where the last step comes from the Weingarten formula Eq. (3.13). The substitution of Eq. (5.66) into Eq. (5.65) gives that

$$\tau_g = -L_{ij}g^{jk}d_i(\mathbf{Y}_{,k} \times \mathbf{Y}_{,s}) \cdot \mathbf{n}/\sin\theta_0$$
$$= -L_{ij}g^{jk}d_id_l(\mathbf{Y}_{,k} \times \mathbf{Y}_{,l}) \cdot \mathbf{n}/\sin^2\theta_0.$$

With the definition of $\mathbf{n}$, Eq. (3.5), direct expansion of the last expression in τ_g into individual terms and with g^{jk} converted into g_{jk} by Eq. (3.4) gives that

$$\begin{aligned}
\tau_g &= [(g_{uu}L_{uv} - g_{uv}L_{uu})d_u^2 + (g_{uu}L_{vv} + g_{uv}L_{uv} - g_{vv}L_{uu} \\
&\quad - g_{uv}L_{uv})d_ud_v + (g_{uv}L_{vv} - g_{vv}L_{uv})d_v^2]/(\sqrt{g}\sin^2\theta_0) \\
&= [(g_{uu}d_u + g_{uv}d_v)(L_{uv}d_u + L_{vv}d_v) \\
&\quad - (g_{uv}d_u + g_{vv}d_v)(L_{uu}d_u + L_{uv}d_v)]/(\sqrt{g}\sin^2\theta_0) \\
&= e^{3mn}(g_{mu}d_u + g_{mv}d_v)(L_{nu}d_u + L_{nv}d_v)/(\sqrt{g}\sin^2\theta_0) \\
&= e^{3mn}g_{km}L_{ln}d_kd_l/(\sqrt{g}\sin^2\theta_0) \quad (k, l, m, n = u, v).
\end{aligned} \tag{5.67}$$

This is practically one of the terms in the expansion of g_{TCLB} in Eq. (5.14).

Another term in Eq. (5.14) is $e^{3mn}g_{kn}(d_{k,m} + \Gamma_{ml}^k d_l)$. With the definition of Γ_{ml}^k, Eq. (3.16), it gives that

$$\begin{aligned}
&e^{3mn}g_{kn}(d_{k,m} + \Gamma_{ml}^k d_l) \\
&= e^{3mn}[(g_{kn}d_k)_{,m} + (1/2)(g_{mn,k} - g_{nk,m} - g_{mk,n})d_k].
\end{aligned}$$

However, $e^{3mn}g_{mn,k} = 0$, and $e^{3mn}(g_{nk,m} + g_{mk,n}) = 0$. These lead to the result that

$$\int e^{3mn}g_{kn}(d_{k,m} + \Gamma^k_{ml}d_l)dudv = \int e^{3mn}(g_{kn}d_k)_{,m}dudv$$

$$= \int \left[\frac{\partial}{\partial u}(g_{kv}d_k) - \frac{\partial}{\partial v}(g_{ku}d_u) \right] dudv.$$

Application of Stokes' theorem[21] (sometimes named Green's theorem) converts this integral into a line integral

$$\int e^{3mn}g_{kn}(d_{k,m} + \Gamma^k_{ml}d_l)dudv = \oint g_{ik}d_k di = \oint (\mathbf{Y}_{,i} \cdot \mathbf{Y}_{,k})d_k di$$

$$= \oint \mathbf{d} \cdot \mathbf{ds} \quad (i,j,k,l,m = u,v), \quad (5.68)$$

where $\mathbf{ds} = \mathbf{Y}_{,i}di$ is the line element of the edge line of TCLB. With Eqs. (5.67) and (5.68), the general formula of free energy of TCLB, Eq. (5.13), now takes the form

$$F = -k_2 t \cos^2 \theta_0 \oint \mathbf{d} \cdot \mathbf{ds} - 2k_2 t \sin^2 \theta_0 \cos \theta_0 \int \tau_g dA, \quad (5.69)$$

where dA is the area element of TCLB. In case of fluid membrane, $\theta_0 = 0$ and $\mathbf{d}$ is orthogonal to $\mathbf{ds}$, Eq. (5.69) gives that $F = 0$. This means that the effect of the chiral curvature elasticity can be displayed only when there is the tilt of the director. The geodesic torsion τ_g can be expressed in the following way.

Equation (3.11) shows that $L_{ij} = \mathbf{n} \cdot \mathbf{Y}_{,ij} = -\mathbf{n}_{,i} \cdot \mathbf{Y}_{,j}$. Therefore, the second fundamental form given by Eq. (2.29) can be written in the form

$$II = \mathbf{n} \cdot \mathbf{Y}_{,ij} = -\mathbf{n}_{,i} \cdot \mathbf{Y}_{,j} = L_{uu}du^2 + 2L_{uv}dudv + L_{vv}dv^2.$$

According to Eq. (2.7) the curvature $c(s)$ of the curve satisfies the relation:

$$c(s)\mathbf{m}(s) = \frac{d\mathbf{t}}{ds} = \mathbf{Y}_{,ss}(s).$$

If θ is the angle between the surface normal $\mathbf{n}$ and the principal normal of the curve $\mathbf{m}(s)$ at a point P on the curve, then

$$c_n(s) = c(s)\cos \theta = c(s)\mathbf{n}(s) \cdot \mathbf{m}(s) = \mathbf{n}(s) \cdot \mathbf{Y}_{,ss}(s),$$

and

$$\mathbf{Y}_{,ss} = \frac{d}{ds}\left(\frac{d\mathbf{Y}}{ds} \right) = \frac{d}{ds}\left(\mathbf{Y}_{,u}\frac{du}{ds} + \mathbf{Y}_{,v}\frac{dv}{ds} \right).$$

If ψ is the angle between the direction of du/dv and the u-curve, Eqs. (2.58) and (2.59) show that

$$\cos\psi = \sqrt{g_{uu}}du/ds, \quad \sin\psi = \sqrt{g_{vv}}dv/ds,$$

and

$$\tau_g = (c_1 - c_2)\cos\psi\sin\psi, \tag{5.70}$$

ψ is simply the angle between one principal direction and the local tilt.

5.2.2   The Effect of Other Elastic Constants

The above discussions concern only the chiral curvature modulus k_2. Besides g_{ch}, the Frank free energy of cholesteric liquid crystal also involves the part of curvature energy density g_N satisfying the $D_{\infty h}$ symmetry[22]:

$$g_N = (1/2)[k_{11}(\nabla\cdot\mathbf{n})^2 + k_{22}(\mathbf{d}\cdot\nabla\times\mathbf{d})^2 + k_{23}(\mathbf{d}\times\nabla\times\mathbf{d})^2]$$
$$= (1/2)[k_{11}(\nabla\cdot\mathbf{n})^2 + k_{22}(\mathbf{d}\cdot\nabla\times\mathbf{d})^2 + k_{23}(\mathbf{d}\cdot\nabla\mathbf{d})^2], \tag{5.71}$$

where k_{11}, k_{22} and k_{33} are the splay, twist and bend elastic constants, respectively. In the case of single elastic constant approximation, Eq. (5.71) reduces to

$$g_N = \frac{1}{2}k_N[(\nabla\cdot\mathbf{d})^2 + (\nabla\times\mathbf{d})^2], \quad k_N \equiv k_{11} = k_{22} = k_{33}. \tag{5.72}$$

The first term $\nabla\cdot\mathbf{d}$ is given by

$$\nabla\cdot\mathbf{d} = (g^{ij}\mathbf{Y}_{,i}\partial/\partial j + \mathbf{n}\partial/\partial n)\cdot(d_l\mathbf{Y}_{,l} + \cos\theta_0\mathbf{n})$$
$$= d_{i,i} + \Gamma_{ij}^j d_i - 2H\cos\theta_0, \tag{5.73}$$

where Eqs. (3.5), (3.13), (3.15), (3.17) and (5.7) have been used to arrive at the final result. The expression $\nabla\times\mathbf{d}$ is given by Eq. (5.11). It is easy to show that

$$\mathbf{d}\cdot\nabla\times\mathbf{d} = g^{-1/2}e^{3mn}[g_{kn}(d_{k,m} + \Gamma_{ml}^k d_l)\cos\theta_0 + 2g_{km}L_{ln}d_k d_l]. \tag{5.74}$$

Since

$$\mathbf{d}\cdot\nabla = (d_l\mathbf{Y}_{,l} + \cos\theta_0\mathbf{n})\cdot[g^{ij}\mathbf{Y}_{,i}(\partial/\partial j) + \mathbf{n}(\partial/\partial n)]$$
$$= d_j\partial/\partial j + \cos\theta_0\partial/\partial n$$

and d_i is not a function of $\mathbf{n}$, it follows that

$$\mathbf{d} \cdot \nabla \mathbf{d} = (d_j \partial/\partial j + \cos\theta_0 \partial/\partial n)(d_i \mathbf{Y}_{,i} + \cos\theta_0 \mathbf{n})$$

$$= (d_j d_{i,j} + d_j d_k \Gamma^i_{kj} - 2\cos\theta_0 d_k L_{jk} g^{ij})\mathbf{Y}_{,i} d_i d_j L_{ij} \mathbf{n}. \quad (5.75)$$

With Eqs. (5.11), (5.73), (5.74) and (5.75), it is not difficult to obtain the free energy g_N for given cases.

In case of wound ribbon, the fundamental quantities are given in Eq. (5.28); d_ϕ and d_h are given by Eq. (5.34). It is easy to show that

$$\nabla \cdot \mathbf{d} = \frac{1}{R}\cos\theta_0, \quad \nabla \times \mathbf{d} = 2d_\phi \mathbf{Y}_{,k}.$$

Therefore, we have

$$g_N = k_N(1 + \sin^2\theta_0)/(2R^2)$$

and

$$F_N = t\cos\theta_0 \int g_N \mathrm{d}A = k_N tA \cos\theta_0(1 + \sin^2\theta_0)/(2R^2), \quad (5.76)$$

where $A = 2n^2 \pi a R$, as given by Eq. (5.38). The total free energy density for the wound ribbon helix is now equal to

$$F_{WT} = F_N + F_W = tA\left[\frac{k_N(1 + \sin^2\theta_0)}{2R^2} - \frac{|k_2|\sin^2\theta_0}{R}\right]\cos\theta_0. \quad (5.77)$$

This shows that the k_N term and the $|k_2|$ term are competitive. For wound ribbon helix with fixed area, minimization of F_{WT} gives the tube radius R as

$$R = \frac{k_N}{|k_2|}\frac{(1 + \sin^2\theta_0)}{\sin^2\theta_0} = \frac{p_{ch}}{\pi}\frac{(1 + \sin^2\theta_0)}{\sin^2\theta_0}, \quad (5.78)$$

where p_{ch} is the pitch of cholesterics and $k_N/|k_2| = p_{ch}/\pi$.[18] Since the gradient angle of the director $\mathbf{d}$ is $45°$, the pitch of the wound ribbon p is given by

$$p = 2\pi|b| = 2\pi R = 2p_{ch}\frac{(1 + \sin^2\theta_0)}{\sin^2\theta_0}. \quad (5.79)$$

This relation demonstrates the effect of chirality p_{ch} and the tilt angle θ_0 on the formation of helical structures and tubes. In the limiting case of either $\theta_0 = 0$ or $p_{ch} \to \infty$, both R and p approach ∞, which means there is no formation of helix and tube at all. Usually, the pitch of cholesterics is of the order 0.1–$100\,\mu m$. Equations (5.78) and (5.79) then give the radius of the tube and the pitch of the helix of the order of magnitude 0.1–$100\,\mu m$. This is in agreement with the experimental observations.[1-7]

In the case of twisted ribbon, the fundamental quantities are given by Eq. (5.46). It is easy to show that in case (1)

$$\nabla \cdot \mathbf{d} = 0 \quad \text{and} \quad \nabla \times \mathbf{d} = \pm \left[\frac{\rho \sin \theta_0}{\rho^2 + b^2} \mathbf{n} + \frac{2b \sin \theta_0}{(\rho^2 + b^2)^{3/2}} \mathbf{Y}_{,\phi} \right], \quad (5.80a)$$

and in case (2)

$$\nabla \cdot \mathbf{d} = \pm \rho \sin \theta_0 / (\rho^2 + b^2) \quad \text{and} \quad \nabla \times \mathbf{d} = \mp 2b \sin \theta_0 \mathbf{Y}_{,\rho} / (\rho^2 + b^2). \tag{5.80b}$$

Both cases lead to

$$g_N = \frac{1}{2} k_N \left[\frac{1}{(\rho^2 + b^2)} + \frac{3b^2}{(\rho^2 + b^2)^2} \right] \sin^2 \theta_0 \tag{5.81}$$

and

$$
\begin{aligned}
F_N &= 2n\pi k_N t \cos\theta_0 \sin^2\theta_0 [\ln(x + \sqrt{1 + x^2}) + 3|b| \arctan x] \\
&= k_N t (A/b^2) \cos\theta_0 \sin^2\theta_0 [\ln(x + \sqrt{1 + x^2}) \\
&\quad + 3|b| \arctan x][x(1 + x^2)^{1/2} + \ln(x + \sqrt{1 + x^2})]^{-1}, \quad (5.82)
\end{aligned}
$$

where A, the area of the strip, is given by Eq. (5.54) and $x = R/|b|$. The total free energy of the twisted strip $F_{TT} = F_N + F_T$ is a rather complicated function of x which renders it difficult to do further calculation.

In the case of spherical vesicle, with fundamental quantities given by Eq. (5.60), it is easy to show that in case (1) where $d_\theta = 0$ and $d_\phi = \sin\theta_0 / (R \sin\theta)$, one has

$$\nabla \cdot \mathbf{d} = 2\cos\theta_0 / R, \quad (\nabla \times \mathbf{d})^2 = \sin^2\theta_0 (1 + 3\sin^2\theta) / (R^2 \sin^2\theta). \tag{5.83}$$

It follows that

$$g_N^{(1)} = k_N (4 + \sin^2\theta_0 \cot^2\theta) / (2R^2). \tag{5.84}$$

Consequently, we have

$$
\begin{aligned}
F_N^{(1)} &= t\cos\theta_0 \int_0^{2\pi} \int_{-\pi/2}^{\pi/2} \frac{1}{2} k_N (4\sin\theta + \sin^2\theta_0 \cos\theta \cot\theta) d\theta d\phi \\
&= \pi k_N t \cos\theta_0 \int_{-\pi/2}^{\pi/2} (4\sin\theta + \sin^2\theta_0 \cos\theta \cot\theta) d\theta.
\end{aligned}
$$

Since both $\sin\theta$ and $\cos\theta\cot\theta$ are odd functions of θ, therefore

$$F_N^{(1)} = 0. \tag{5.85}$$

In case (2) where $d_\theta = \sin\theta_0/R, d_\phi = 0$, it gives that

$$\nabla \cdot \mathbf{d} = (\sin\theta_0\cos\theta + 2\cos\theta_0\sin\theta)/(R\sin\theta),$$

$$(\nabla \times \mathbf{d})^2 = 4\sin^2\theta_0/R^2. \tag{5.86}$$

It follows that

$$g_N^{(2)} = k_N(\sin^2\theta_0\cos^2\theta + 4\sin^2\theta + 4\sin\theta_0\cos\theta_0\sin\theta\cos\theta)/(2R^2\sin^2\theta) \tag{5.87}$$

and

$$F_N^{(2)} = \frac{t\cos\theta_0}{2} \int_0^{2\pi} \int_{-\pi/2}^{\pi/2} k_N$$
$$\cdot (4\sin\theta + \sin^2\theta_0\cos\theta\cot\theta + 4\sin\theta_0\cos\theta_0\cos\theta)\mathrm{d}\theta\mathrm{d}\phi$$
$$= 8\pi k_N t\sin\theta_0\cos^2\theta_0. \tag{5.88}$$

Therefore there are two different cases:

$$(1) \quad F_{ST}^{\text{latitude}} = 0; \tag{5.89a}$$

$$(2) \quad F_{ST}^{\text{longitude}} = 8\pi k_N t\sin\theta_0\cos^2\theta_0. \tag{5.89b}$$

The latitudinal case is the favorable state.

5.2.3   High- and Low-Pitch Helical Structures of TCLBs

Tubular and helical structures are rather commonly observed in nature.[23] For example, many authors found various helical structures of chiral lipid bilayers in recent experiments.[1-6] Among several theoretical approaches, a fundamental basis which is consistent with the experiments is that the molecular packing interaction is chiral as in cholesteric liquid crystals.[11-13, 24] According to the early theories, the pitch angle of the helix (ϕ_0 in Fig. 5.6(a)) was predicted to be $45°$ which was in good agreement with the previous experimental observations.[1-6]

Although the concept of molecular packing interaction in cholesteric liquid crystals has been experimentally confirmed[25] and theoretically extended,[26, 27] a recent challenge is to understand the new findings in cholesterol crystallization in native and model biles.[28, 29] Micellar model

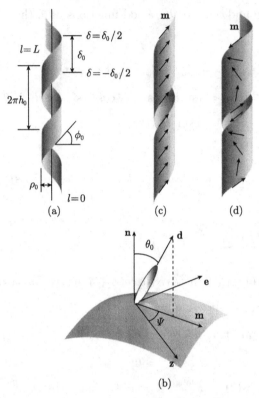

Fig. 5.6. Schematic illustration of the helical ribbon: (a) the geometric parameters characterizing the ribbon, (b) local coordinate system of TCLB, (c) helical ribbon with "parallel packing" of molecules (P-helix), (d) helical ribbon with "antiparallel packing" of molecules (A-helix).

biles composed of bile salt sodium taurocholate, lecithin, and cholesterol in a molar ratio of 97.5:0.8:1.7 were initially prepared and contained both micelles and vesicles. Within 2–4 hours of dilution, filamentous structures were observed. A few days later, the filaments were bent to form high-pitch helices ($\approx 54°$, Ref. [29, Fig. 2(a)]). These helices grew laterally while maintaining the pitch angle to form tubules (Ref. [29, Fig. 2(b)]). Within a few weeks, high-pitch helices and tubules disappeared, while new helices with low-pitch angle ($\approx 11°$) appeared and grew to form new tubules (Ref. [29, Figs. 2(c) and 2(d)]). Eventually, only plate-like cholesterol monohydrate crystals remained (Ref. [29, Figs. 2(e) and 2(f)]). One of their interesting findings is the existence of helical ribbon structures with two distinctive pitch angles: high-pitch (54°) and low-pitch (11°) helices. To explain these, Chung et al.[29] used the previous theory[30] appropriate for anisotropic bilayers of tilted chiral amphiphiles case and derived a free energy per unit area

$$g = \frac{1}{2\rho_0^2} \left(k^{cc} \cos^4 \phi_0 + k^{pp} \sin^4 \phi_0 + \frac{k^{cp}}{2} \sin^2 2\phi_0 \right) - \frac{k^*}{2\rho_0} \sin 2\phi_0,$$

(5.90)

where ρ_0 and ϕ_0 are the radius and pitch angle of the helical ribbon, respectively (see Fig. 5.6); k^{cc}, k^{pp}, k^{cp} and k^* are elastic constants. For the equilibrium helix, they obtained $\tan \phi_0 = (k^{cc}/k^{pp})^{1/4}$. Corresponding to $\phi_0 = 54°$ and $11°$, the calculated ratio is $k^{cc}/k^{pp} = 3.4$ and 0.0015, respectively. However, as commented very recently by Selinger et al.,[31] the ratio of 3.4 is reasonable, but that of 0.0015 is surprising. For this reason, Selinger et al. assumed that the molecular tilt direction on the ribbons (**m** in Fig. 5.6) varies linearly across the ribbon.[31] Although the observed low-pitch angle $11°$ has been attributed to the difference between the helical direction and the tilt direction in their paper, there is no evidence for the linear ansatz and the theory has not yet reached a conclusive explanation on the existence of two types of helices. Thus, two general questions can be posed as follows: What is the nature of the existence of two types of helices? Can we theoretically discuss the above sequence of transition as well as each structure from the energetic point of view?

In a recent paper, Komura and Ou-Yang[14] investigated the molecular packing on a helical ribbon of chiral lipid bilayers as cholesteric liquid crystals by generalizing a previous theory.[12,13] In Refs. [12, 13], only the chiral free energy term $g_{ch} = -k_2 \mathbf{d} \cdot \nabla \times \mathbf{d}$ is employed, where **d** is a unit vector parallel to the long axis of the molecules (see Fig. 5.6(b)) and k_2 characterizes the chirality of cholesteric liquid crystal. With this simplified energy, the pitch angle was predicted to be $45°$. A calculation is given in this section[14] by using the complete free energy density per unit area of cholesteric liquid crystal[18]

$$g_{LC} = \frac{k_{11}}{2}(\nabla \cdot \mathbf{d})^2 + \frac{k_{22}}{2} \left(\mathbf{d} \cdot \nabla \times \mathbf{d} - \frac{k_2}{k_{22}} \right)^2 + \frac{k_{33}}{2}(\mathbf{d} \times \nabla \times \mathbf{d})^2,$$

(5.91)

where k_{ii} ($i = 1, 2$ and 3) are the splay, twist, and bending elastic constants, respectively, and ∇ is the 3D gradient operator. To view Eq. (5.91) as an energy per unit area, we regard k_{ii} and k_2 as the corresponding elastic constants of cholesteric liquid crystal multiplied by the thickness of

the chiral lipid bilayer, and hence the total free energy of the chiral lipid
bilayer is given by $F = \int g_{LC} dA$, where dA is the area element of the
chiral lipid bilayer surface. We allow the molecular tilt direction, vector
$\mathbf{m}$, to vary on the bilayer as proposed by Selinger et al.,[24,31] but keep the
boundary condition where the tilt direction is aligned with the helical direction at the ribbon edges. The latter condition has also been considered in
the previous theories[11, 26, 29] and confirmed by X-ray diffraction studies.[32]
Especially, instead of assuming a simple linear ansatz for the tilt direction field,[31] we obtain it exactly by solving the Euler–Lagrange equation
derived from g_{LC}. With these refinements, we will reveal that for the high-
pitch helices, molecules at the two edges of bilayers tilt parallel to each other
(Fig. 5.6(c)), whereas for the low-pitch helices, they tilt antiparallel to each
other (Fig. 5.6(d)). Helix formation takes place through a transition from
a fluid chiral lipid bilayer (chain-melted L_α) phase to a tilted chiral lipid
bilayer (TCLB, chain-frozen $L_{\beta*}$) phase under cooling or dilution.[25] The
observed sequence of structures from filaments to cholesterol monohydrate
crystals by way of two types of helices can be shown as a decreasing succes-
sion of free energy of the metastable intermediate structures and regarded
as a quench-like cooling process of the molecular tilt.

Mathematically, a surface of a helical ribbon wound by a cylinder of
radius ρ_0 can be represented by a vector function with two variables as

$$\mathbf{Y}(\ell, \delta) = (\rho_0 \cos \omega_0 \ell, \rho_0 \sin \omega_0 \ell, h_0 \omega_0 \ell + \delta),$$

where $\omega_0 = 1/\sqrt{\rho_0^2 + h_0^2}$ (see Fig. 5.6(a)). The pitch angle and the pitch of
the helix are given by $\phi_0 = \arctan(h_0/\rho_0)$ and $2\pi|h_0|$, respectively. More-
over, ℓ and δ satisfy $0 \leq \ell \leq L$ and $-\delta_0/2 \leq \delta \leq \delta_0/2$, respectively, where
δ_0 is the ribbon width along z-axis (the central axis of the cylinder), and L
is the arc length along the edge of the ribbon. We introduce the tilt of the
director field in $\mathbf{d} = \cos \theta_0 \mathbf{n} + \sin \theta_0 \mathbf{m}$, where $\mathbf{n}$ is the unit normal vector of
the ribbon surface, $\mathbf{m}$ is the unit vector parallel to the projection of $\mathbf{d}$ on the
surface, and the tilt angle θ_0 is the angle between $\mathbf{n}$ and $\mathbf{d}$ (see Fig. 5.6(b)).[30]
θ_0 characterizes the transition from the fluid (L_α) phase with $\theta_0 = 0$ to the
TCLB $(L_{\beta*})$ phase with a constant $\theta_0 > 0$. Generally, $\mathbf{m}$ can be written in
terms of the single angle $\psi(\ell, \delta)$ as $\mathbf{m} = \cos \psi(\ell, \delta)\mathbf{z} + \sin \psi(\ell, \delta)\mathbf{e}$, where
$\mathbf{z}$ is the unit vector along the z-axis and $\mathbf{e}$ is the unit vector along the
circumference of the cylinder.

For simplicity but without loss of generality, we consider the case
where ψ is a function of only δ, and use the one-constant approximation,

i.e., $k_{11} = k_{22} = k_{33} \equiv k$. Then Eq. (5.91) becomes

$$g_{\mathrm{LC}} = [k/(2\rho_0^2)][\cos^2\theta_0 + \sin^2\theta_0(4\sin^2\psi + \omega_0^{-2}\psi_\delta^2)$$
$$- \sin 2\theta_0(h_0\cos\psi + \rho_0\sin\psi)\psi_\delta]$$
$$- (k_2/2\rho_0)[\sin 2\theta_0(h_0\sin\psi - \rho_0\cos\psi)\psi_\delta$$
$$+ 2\sin^2\theta_0\sin 2\psi] + k\mu^2/2, \tag{5.92}$$

where $\psi_\delta = d\psi/d\delta$, $\mu = k_2/k$ is the inverse of the pitch of cholesteric liquid crystal.[12, 13, 18] For constant ψ, one can show that Eq. (5.92) coincides with Eq. (5.90) when $k^{pp} = k\cos^2\theta_0$, $k^{cc} = k(\cos^2\theta_0 + 4\sin^2\theta_0)$, $k^{cp} = k\cos^2\theta_0$ and $k^* = 2k_2\sin^2\theta_0$ except the constant term. One term of progress in the present work is to extend to the case without the constraint of constant ψ. We therefore derive the Euler–Lagrange equation from Eq. (5.92) as

$$\psi_{\delta\delta} = 2\omega_0^2\sin(2\psi - \alpha_0)/\cos\alpha_0, \tag{5.93}$$

where $\psi_{\delta\delta} = d^2\psi/d\delta^2$ and $\alpha_0 = \arctan(\mu\rho_0)$. The general form of the first integral of Eq. (5.93) is

$$\psi_\delta = (2\omega_0/\sqrt{\cos\alpha_0})\sqrt{\cosh^2 C - \cos^2(\psi - \alpha_0/2)}, \tag{5.94}$$

where C is an integral constant.

Obviously, Eq. (5.93) has a constant solution of $\psi = \psi_0 = \alpha_0/2$ and we let $\phi_0 = \pi/2 - \psi_0 = \pi/2 - \alpha_0/2$ to satisfy the boundary condition such that the molecules at the two edges of the ribbon tilt parallel to each other. We call this configuration "parallel packing" (P-helix) as shown in Fig. 5.6(c). It is necessary to check whether such a helix is in mechanical equilibrium or not. We define the curvature κ_0 and the torsion τ_0 of the helix by $\kappa_0 = \rho_0\omega_0^2$ and $\tau_0 = h_0\omega_0^2$, respectively. Then the elastic energy of the P-helix is expressed from Eq. (5.92) in terms of κ_0 and τ_0 as

$$g_P = \frac{k}{2}[4\sin^2\theta_0(\kappa_0^2 + \tau_0^2) + \cos^2\theta_0(\kappa_0 + \tau_0^2/\kappa_0)^2] - 2k_2\tau_0\sin^2\theta_0 + \frac{k\mu^2}{2}.$$

Minimizing this with respect to κ_0 and τ_0, i.e., $\partial g_P/\partial\kappa_0 = \partial g_P/\partial\tau_0 = 0$, we obtain the optimal geometry of the helical ribbon characterized by

$$\tan\phi_0 = (1 + 4\tan^2\theta_0)^{1/4}, \tag{5.95}$$

$$\mu\rho_0 = -\tan 2\phi_0 = \frac{2(1 + 4\tan^2\theta_0)^{1/4}}{(1 + 4\tan^2\theta_0)^{1/2} - 1}, \tag{5.96}$$

and the corresponding energy per unit area is $g_P = k\mu^2/2$. These results reveal some important features of the P-helix. Firstly, we note that Eq. (5.96) is identical to the assumed boundary condition, $\phi_0 = \pi/2 - \alpha_0/2$ with $\alpha_0 = \arctan(\mu\rho_0)$, and hence the calculation is self-consistent. This confirms that the helix is in mechanical equilibrium since the minimization condition requires the state of both free external force and moment. Secondly, the result of Eq. (5.95) essentially corresponds to that by Chung et al. as mentioned before (note also the relations between k^{cc}, k^{pp}, and k), and predicts a high-pitch angle of $\phi_0 > \pi/4$. We have also performed the minimizing calculation without using the one-constant approximation and obtained an equation as $\cos^2 2\phi_0 + [2 + (k_{11}/k_{33})\cot^2\theta_0]\cos 2\phi_0 + 1 = 0$. Since this equation possesses a negative solution for $\cos 2\phi_0$, we can show $\phi_0 > \pi/4$ as before, and it reduces Eq. (5.95) when $k_{11} = k_{33}$. Thirdly, since the radius ρ_0 is inversely proportional to the chiral elastic constant k_2, we see that the origin of the cylindrical curvature is due to the chirality of molecules. In fact, Eq. (5.96) reveals that the pitch (μ^{-1}) should be proportional to the radius. This is one of the remarkable facts observed in experiments (see Figs. 4–6 in Ref. [29]).

According to the experiments, a straight filament appears at the initial stage of the crystallization process in model biles and can be identified as $\rho_0 = 0$.[28] This fact indicates that the initial stage of the TCLB, the helix is quenched to the structure with $\phi_0 = \pi/2$ (note Eq. (5.96) and $\phi_0 > \pi/4$). From Eq. (5.95), one can then expect that the tilt angle takes the maximum value $\theta_0 = \pi/2$. After the quench, θ_0 will decrease from $\pi/2$ to its stable value and the straight filaments will bend and twist by decreasing ϕ_0 and increasing ρ_0.[28, 29] This is the expected mechanism underlying the transition from the straight filaments to the high-pitch helices.

We now turn to consider the case where the tilt direction can vary on the bilayer across the ribbon. Since the tilt direction coincides with the helical direction at the ribbon edges, the allowed boundary condition for the nonuniform solution is $\psi(-\delta_0/2) = \pi/2 - \phi_0$ and $\psi(\delta_0/2) = \psi(-\delta_0/2) + \pi = 3\pi/2 - \phi_0$. The molecular configuration subjected to this boundary condition is called "antiparallel packing" (A-helix) and is depicted in Fig. 5.6(d). We will show below that the helical structure produced from the anti-parallel packing of molecules indeed results in the low-pitch angle. The integral constant C in Eq. (5.94) is now determined by the boundary condition as $\delta_0 = \sqrt{\cos\alpha_0}qK(q)/\omega_0$ with $q = 1/\cosh C$, whereas the energy per unit area can be calculated from

Eq. (5.92) as

$$g_A = (k\mu^2/2)\{1 + (1 + \sin^2\theta_0)x^2 + 2\sin^2\theta_0 x\sqrt{1+x^2}(1 - 2/q^2)$$
$$+ 8\sin^2\theta_0 x\sqrt{1+x^2}E(q)/[q^2 K(q)]$$
$$- 2\sin 2\theta_0\sqrt{x}(1+x^2)^{1/4}/[qK(q)]\}, \tag{5.97}$$

where $x = \cot\alpha_0$, $K(q)$ and $E(q)$ are the first and the second complete elliptic integrals given by $K(q) = \int_0^{\pi/2} d\varphi/(1 - q^2\sin^2\varphi)^{1/2}$ and $E(q) = \int_0^{\pi/2}(1 - q^2\sin^2\varphi)^{1/2}d\varphi$ respectively. To ensure that the helix is in mechanical equilibrium, we must minimize g_A. Because q is a function of δ_0, ϕ_0 and α_0, minimization of g_A over these quantities leads simultaneously to $\partial g_A/\partial q = 0$. From this and with the help of differential relations, $dK(q)/dq = [E(q)/q(1 - q^2)] - K(q)/q$ and $dE(q)/dq = [E(q) - K(q)]/q$, we can obtain the relation between α_0 and q as $\sqrt{\cos\alpha_0}/\sin\alpha_0 = q\cot\theta_0/[2E(q)]$. By substituting this into Eq. (5.97), g_A then becomes

$$g_A = (k\mu^2/2)[1 + (1 + \sin^2\theta_0)(\sqrt{1 + 4y^4} - 1)/2$$
$$+ 2y^2\sin^2\theta_0 - \cos^2\theta_0/E^2(q)], \tag{5.98}$$

where $y = \sqrt{\cos\alpha_0}/\sin\alpha_0$ and $\alpha_0 = \arctan(\mu\rho_0)$ as before.

In Fig. 5.7(a), we have plotted scaled $g_A - g_P$ as a function of q for several values of $t = \tan\theta_0$. There exists a critical value of θ_0 (denoted as θ_0^*)

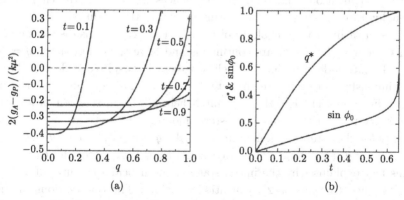

Fig. 5.7. (a) $2(g_A - g_P)/(k\mu^2)$ versus q for several values of $t = \tan\theta_0$, (b) q^* and $\sin\phi_0$ versus $t = \tan\theta_0$ for $0 < \theta_0 < \theta_0^*$ ($\tan\theta_0^* \approx 0.654$), respectively.

such that for $0 < \theta_0 < \theta_0^*$, there exists a critical value of q (denoted as q^*) at which g_P is equal to g_A, whereas for $\theta_0^* < \theta_0 < \pi/2$, g_A is always smaller than g_P predicting the transition from P-helix to A-helix. In Fig. 5.7(b), q^* is plotted against $t = \tan \theta_0$ for $0 < \theta_0 < \theta_0^*$. θ_0^* can be obtained by solving $g_P = g_A$ at $q = 1$ and is $\tan \theta_0^* \approx 0.654$. If we consider the case when the helix closes to form a tube ($\delta_0 = 2\pi h_0$), then $\delta_0 = (2\pi/\mu) \tan \alpha_0 \tan \phi_0$ holds. With this relation, we can calculate the pitch angle of the helix as

$$\sin \phi_0 = \frac{K(q)}{2\pi} \left[\sqrt{q^4 + 4\tan^4 \theta_0 E^4(q)} - 2\tan^2 \theta_0 E^2(q) \right]^{1/2}. \tag{5.99}$$

In Fig. 5.7(b), we have also plotted $\sin \phi_0$ versus $t = \tan \theta_0$ with $q = q^*$. Except at the singular point $\theta_0 = \theta_0^*$, this graph generally shows $\phi_0 < 30°$ indicating that the antiparallel packing results in the low-pitch angle.

Another interesting feature is that for each curve in Fig. 5.7(a), we see that $q = 0$ gives the minimum of g_A and hence it is the unique stable state. The state $q = 0$ leads to $\alpha_0 = \pi/2$ which corresponds to $\rho_0 \to \infty$ since $\alpha_0 = \arctan(\mu \rho_0)$. In fact, this result confirms the observation in native and model biles. At the final stage of the crystallization process, the tubes fracture at their ends to produce plate-like cholesterol monohydrate crystals (see Ref. [29, Fig. 2]). In other words, P-helix or A-helix are metastable intermediates although they are static solutions of the Euler–Lagrange equation (5.93).

To progress the previous theories,[29, 31] Fig. 5.7(a) predicts an important fact: the transition from the low-pitch helices (A-helix) to the cholesterol monohydrate crystals can be continuous while that from the high-pitch helices (P-helix) to the cholesterol monohydrate crystals is discontinuous. All the curves for $g_A(q)$ are continuous and have their minima at $q = 0$. This feature indicates that the helical tubes of low-pitch angle transform continuously into plate-like cholesterol monohydrate crystals as shown in Ref. [29, Fig. 2(f)]. For the high-pitch helices, the dissolution process is necessary to transform into cholesterol monohydrate crystals.[29] Since the energy for cholesterol monohydrate crystals ($q = 0$) is $g_{ChM} = (k\mu^2/2)[1 - (2/\pi)^2 \cos^2 \theta_0]$, $g_P = k\mu^2/2$ is always larger than g_{ChM} except for $\theta_0 = \pi/2$. This fact explains why the initial stage of the helical structures of TCLB is quenched to $\theta_0 = \pi/2$ as mentioned before. The observations in the experiment,[29] as mentioned at the beginning, confirm very well the above predictions.

These considerations now lead us to determine the optimal high-pitch angle. For the transition from the high-pitch helices to the cholesterol monohydrate crystals, θ_0 must reduce to θ_0^* in order to satisfy the condition $g_P = g_A$ from which the branch of low-pitch helix starts in Fig. 5.7(a). Hence, the optimal high-pitch angle is precisely obtained from Eq. (5.95) as

$$\phi_0 = \arctan \left[\frac{8}{3} \cos \left(\frac{1}{3} \arccos \frac{5}{32} \right) + \frac{1}{3} \right]^{1/4} \approx 52.1^\circ, \qquad (5.100)$$

which is in excellent agreement with the experimentally observed value $\phi_0 = 53.7^\circ \pm 0.8^\circ$.[29]

5.3 Concise Theory for Chiral Lipid Membranes

Based on symmetric argument or Frank energy in the theory of liquid crystal, many theoretical models and results were achieved.[11–14, 27, 31] These theoretical models contain much complicated terms and many parameters, which make it is impossible to derive the exact governing equations for describing equilibrium configurations of chiral lipid membranes. So, these theoretical models cannot give good explanations to some recent experimental results. Schnur et al. have observed that the spherical vesicles in solution have very weak circular dichroism signals while tubules have strong ones.[33] Fang's group has carefully resolved the molecular tilting order in the tubules and concluded that the projected direction of the molecules on the tubular surfaces departs 45° from the equator of the tubules at the uniform tilting state.[34] Helical ripples in lipid tubules are also observed by the same group with atomic force microscopy.[35] Their pitch angles are found to be concentrated on about 5° and 28°.[35] Oda et al. have reported twisted ribbons of achiral cationic amphiphiles interacting with chiral tartrate counterions.[36, 37] It is found that the twisted ribbons can be tuned by the introduction of opposite-handed chiral counterions in various proportions.[36] From the experimental data,[36] we see that the ratio between the width and pitch of the ribbons is proportional to the relative concentration difference of left- and right-handed enantiomers in the low relative concentration difference region. Here, we will introduce a simplified theory of chiral lipid membranes (CLMs) proposed by Tu and Seifert.[15] Interestingly, this concise theory is consistent with these experimental results. A torus with the ratio between its two generated radii larger than $\sqrt{2}$ is predicted within this concise theory.

5.3.1 Brief Introduction to Moving Frame Method and Exterior Differential Forms

First, we introduce some new geometry conceptions that will be used in the following contents. If we take a frame $\{e_1, e_2, e_3\}$ at any point r on a surface as shown in Fig. 5.8, then the infinitesimal tangential vector at r is expressed as

$$d\mathbf{r} = \omega_1 \mathbf{e}_1 + \omega_2 \mathbf{e}_2, \qquad (5.101)$$

and the difference of frame between at points $\mathbf{r} + d\mathbf{r}$ and $\mathbf{r}$ is denoted as

$$d\mathbf{e}_i = \omega_{ij} \mathbf{e}_j \quad (i = 1, 2, 3), \qquad (5.102)$$

where ω_1, ω_2 and $\omega_{ij} = -\omega_{ji}(i, j = 1, 2, 3)$ are 1-forms.[38, 39] The repeated subscripts in Eq. (5.102) and the following contents represent the Einstein summation convention. With these 1-forms, the structure equations of a surface can be expressed as[38, 39]

$$\begin{cases} d\omega_1 = \omega_{12} \wedge \omega_2, \\ d\omega_2 = \omega_{21} \wedge \omega_1, \\ d\omega_{ij} = \omega_{ik} \wedge \omega_{kj} \quad (i, j = 1, 2, 3), \end{cases} \qquad (5.103)$$

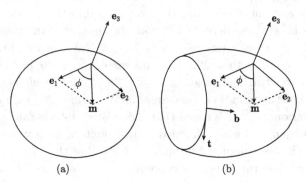

(a)   (b)

Fig. 5.8. Right-handed orthonormal frame $\{e_1, e_2, e_3\}$ at any point in a surface where e_3 is the normal vector of the surface. (a) Surface without boundary curve. (b) Surface with boundary curve where t is the tangent vector of the boundary curve, and b, in the tangent plane of the surface, is perpendicular to t. (Reprint from Ref. [15]).

and

$$\begin{pmatrix} \omega_{13} \\ \omega_{23} \end{pmatrix} = \begin{pmatrix} a & b \\ b & c \end{pmatrix} \begin{pmatrix} \omega_1 \\ \omega_2 \end{pmatrix}, \tag{5.104}$$

where the symbol "$\wedge$" expresses the wedge product between differential forms and "d" is the exterior differential operator.[38, 39] The matrix $\begin{pmatrix} a & b \\ b & c \end{pmatrix}$ is called the curvature matrix which is related to the mean and Gaussian curvature by

$$2H = a + c, \quad K = ac - b^2. \tag{5.105}$$

For a unit vector $\mathbf{m} = \cos\phi\mathbf{e}_1 + \sin\phi\mathbf{e}_2$, the normal curvature and the geodesic torsion along the direction of $\mathbf{m}$ can be expressed as[38]

$$\kappa_{\mathbf{m}} = a\cos^2\phi + 2b\cos\phi\sin\phi + c\sin^2\phi \tag{5.106}$$

and

$$\tau_{\mathbf{m}} = b(\cos^2\phi - \sin^2\phi) + (c - a)\cos\phi\sin\phi, \tag{5.107}$$

respectively.

The normal curvature along the direction of an arbitrary vector $\mathbf{u}$ can be expressed as

$$\kappa_{\mathbf{u}} = (au_1^2 + 2bu_1u_2 + cu_2^2)/\mathbf{u}^2, \tag{5.108}$$

where u_1 and u_2 are the components of $\mathbf{u}$ in the directions of $\mathbf{e}_1$ and $\mathbf{e}_2$.

In our following derivations, several relations between vector forms and differential forms are used frequently. For convenience, we list them below:

$$(\nabla \times \mathbf{m})dA = d(\mathbf{m} \cdot d\mathbf{r}), \tag{5.109}$$

$$(\nabla \cdot \mathbf{m})dA = d(*\mathbf{m} \cdot d\mathbf{r}), \tag{5.110}$$

$$\tau_{\mathbf{m}}dA = \mathbf{m} \cdot d\mathbf{e}_3 \wedge \mathbf{m} \cdot d\mathbf{r}, \tag{5.111}$$

$$\kappa_{\mathbf{m}}dA = -\mathbf{m} \cdot d\mathbf{e}_3 \wedge *\mathbf{m} \cdot d\mathbf{r}, \tag{5.112}$$

$$\nabla\phi \cdot d\mathbf{r} = d\phi, \tag{5.113}$$

$$(\nabla^2\phi)dA = d * d\phi, \tag{5.114}$$

$$\mathbf{S} \cdot d\mathbf{r} = -\omega_{12}, \tag{5.115}$$

$$(\nabla \cdot \mathbf{S})dA = -d * \omega_{12}, \tag{5.116}$$

$$(\nabla \times \mathbf{S})dA = -d\omega_{12} = KdA, \qquad (5.117)$$

$$\mathbf{v} \cdot d\mathbf{r} = d\phi + \omega_{12}, \qquad (5.118)$$

$$\mathbf{v}^2 dA = \mathbf{v} \cdot d\mathbf{r} \wedge *\mathbf{v} \cdot d\mathbf{r}, \qquad (5.119)$$

$$(\nabla\mathbf{v} : \nabla\mathbf{e}_3)dA = d(\mathbf{v} \cdot d\mathbf{e}_3), \qquad (5.120)$$

where $dA = \omega_1 \wedge \omega_2$ is the area element, $\mathbf{m}$ is a unit vector and ϕ is the angle between $\mathbf{m}$ and $\mathbf{e}_1$, $\mathbf{S}$ is the spin connection and $\mathbf{v} \equiv \nabla\phi - \mathbf{S}$, $*$ is the Hodge star operator[38, 40] which satisfies $*\omega_1 = \omega_2$ and $*\omega_2 = -\omega_1$. The operator ∇ and ∇^2 are 2D gradient and Laplace operators on the surface, respectively. In particular, "$\times$" represents the 2D cross product which gives a scalar.

Using Eqs. (5.105), (5.108), (5.109), (5.110) and (5.118), we can prove

$$(\kappa_\mathbf{v} - H)\mathbf{v}^2 = (v_1^2 - v_2^2)(a - c)/2 + 2bv_1v_2 \qquad (5.121)$$

and

$$[\nabla \cdot (\mathbf{m}\nabla \times \mathbf{m}) + \nabla \times (\mathbf{m}\nabla \cdot \mathbf{m})]dA$$
$$= d[(v_2\cos 2\phi - v_1\sin 2\phi)\omega_1 + (v_1\cos 2\phi + v_2\sin 2\phi)\omega_2], \qquad (5.122)$$

where v_1 and v_2 are the components of $\mathbf{v}$ in the directions of $\mathbf{e}_1$ and $\mathbf{e}_2$.

5.3.2 Constructing the Free Energy

The free energy density for a chiral lipid membrane is supposed to consist of the following contributions:

(1) The bending energy per area is still taken as Helfrich's form:

$$f_H = (k_c/2)(2H + c_0)^2 - \bar{k}K + \lambda, \qquad (5.123)$$

where k_c and $\bar{k}$ are bending rigidities, λ is the surface tension, c_0 is the spontaneous curvature reflecting the asymmetrical factors between two sides of the membrane, H and K are the mean curvature and Gaussian curvature of the membrane, respectively, which can be expressed as $2H = -(1/R_1 + 1/R_2)$, $K = 1/(R_1 R_2)$ by two principal curvature radii R_1 and R_2. The curvature energy in Eq. (5.123) is invariant under the coordinate rotation around the normal of the membrane surface, but will change under the inversion of the normal if $c_0 \neq 0$. That is, we neglect the anisotropic effect of lipid molecules' tilting on the bending moduli.

(2) The energy per area originating from the chirality of tilting molecules has the form[12]:

$$f_{ch} = -h\tau_m, \qquad (5.124)$$

where h reflects the strength of molecular chirality. Without losing the generality, here, we only discuss the case of $h > 0$. τ_m is the geodesic torsion along the unit vector m at each point. Here, m represents the projected direction of the lipid molecules on the membrane surface. If we take a right-handed orthonormal frame $\{e_1, e_2, e_3\}$ as shown in Fig. 5.8, m can be expressed as $m = \cos\phi e_1 + \sin\phi e_2$, where ϕ is the angle between m and e_1. At this frame, the geodesic torsion along m can be expressed as Eq. (5.107). In the principal frame, this equation is simplified as

$$\tau_m = (1/R_1 - 1/R_2)\cos\phi\sin\phi. \qquad (5.125)$$

We will confine ϕ to the region $(-\pi/2, \pi/2]$ because of the relation $\tau_m(\phi + \pi) = \tau_m(\phi)$. Moreover, it is easy to see that the geodesic torsion along the mirror image of m with respect to e_1 changes its sign because $\phi \mapsto -\phi$ under the reflection with respect to e_1. Thus, this term breaks the inversion symmetry in the tangent plane at each point of the membrane, which allows us to distinguish the handedness. Additionally, from Eq. (5.125), we can see that the minimum of Eq. (5.124),

$$f_{ch}^{min} = -(h/2)|1/R_1 - 1/R_2|, \qquad (5.126)$$

is reached when m departs from the principal direction at an angle $+\pi/4$ or $-\pi/4$ for a given shape of the membrane. Here, the sign in front of $\pi/4$ depends on the sign of $(1/R_1 - 1/R_2)$. The larger difference between R_1 and R_2 is that the larger absolute value of G_{ch}^{min} is reached. In other words, the chiral term favors saddle surfaces (for example, the twisted ribbons shown in Fig. 5.3) whose two principal curvature radii R_1 and R_2 have opposite signs.

(3) The energy per area due to the orientational variation is taken as[41]

$$f_{ov} = (k_f/2)[(\nabla \times m)^2 + (\nabla \cdot m)^2], \qquad (5.127)$$

where k_f is a constant in the dimension of energy. This is the simplest term of energy cost due to tilting order invariance under the coordinate rotation around the normal of the membrane surface. By defining a spin connection field S such that $\nabla \times S = K$,[42] one can derive $(\nabla \times m)^2 + (\nabla \cdot m)^2 = (\nabla\phi - S)^2$ through simple calculations.[42]

The total free energy density adopted in the present paper, $G = f_H + f_{ch} + f_{ov}$, has the following concise form:

$$G = \frac{k_c}{2}(2H + c_0)^2 + \bar{k}K + \lambda - h\tau_{\mathbf{m}} + \frac{k_f}{2}\mathbf{v}^2, \qquad (5.128)$$

with $\mathbf{v} \equiv \nabla\phi - \mathbf{S}$. This special form might arguably be the most natural and concise construction including the bending, chirality and tilting order, for the given vector field $\mathbf{m}$ and normal vector field $\mathbf{e}_3$.

5.3.3 Governing Equations to Describe Equilibrium Configurations of CLMs Without Free Edges

CLMs without free edges usually correspond to closed vesicles. Here, a long enough tubule, where the end effect is neglected, is also regarded as a CLM without free edges. The free energy for a closed chiral lipid vesicle may be expressed as

$$F = \int G\mathrm{d}A + p\int \mathrm{d}V, \qquad (5.129)$$

where $\mathrm{d}A$ is the area element of the membrane, $\mathrm{d}V$ is the volume element enclosed by the vesicle, p is either the pressure difference between the outer and inner sides of the vesicle or used to implement a volume constraint. For tubular configuration, we usually take $p = 0$.

Now, we will derive Euler–Lagrange equations of CLMS without free edges through the variational method developed in Ref. [38] with the aid of the moving frame method and exterior differential forms.

First, assume $\delta\phi \equiv \Xi$. Through simple calculations, we arrive at

$$\delta\tau_{\mathbf{m}} = 2(H - \kappa_{\mathbf{m}})\Xi \qquad (5.130)$$

and

$$\delta(\mathbf{v}^2\mathrm{d}A) = 2\mathrm{d}\Xi \wedge *(\mathrm{d}\phi + \omega_{12}). \qquad (5.131)$$

Combining with the above two equations and using the integral by parts and Stokes' theorem, we derive

$$\delta F = \int [2h(\kappa_{\mathbf{m}} - H) - k_f\nabla^2\phi]\Xi\mathrm{d}A + k_f \oint \Xi * \mathbf{v} \cdot \mathrm{d}\mathbf{r}, \qquad (5.132)$$

where the second term is the integral along the boundary curve, which vanishes for CLMs without free edges. Since Ξ is an arbitrary function, from

$\delta F = 0$, we derive the first Euler–Lagrange equation for a CLM without free edges:

$$2h(\kappa_m - H) - k_f \nabla^2 \phi = 0. \tag{5.133}$$

It should be noted that we have used the assumption $\nabla \cdot \mathbf{S} = 0$ when we write Eq. (5.132). Thus, $\nabla^2 \phi$ should be replaced with $\nabla^2 \phi - \nabla \cdot \mathbf{S}$ in Eq. (5.132) as well as Eq. (5.133) when this condition does not hold.

Nextly, let us denote F_0 as the functional (5.129) with vanishing h and k_f, and define the additional functional

$$F_{ad} = \int \left(\frac{k_f}{2} \mathbf{v}^2 - h\tau_m \right) dA. \tag{5.134}$$

Any small deformation of a CLM without free edges can always be achieved from small normal displacement Ω_3 at each point $\mathbf{r}$ in the surface. That is, $\delta \mathbf{r} = \Omega_3 \mathbf{e}_3$. The frame is also changed because of the deformation of the surface, which is denoted as

$$\delta \mathbf{e}_i = \Omega_{ij} \mathbf{e}_j \quad (i = 1, 2, 3), \tag{5.135}$$

where $\Omega_{ij} = -\Omega_{ji}$ $(i, j = 1, 2, 3)$ corresponds to the rotation of the frame due to the deformation of the surface.

The variation of F_0 is

$$\delta F_0 = \int [2k_c \nabla^2 H - 2\lambda H + k_c(2H + c_0)(2H^2 - c_0 H - 2K) + p]\Omega_3 dA. \tag{5.136}$$

Following Ref. [38], and considering Eqs. (5.111), (5.118) and (5.119), through somewhat involved calculations, we can derive

$$\delta(\tau_m dA) = 2\Omega_{12}(H - \kappa_m)dA + \nabla \cdot \mathbf{m}d\Omega_3 \wedge \mathbf{m} \cdot d\mathbf{r}$$
$$+ \mathbf{m} \cdot d\mathbf{r} \wedge d(d\Omega_3 \wedge *\mathbf{m} \cdot d\mathbf{r}/dA), \tag{5.137}$$

$$\delta(\mathbf{v}^2 dA) = 2d\Omega_{12} \wedge *(d\phi + \omega_{12}) + 2d\Omega_3 \wedge \mathbf{v} \cdot d\mathbf{e}_3$$
$$+ 2(\kappa_v - H)\mathbf{v}^2\Omega_3 dA. \tag{5.138}$$

Combining with the above two equations and using the integral by parts and Stokes' theorem, we derive

$$\delta F_{ad} = \int h[\nabla \cdot (\mathbf{m}\nabla \times \mathbf{m}) + \nabla \times (\mathbf{m}\nabla \cdot \mathbf{m})]\Omega_3 dA$$

$$+ \int k_f[(\kappa_v - H)\mathbf{v}^2 - \nabla \mathbf{v} : \nabla \mathbf{e}_3]\Omega_3 dA. \tag{5.139}$$

Since Ω_3 is an arbitrary function, $\delta F = \delta F_0 + \delta F_{\text{ad}} = 0$ follows the second Euler–Lagrange equation for a CLM without free edges:

$$2\nabla^2 H + (2H + c_0)(2H^2 - c_0 H - 2K) - 2\tilde{\lambda} H + \tilde{p}$$
$$+ \tilde{h}[\nabla \cdot (\mathbf{m}\nabla \times \mathbf{m}) + \nabla \times (\mathbf{m}\nabla \cdot \mathbf{m})]$$
$$+ \tilde{k}_f[(\kappa_{\mathbf{v}} - H)\mathbf{v}^2 - \nabla\mathbf{v} : \nabla\mathbf{e}_3] = 0, \tag{5.140}$$

with reduced parameters $\tilde{h} = h/k_c$, $\tilde{k}_f = k_f/k_c$, $\tilde{p} = p/k_c$, and $\tilde{\lambda} = \lambda/k_c$. $\kappa_{\mathbf{m}}$ and $\kappa_{\mathbf{v}}$ are the normal curvature along the directions of $\mathbf{m}$ and $\mathbf{v}$, respectively.

5.3.4   Governing Equations to Describe Equilibrium Configurations of CLMs with Free Edges

Consider a chiral lipid membrane with a free edge as shown in Fig. 5.8(b). Its free energy can be expressed as

$$F = \int G \mathrm{d}A + \gamma \oint \mathrm{d}s, \tag{5.141}$$

where $\mathrm{d}s$ is the arc length element of the edge, γ represents the line tension of the edge.

Let $\delta\phi = \Xi$, we still have Eq. (5.132) which can be further transformed into

$$\delta F = \int [2h(\kappa_{\mathbf{m}} - H) - k_f\nabla^2\phi]\Xi \mathrm{d}A - k_f \oint v_b\Xi \mathrm{d}s \tag{5.142}$$

with $v_b = \mathbf{v} \cdot \mathbf{b}$. Thus, from $\delta F = 0$, we can derive the first Euler–Lagrange equation

$$2\tilde{h}(\kappa_{\mathbf{m}} - H) - \tilde{k}_f\nabla^2\phi = 0, \tag{5.143}$$

and the first boundary condition

$$v_b = 0. \tag{5.144}$$

It should also be noted that we have used the assumption $\nabla \cdot \mathbf{S} = 0$ when we write Eq. (5.142). Thus, $\nabla^2\phi$ should be replaced with $\nabla^2\phi - \nabla \cdot \mathbf{S}$ in Eq. (5.142) as well as Eq. (5.143) when this condition is broken.

Because one can select an arbitrary frame $\{\mathbf{r}; \mathbf{e}_1, \mathbf{e}_2, \mathbf{e}_3\}$ here, we take it such that $\mathbf{e}_1$ and $\mathbf{e}_2$ align with $\mathbf{t}$ and $\mathbf{b}$ in the boundary curve. Then any small deformation of a CLM with free edge can always be expressed as the linear superposition of a small normal displacement Ω_3 and tangent displacement Ω_2 along $\mathbf{e}_2$ at each point $\mathbf{r}$ in the surface.

First, we consider the in-plane deformation mode $\delta \mathbf{r} = \Omega_2 \mathbf{e}_2$. The change of the frame is still denoted as Eq. (5.135). In terms of Ref. [43], we have $\delta \oint ds = - \oint \kappa_g \Omega_2 ds$, where κ_g is the geodesic curvature of the boundary curve. Additionally, we can derive

$$\delta \int G dA = - \oint G \Omega_2 ds. \tag{5.145}$$

Although the derivation of the above equation is somewhat involved, its physical meaning is quite clear. Under the displacement Ω_2 along $\mathbf{e}_2$, G is similar to the surface tension and the area element near the boundary decreases by $\Omega_2 ds$. Thus, the variation of $\int G dA$ gives Eq. (5.145) and $\delta F = 0$ results in the second boundary condition

$$(1/2)(2H + c_0)^2 + \tilde{k} K - \tilde{h} \tau_{\mathbf{m}} + (\tilde{k}_f/2)\mathbf{v}^2 + \tilde{\lambda} + \tilde{\gamma}\kappa_g = 0, \tag{5.146}$$

with $\tilde{k} = \bar{k}/k_c$ and $\tilde{\gamma} = \gamma/k_c$.

Next, we consider the out-of-plane deformation mode $\delta \mathbf{r} = \Omega_3 \mathbf{e}_3$. Let us denote F_0 as the functional (5.141) with h and k_f vanishing, and define the additional functional as Eq. (5.134). δF_0 is fully discussed in Ref. [43] as

$$\delta F_0 = \int [k_c(2H + c_0)(2H^2 - c_0 H - 2K) - 2\lambda H]\Omega_3 dA$$

$$+ \int 2k_c(\nabla^2 H)\Omega_3 dA - \oint [k_c(2H + c_0) + \bar{k}\kappa_n]\Omega_{23} ds$$

$$- \oint \left[-2k_c \frac{\partial H}{\partial \mathbf{b}} + \gamma \kappa_n + \bar{k}\dot{\tau}_g \right] \Omega_3 ds, \tag{5.147}$$

where the "dot" represents the derivative with respect to arc length parameter s. Similar to the derivation of Eq. (5.139) from Eqs. (5.137) and (5.138), by using the integral by parts and Stokes' theorem, we have

$$\delta F_{\mathrm{ad}} = \int h[\nabla \cdot (\mathbf{m}\nabla \times \mathbf{m}) + \nabla \times (\mathbf{m}\nabla \cdot \mathbf{m})]\Omega_3 dA$$

$$+ \int k_f[(\kappa_{\mathbf{v}} - H)\mathbf{v}^2 - \nabla \mathbf{v} : \nabla \mathbf{e}_3]\Omega_3 dA$$

$$+ \oint [h(v_t + \dot{\phi})\sin 2\bar{\phi} - k_f \kappa_n v_t]\Omega_3 ds$$

$$+ \oint \frac{h}{2} \sin 2\bar{\phi}\Omega_{23} ds, \tag{5.148}$$

where $v_t = \mathbf{v} \cdot \mathbf{t}$. κ_n, τ_g and κ_g are the normal curvature, geodesic torsion, and geodesic curvature of the boundary curve (i.e., the edge), respectively, $\bar{\phi}$ is the angle between $\mathbf{m}$ and $\mathbf{t}$ at the boundary curve. In Eqs. (5.147) and (5.148), Ω_3 represents the arbitrary small displacement of a point on the surface along $\mathbf{e}_3$ and Ω_{23} is the arbitrary small rotation of $\mathbf{e}_3$ around $\mathbf{t}$ at the edge. Thus, $\delta F = 0$ will give the second Euler–Lagrange equation

$$2\nabla^2 H + (2H + c_0)(2H^2 - c_0 H - 2K) - 2\tilde{\lambda}H$$

$$+ \tilde{h}[\nabla \cdot (\mathbf{m}\nabla \times \mathbf{m}) + \nabla \times (\mathbf{m}\nabla \cdot \mathbf{m})]$$

$$+ \tilde{k}_f[(\kappa_\mathbf{v} - H)\mathbf{v}^2 - \nabla\mathbf{v} : \nabla\mathbf{e}_3] = 0, \tag{5.149}$$

and the remaining two boundary conditions

$$(2H + c_0) + \tilde{k}\kappa_n - (\tilde{h}/2)\sin 2\bar{\phi} = 0, \tag{5.150}$$

$$\tilde{\gamma}\kappa_n + \tilde{k}\dot{\tau}_g - 2\partial H/\partial \mathbf{b} - \tilde{h}(v_t + \dot{\bar{\phi}})\sin 2\bar{\phi} + \tilde{k}_f\kappa_n v_t = 0. \tag{5.151}$$

We emphasize that Eqs. (5.144)–(5.151) describe the force and moment balance relations in the edge. Thus, they are also available for a chiral lipid membrane with several edges.

5.3.5 Solutions to Governing Equations of CLMs Without Free Edges

Now, we will present some analytic solutions to the governing equations of CLMs without free edges.

5.3.5.1 *Sphere*

For spherical vesicles of chiral lipid molecules with radius R, $\tau_\mathbf{m}$ is always vanishing because $a = c = 1/R$ and $b = 0$. Thus, the free energy (5.129) is independent of the molecular chirality and permits the same existence probability of left- and right-handed spherical vesicles. This is an uninteresting case in practice. Naturally, no evident circular dichroism signal would be observed, which is consistent with the experiment.[33]

5.3.5.2 *Cylinder*

Here, we consider a long enough cylinder with radius ρ such that its two ends can be neglected. The cylinder can be parameterized by two variables s and z which are the arc length along the circumferential direction and coordinate along axial direction, respectively. Let ϕ be the angle between m and the circumferential direction. We take a frame such that $\mathbf{e}_1$, $\mathbf{e}_2$ and $\mathbf{e}_3$ are along the circumferential, axial and radial directions, respectively. Then we have

$$\omega_1 = \mathrm{d}s, \quad \omega_2 = \mathrm{d}z \qquad (5.152)$$

$$a = -1/\rho, \quad b = c = 0. \qquad (5.153)$$

Thus,

$$2H = a + c = -1/\rho, \quad K = ac - b^2 = 0. \qquad (5.154)$$

Because $\mathrm{dd}s = 0$ and $\mathrm{dd}z = 0$, using the structure equation (5.103), we have $\omega_{12} = 0$. Considering Eq. (5.115), we obtain the spin connection

$$\mathbf{S} = 0, \qquad (5.155)$$

and then

$$\mathbf{v} = \nabla\phi = \phi_s\mathbf{e}_1 + \phi_z\mathbf{e}_2. \qquad (5.156)$$

From Eqs. (5.106) and (5.114), we have

$$\kappa_\mathbf{m} = -\cos^2\phi/\rho, \qquad (5.157)$$

$$\nabla^2\phi = \phi_{ss} + \phi_{zz}. \qquad (5.158)$$

Substituting the above two equations and Eq. (5.154) into the Euler–Lagrange equation (5.133), we can derive

$$\tilde{k}_f(\phi_{ss} + \phi_{zz}) + (\tilde{h}/\rho)\cos 2\phi = 0, \qquad (5.159)$$

where the subscripts s and z represent the partial derivatives with respect to s and z, respectively.

Additionally, using Eqs. (5.102), (5.104) and (5.156), we derive

$$\mathbf{v} \cdot \mathrm{d}\mathbf{e}_3 = (\phi_s/\rho)\mathrm{d}s. \qquad (5.160)$$

From Eq. (5.120), we can derive

$$\nabla\mathbf{v} : \nabla\mathbf{e}_3 = -\phi_{sz}/\rho. \qquad (5.161)$$

In this derivation, one should note that $\mathrm{d}A = \mathrm{d}s \wedge \mathrm{d}z = -\mathrm{d}z \wedge \mathrm{d}s$. From Eqs. (5.121), (5.122), (5.153) and (5.156), we can obtain

$$(\kappa_\mathbf{v} - H)\mathbf{v}^2 = (\phi_z^2 - \phi_s^2)/(2\rho) \qquad (5.162)$$

and

$$\nabla \cdot (\mathbf{m}\nabla \times \mathbf{m}) + \nabla \times (\mathbf{m}\nabla \cdot \mathbf{m})$$
$$= 2\sin 2\phi(\phi_z^2 - \phi_s^2 + \phi_{sz}) + \cos 2\phi(\phi_{ss} - \phi_{zz} + 4\phi_z\phi_s). \quad (5.163)$$

Substituting Eqs. (5.154) and (5.161)–(5.163) into the Euler–Lagrange equation (5.140) with vanishing p, we can obtain

$$\tilde{h}[2(\phi_z^2 - \phi_s^2 + \phi_{sz})\sin 2\phi + (\phi_{ss} - \phi_{zz} + 4\phi_z\phi_s)\cos 2\phi]$$
$$+ \tilde{\lambda}/\rho + (c_0^2 - 1/\rho^2)/(2\rho) + \tilde{k}_f[(\phi_z^2 - \phi_s^2)/(2\rho) + \phi_{sz}/\rho] = 0, \quad (5.164)$$

where the subscripts s and z represent the partial derivatives with respect to s and z, respectively.

It is not hard to see that $\phi = \pi/4$ and $2\tilde{\lambda}\rho^2 - 1 + c_0^2\rho^2 = 0$ can satisfy the above two equations. Thus, the cylinder shown in Fig. 5.9(a) with uniform tilting state (tilting angle $\phi = \pi/4$) is a solution, which is in good agreement with the experiment by Fang's group[34] where the projected direction of the molecules on the tubular surfaces indeed departs 45° from the equator of the tubules with the uniform tilting state.

Another possible tubule with helically modulated tilting state was proposed by Selinger *et al.*[31] The orientational variation of the tilting molecules

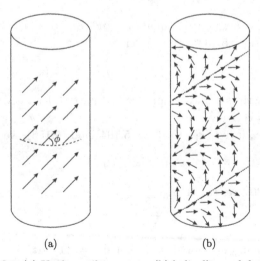

(a)   (b)

Fig. 5.9. Tubules. (a) Uniform tilting state, (b) helically modulated tilting state. Arrows represent the projected directions $\{\mathbf{m}\}$ of the tilting molecules on the tubules. (Reprint from Ref. [15]).

is assumed to be linear and confined to a small range. Here, we will directly investigate the orientational variation by using the Euler–Lagrange equation (5.133) without this assumption. Additionally, whether the tubules with helically modulated titling state are equilibrium configurations or not is not addressed in Ref. [31]. Here, our theory will unambiguously reveal that they are not equilibrium configurations.

At the helically modulated tilting state, ϕ is invariant along the direction of a fictitious helix enwinding around the tubule as shown in Fig. 5.9(b). Let ψ be the pitch angle of that helix. Apply a coordinate transformation $(s, z) \rightarrow (\zeta, \eta)$ via $\zeta = s \cos\psi + z \sin\psi, \eta = -s \sin\psi + z \cos\psi$, where ζ is the coordinate along the helix and η is the coordinate orthogonal to ζ. In the new coordinates, ϕ depends only on η. Changing variable $\Theta = \phi - \psi$ and introducing the dimensionless parameters $\chi = \eta/\rho$ and $\bar{h} = \tilde{h}\rho/\tilde{k}_f$, we transform Eqs. (5.159) and (5.164), respectively, into

$$\Theta_{\chi\chi} = -\bar{h}\cos 2(\Theta + \psi), \qquad (5.165)$$

and

$$(1 - c_0^2\rho^2) - 2\tilde{\lambda}\rho^2 = \tilde{k}_f[\Theta_\chi^2 \cos 2\psi - \Theta_{\chi\chi} \sin 2\psi$$
$$+ 2\bar{h}(2\Theta_\chi^2 \sin 2\Theta - \Theta_{\chi\chi}\cos 2\Theta)]. \qquad (5.166)$$

The first integral of Eq. (5.165) is

$$\Theta_\chi^2 = \mu^2 - \bar{h}\sin 2(\Theta + \psi), \qquad (5.167)$$

with an unknown constant μ. Substituting Eqs. (5.165) and (5.167) into Eq. (5.166), we obtain

$$(1 - c_0^2\rho^2) - 2\tilde{\lambda}\rho^2 = \tilde{k}_f[\mu^2 \cos 2\psi + \bar{h}(4\mu^2 - 1)\sin 2\Theta]$$
$$+ 2\tilde{k}_f\bar{h}^2[\cos(4\Theta + 2\psi) - \sin 2\Theta \sin 2(\Theta + \psi)]. \qquad (5.168)$$

The necessary condition for validity of Eq. (5.168) is that its right-hand side term is constant, which holds if and only if $\bar{h} = 0$ for varying Θ. Therefore, tubules with helically modulated tilting state are not admitted by the Euler–Lagrange equation (5.140) if $h \neq 0$. In other words, the orientational variation of the tilting lipid molecules breaks the force balance along the normal direction of the tubule at this state, which might rather induce helical ripples in tubules.

5.3.5.3 Helical Ripples in a Tubule

Assume now that a tubule with radius ρ undergoes small out-of-plane deformations and reaches a new configuration expressed as a vector $\{\rho(1 + y)\cos(s/\rho), \rho(1 + y)\sin(s/\rho), z\}$, where $|y| \ll 1$ is a function of s and z. The first and second fundamental forms of the surface can be calculated as

$$I = (1 + 2y)\mathrm{d}s^2 + \mathrm{d}z^2, \tag{5.169}$$

$$II = -\frac{1}{\rho}(1 + y - \rho^2 y_{ss})\mathrm{d}s^2 + 2\rho y_{sz}\mathrm{d}s\mathrm{d}z + \rho y_{zz}\mathrm{d}z^2 \tag{5.170}$$

up to the order of $O(y)$, respectively. The same order is kept in the following expressions in this section. In terms of the correspondence relations $I = \omega_1^2 + \omega_2^2$ and $II = a\omega_1^2 + 2b\omega_1\omega_2 + \omega_2^2$, we have

$$\omega_1 = (1 + y)\mathrm{d}s, \quad \omega_2 = \mathrm{d}z, \tag{5.171}$$

$$a = -(1 - y - \rho^2 y_{ss})/\rho, \quad b = \rho y_{sz}, \quad c = \rho y_{zz}, \tag{5.172}$$

$$2H = -(1 - y - \rho^2 y_{ss} - \rho^2 y_{zz})/\rho, \quad K = -y_{zz}. \tag{5.173}$$

Since $2H$ is a function of s and z, using $\nabla^2(2H)\mathrm{d}A = \mathrm{d} * \mathrm{d}(2H)$, we can derive

$$\nabla^2(2H) = (y_{ss} + y_{zz} + \rho^2 y_{ssss} + 2\rho^2 y_{zzss} + \rho^2 y_{zzzz})/\rho. \tag{5.174}$$

Because $\mathrm{d}\omega_1 = y_z\mathrm{d}z \wedge \mathrm{d}s$ and $\mathrm{d}\omega_2 = \mathrm{d}\mathrm{d}z = 0$, using the structure equation (5.103), we have $\omega_{12} = -y_z\omega_1$. Considering Eqs. (5.115) and (5.116), we have

$$\mathbf{S} = y_z\mathbf{e}_1, \quad \nabla \cdot \mathbf{S} = y_{zs}. \tag{5.175}$$

Let ϕ be the angle between $\mathbf{m}$ and the equator of the tubule. Using Eqs. (5.113) and (5.114), we obtain

$$\nabla\phi = \phi_s(1 - y)\mathbf{e}_1 + \phi_z\mathbf{e}_2, \tag{5.176}$$

$$\nabla^2\phi = \phi_{ss} + \phi_{zz} + \phi_z y_z - \phi_s y_s - 2y\phi_{ss}, \tag{5.177}$$

and then

$$\mathbf{v} = \nabla\phi - \mathbf{S} = [\phi_s(1 - y) - y_z]\mathbf{e}_1 + \phi_z\mathbf{e}_2. \tag{5.178}$$

From Eqs. (5.102), (5.104), (5.120), (5.121), and (5.178), we can derive

$$\nabla \mathbf{v} \colon \nabla \mathbf{e}_3 = -\phi_{sz}/\rho + (y_z \phi_s + 2y\phi_{sz} + y_{zz})/\rho, \tag{5.179}$$

$$(\kappa_{\mathbf{v}} - H)\mathbf{v}^2 = (\phi_z^2 - \phi_s^2)/(2\rho) + \phi_s(y\phi_s + y_z)/\rho$$
$$+2\rho y_{sz}\phi_s\phi_z + (y + \rho^2 y_{ss} - \rho^2 y_{zz})(\phi_s^2 - \phi_z^2)/(2\rho). \tag{5.180}$$

Since we only consider the case of $\bar{h} \ll 1$, up to the order $O(\bar{h})$, we obtain

$$\bar{h}(\kappa_{\mathbf{m}} - H) = -[\bar{h}/(2\rho)]\cos 2\phi, \tag{5.181}$$

$$\bar{h}[\nabla \cdot (\mathbf{m}\nabla \times \mathbf{m}) + \nabla \times (\mathbf{m}\nabla \cdot \mathbf{m})]$$
$$= \bar{h}[2(\phi_z^2 - \phi_s^2 + \phi_{sz})\sin 2\phi + (\phi_{ss} - \phi_{zz} + 4\phi_z\phi_s)\cos 2\phi] \tag{5.182}$$

by using Eqs. (5.106) and (5.163).

For simplicity, we take $c_0 = 0$, $\lambda = 0$, $k_f \simeq k_c$ such that $\tilde{k}_f \simeq 1$. Substituting the above Eqs. (5.173)–(5.182) into the Euler–Lagrange equations (5.133) and (5.140), we can derive

$$\rho^2(\phi_{ss} + \phi_{zz} + \phi_z y_z - \phi_s y_s - 2y\phi_{ss} - y_{zs}) = -\bar{h}\cos 2\phi \tag{5.183}$$

and

$$\rho^2[(1 + \rho^2\partial_{ss} + \rho^2\partial_{zz})^2 y - 1/2] + \rho^2[(1 + \rho^2\partial_{ss} + \rho^2\partial_{zz})y/2 - 2\rho^2 y_{zz}]$$
$$+ \bar{h}[2(\phi_z^2 - \phi_s^2 + \phi_{sz})\sin 2\phi + (\phi_{ss} - \phi_{zz} + 4\phi_z\phi_s)\cos 2\phi]$$
$$+(\phi_s^2 - \phi_z^2)(y + \rho^2 y_{ss} - \rho^2 y_{zz} - 1)/2 + \phi_{sz}$$
$$+ y\phi_s^2 - 2y\phi_{sz} - y_{zz} + 2\rho^2 y_{sz}\phi_s\phi_z = 0. \tag{5.184}$$

Now, we consider helical ripples in the tubule where ϕ and y are invariant along the direction of a fictitious helix enwinding around the tubule as shown in Fig. 5.10. We adopt the same coordinate transformation as the above subsection, and let $\vartheta = \phi - (\Theta + \psi)$, where ψ is the pitch angle of the fictitious helix and Θ is governed by Eq. (5.165). Then Eqs. (5.183) and (5.184) are reduced to a matrix equation

$$\mathcal{L}\Psi = \Phi, \tag{5.185}$$

with $\Psi \equiv \{\vartheta, y\}^T$ and $\Phi \equiv \{0, (\bar{h}/2)(1 - 4\mu^2)\sin 2\Theta\}^T$, where the superscript "T" represents the transpose. The differential operator $\mathcal{L}$ has four

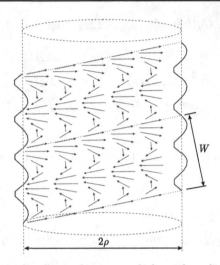

Fig. 5.10. Small amplitude ripples in a tubule with radius ρ. Arrows represent the projected directions $\{\mathbf{m}\}$ of the tilting molecules on the ripples' surface. (Reprint from Ref. [15]).

matrix elements:

$$\mathcal{L}_{11} \equiv -\mathrm{d}^2/\mathrm{d}\chi^2,$$

$$\mathcal{L}_{12} \equiv -\mu\cos 2\psi \mathrm{d}/\mathrm{d}\chi - \sin\psi\cos\psi \mathrm{d}^2/\mathrm{d}\chi^2,$$

$$\mathcal{L}_{21} \equiv \mu\cos 2\psi \mathrm{d}/\mathrm{d}\chi - \sin\psi\cos\psi \mathrm{d}^2/\mathrm{d}\chi^2,$$

$$\mathcal{L}_{22} \equiv \mu^2\cos^2\psi + [2\mu^2\cos 2\psi\sin^2\psi + (\mu^2-1)\cos^2\psi]\mathrm{d}^2/\mathrm{d}\chi^2$$
$$+ \mu^2\cos 2\psi \mathrm{d}^4/\mathrm{d}\chi^4,$$

where μ satisfies

$$\mu^2\cos 2\psi \simeq 1 \quad (\simeq 1/\tilde{k}_f). \tag{5.186}$$

It is not hard to find a special solution of Eq. (5.185) as

$$\tilde{\Psi} = \frac{\bar{h}(1-4\mu^2)}{4\Gamma}\begin{pmatrix} \cos 2(\Theta+\psi) \\ 2\sin 2\Theta \end{pmatrix} \tag{5.187}$$

with $\Gamma \equiv \mu^2(3\cos^2\psi + \cos 2\psi - 8\mu^2\cos 2\psi\sin^2\psi - 4\mu^2\cos^2\psi + 16\mu^4\cos 2\psi)$ which is positive.

Note that Eq. (5.185) contains only the first-order terms of ϑ and y, which can also be obtained from the variation of the free energy F expanded up to the second-order terms of ϑ and y. The dimensionless mean energy difference between the tubule with helical ripples and that with helically modulated tilting state is expressed as

$$\Delta F = \int_0^{W/\rho} \left(\frac{1}{2} \Psi^T \mathcal{L} \Psi - \Psi^T \Phi \right) d\chi, \tag{5.188}$$

where W is the period of the ripples along η direction as shown in Fig. 5.10. Substituting the solution (5.187) into the above equation, we have

$$\Delta F = -\frac{\bar{h}^2 (1 - 4\mu^2)^2}{8\Gamma} \int_0^{W/\rho} \sin^2 2\Theta d\chi < 0, \tag{5.189}$$

which reveals that a tubule with ripples is energetically more favorable than that in a helically modulated tilting state. Additionally, we can easily prove from Eq. (5.186) that the pitch angle obeys $\psi < 45°$. All ripples in tubules resolved in the recent experiment[35] with atomic force microscope indeed have pitch angles smaller than $45°$.

Considering $\bar{h} \ll 1$, we have $\Theta_\chi \approx \mu$ from Eq. (5.167) and then

$$\mu W/\rho \approx 2\pi, \tag{5.190}$$

due to W being the period of the ripples along η direction. If the pitch angle $\psi = 0$, the fictitious helix is a circle, and Eqs. (5.186) and (5.190) require $2\pi\rho/W \approx 1$ ($\simeq \sqrt{1/\tilde{k}_f}$). If $\psi \neq 0$, the fictitious helix is indeed a helix satisfying $W = 2\pi\rho \sin\psi$. It is then easy to derive

$$\cos 2\psi / \sin^2 \psi \approx 1 \quad (\simeq 1/\tilde{k}_f) \tag{5.191}$$

from Eqs. (5.186) and (5.190). Therefore, two kinds of ripples are permitted in our theory: one has a pitch angle about $0°$; another satisfies Eq. (5.191), which gives $\psi \simeq 35°$ for $\tilde{k}_f \simeq 1$ (i.e., $k_c \simeq k_f$). The theoretical results are thus close to the most frequent pitch angles (about $5°$ and $28°$) of the ripples in tubules observed by Fang's group.[35]

5.3.5.4 Torus

A torus is a revolution surface generated by a cycle with radius ρ rotating around an axis in the same plane of the cycle as shown in Fig. 5.11(a). The revolution radius r should be larger than ρ. A point in the torus can be expressed as a vector

$$\mathbf{r} = \{(r + \rho\cos\varphi)\cos\theta, (r + \rho\cos\varphi)\sin\theta, \rho\sin\varphi\} \tag{5.192}$$

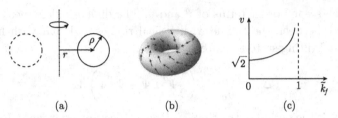

(a)   (b)   (c)

Fig. 5.11.   (a) Torus generated by a cycle rotating around an axis in the same plane of the cycle, (b) uniform tilting state ($\phi = -\pi/4$), (c) ratio of the generated radii (ν), Eq. (5.213), as a function of $\tilde{k}_f \equiv k_f/k_c$.

with $r > \rho$. A frame $\{e_1, e_2, e_3\}$ can be taken as

$$\begin{cases} e_1 = \{-\sin\theta, \cos\theta, 0\}, \\ e_2 = \{-\sin\varphi\cos\theta, -\sin\varphi\sin\theta, \cos\varphi\}, \\ e_3 = \{\cos\theta\cos\varphi, \sin\theta\cos\varphi, \sin\varphi\}. \end{cases} \tag{5.193}$$

By using Eqs. (5.101), (5.192) and (5.193), we can derive

$$\omega_1 = (r + \rho\cos\varphi)d\theta, \quad \omega_2 = \rho d\varphi. \tag{5.194}$$

From Eqs. (5.102) and (5.193), we obtain

$$\omega_{12} = [\sin\varphi/(r + \rho\cos\varphi)]\omega_1, \tag{5.195}$$

$$\omega_{13} = -[\cos\varphi/(r + \rho\cos\varphi)]\omega_1, \tag{5.196}$$

$$\omega_{23} = -(1/\rho)\omega_2. \tag{5.197}$$

Considering Eqs. (5.115), (5.116) and (5.195), we have

$$\mathbf{S} = -[\sin\varphi/(r + \rho\cos\varphi)]e_1, \tag{5.198}$$

$$\nabla \cdot \mathbf{S} = 0. \tag{5.199}$$

Additionally, Eqs. (5.104), (5.196) and (5.197) give

$$a = -\cos\varphi/(r + \rho\cos\varphi), \quad b = 0, \quad c = -1/\rho. \tag{5.200}$$

Thus, by considering Eqs. (5.105), (5.106), (5.107), and (5.114), we obtain

$$2H = -\frac{r + 2\rho\cos\varphi}{\rho(r + \rho\cos\varphi)}, \quad K = \frac{\cos\varphi}{\rho(r + \rho\cos\varphi)}, \tag{5.201}$$

$$\kappa_m = -\frac{\cos^2\phi\cos\varphi}{r + \rho\cos\varphi} - \frac{\sin^2\phi}{\rho}, \tag{5.202}$$

$$\tau_{\mathbf{m}} = -\frac{r\sin 2\phi}{2\rho(r + \rho\cos\varphi)}, \tag{5.203}$$

$$\nabla^2\phi = \frac{\phi_{\theta\theta}}{(r + \rho\cos\varphi)^2} + \frac{\phi_{\varphi\varphi}}{\rho^2} - \frac{\sin\varphi\,\phi_{\varphi}}{\rho(r + \rho\cos\varphi)}. \tag{5.204}$$

Substituting the above four equations into Eq. (5.133), we can derive

$$\frac{1}{\nu + \cos\varphi}\frac{\partial^2\phi}{\partial\theta^2} + \frac{\partial}{\partial\varphi}\left[(\nu + \cos\varphi)\frac{\partial\phi}{\partial\varphi}\right] - \frac{\nu\tilde{h}\rho}{\tilde{k}_f}\cos 2\phi = 0, \tag{5.205}$$

where ϕ is the angle between $\mathbf{m}$ and the latitude of the torus, while $\nu \equiv r/\rho$ is the ratio between two generated radii of the torus.

The uniform tilting state ($\phi = -\pi/4$) satisfies Eq. (5.205) and makes $-\int h\tau_{\mathbf{m}}dA$ take the minimum. At this state, $\nabla\phi = 0$, and

$$\mathbf{v} \equiv \nabla\phi - \mathbf{S} = [\sin\varphi/(r + \rho\cos\varphi)]\mathbf{e}_1, \tag{5.206}$$

$$\mathbf{v}\cdot d\mathbf{e}_3 = v_1\omega_{31} = \{\sin 2\varphi/[2(r + \rho\cos\varphi)]\}d\theta, \tag{5.207}$$

where v_1 is the component of $\mathbf{v}$ in the direction of $\mathbf{e}_1$. Thus, from Eq. (5.120), we can derive

$$\nabla\mathbf{v}:\nabla\mathbf{e}_3 = -(r\cos 2\varphi + \rho\cos^3\varphi)/[\rho(r + \rho\cos\varphi)^3]. \tag{5.208}$$

Using Eqs. (5.121), (5.122), (5.200) and (5.206), we can obtain

$$(\kappa_{\mathbf{v}} - H)\mathbf{v}^2 = r\sin^2\varphi/[2\rho(r + \rho\cos\varphi)^3] \tag{5.209}$$

and

$$\nabla\cdot(\mathbf{m}\nabla\times\mathbf{m}) + \nabla\times(\mathbf{m}\nabla\cdot\mathbf{m}) = -\cos\varphi/[\rho(r + \rho\cos\varphi)]. \tag{5.210}$$

Additionally, from $\nabla^2(2H)dA = d*d(2H)$, we have

$$\nabla^2(2H) = r(r\cos\varphi + \rho)/[\rho^2(r + \rho\cos\varphi)^3]. \tag{5.211}$$

Substituting the above Eqs. (5.201) and (5.208)–(5.211) into Eq. (5.140), we can derive

$$\begin{aligned}
(2 - \tilde{k}_f)/\nu^2 &+ (c_0^2\rho^2 - 1) + 2(\tilde{p}\rho + \tilde{\lambda})\rho^2 \\
&+ [(4c_0^2\rho^2 - 4c_0\rho - 2\tilde{h}\rho + 8\tilde{\lambda}\rho^2 + 6\tilde{p}\rho^3)/\nu]\cos\varphi \\
&+ [(5c_0^2\rho^2 - 8c_0\rho - 4\tilde{h}\rho + 10\tilde{\lambda}\rho^2 + 3\tilde{k}_f + 6\tilde{p}\rho^3)/\nu^2]\cos^2\varphi \\
&+ [(2c_0^2\rho^2 - 4c_0\rho - 2\tilde{h}\rho + 4\tilde{\lambda}\rho^2 + 2\tilde{k}_f + 2\tilde{p}\rho^3)/\nu^3]\cos^3\varphi = 0.
\end{aligned} \tag{5.212}$$

Because ν is finite for a torus, then the above equation holds if and only if the coefficients of $\{1, \cos\varphi, \cos^2\varphi, \cos^3\varphi\}$ vanish. It follows that $2\tilde{\lambda}\rho^2 = (4\rho c_0 - \rho^2 c_0^2) - 3\tilde{k}_f + 2\tilde{h}\rho$, $\tilde{p}\rho^3 = 2\tilde{k}_f - 2\rho c_0 - \tilde{h}\rho$ and

$$\nu = \sqrt{(2 - \tilde{k}_f)/(1 - \tilde{k}_f)}. \tag{5.213}$$

Thus, a torus with uniform tilting state as shown in Fig. 5.11(b) is an exact solution to governing equations of chiral lipid vesicles. The ratio of two generation radii satisfies Eq. (5.213), which increases with $\tilde{k}_f$ as shown in Fig. 5.11(c). Especially, $\nu = \sqrt{2}$ for $\tilde{k}_f = 0$, which leads to the $\sqrt{2}$ torus of non-tilting lipid molecules.[44] Since this kind of torus was observed in the experiment,[45] tori with $\nu > \sqrt{2}$ for $0 < \tilde{k}_f < 1$ might also be observed in some experiments on chiral lipid membranes.

5.3.5.5   Twisted Ribbons

Now, we consider a solution to the governing equations of chiral lipid membranes with free edges. Two long enough twisted ribbons with lipid molecules in different tilting states are shown in Fig. 5.3. A twisted ribbon can be expressed as a vector form

$$\mathbf{r} = \{u\cos\varphi, u\sin\varphi, \alpha\varphi\}. \tag{5.214}$$

A frame $\{\mathbf{e}_1, \mathbf{e}_2, \mathbf{e}_3\}$ can be taken as

$$\begin{cases} \mathbf{e}_1 = \{\cos\varphi, \sin\varphi, 0\}, \\ \mathbf{e}_2 = \{-u\sin\varphi, u\cos\varphi, \alpha\}/\sqrt{u^2 + \alpha^2}, \\ \mathbf{e}_3 = \{\alpha\sin\varphi, -\alpha\cos\varphi, u\}/\sqrt{u^2 + \alpha^2}. \end{cases} \tag{5.215}$$

By using Eqs. (5.101), (5.214) and (5.215), we can derive

$$\omega_1 = du, \quad \omega_2 = \sqrt{u^2 + \alpha^2}\,d\varphi. \tag{5.216}$$

From Eq. (5.102) and (5.215), we obtain

$$\omega_{12} = [u/(u^2 + \alpha^2)]\omega_2, \tag{5.217}$$

$$\omega_{13} = -[\alpha/(u^2 + \alpha^2)]\omega_2, \tag{5.218}$$

$$\omega_{23} = -[\alpha/(u^2 + \alpha^2)]\omega_1. \tag{5.219}$$

Considering Eqs. (5.115), (5.116) and (5.217), we have

$$\mathbf{S} = -[u/(u^2 + \alpha^2)]\mathbf{e}_2, \tag{5.220}$$

$$\nabla \cdot \mathbf{S} = 0. \tag{5.221}$$

Additionally, Eqs. (5.104), (5.218) and (5.219) give

$$a = c = 0, \quad b = -\alpha/(u^2 + \alpha^2). \tag{5.222}$$

Thus, by considering Eqs. (5.107), (5.105), (5.106), and (5.114), we obtain

$$H = 0, \quad K = -\alpha^2/(u^2 + \alpha^2)^2, \tag{5.223}$$

$$\kappa_{\mathbf{m}} = -\alpha \sin 2\phi/(u^2 + \alpha^2), \tag{5.224}$$

$$\tau_{\mathbf{m}} = -\alpha \cos 2\phi/(u^2 + \alpha^2), \tag{5.225}$$

$$\nabla^2 \phi = \phi_{uu} + (u\phi_u + \phi_{\varphi\varphi})/(u^2 + \alpha^2). \tag{5.226}$$

Substituting the above four equations into Eq. (5.143), we can derive

$$\tilde{k}_f \left(\phi_{uu} + \frac{u\phi_u + \phi_{\varphi\varphi}}{u^2 + \alpha^2} \right) + \frac{2\tilde{h}\alpha \sin 2\phi}{u^2 + \alpha^2} = 0, \tag{5.227}$$

where ϕ is the angle between $\mathbf{m}$ and the horizontal, ϕ_u represents the first-order derivative with respect to u while ϕ_{uu} and $\phi_{\varphi\varphi}$ are the second-order derivatives with respect to u and φ, respectively.

If we only consider the uniform tilting state, the above equation requires $\phi = 0$ or $\pi/2$. It is easy to see that $\phi = 0$ minimizes $-h \int \tau_{\mathbf{m}} dA$ for $\alpha < 0$ while $\phi = \pi/2$ minimizes $-h \int \tau_{\mathbf{m}} dA$ for $\alpha > 0$ because $\tau_{\mathbf{m}} = -\alpha \cos 2\phi/(u^2 + \alpha^2)$.[15] Thus, we should take $\phi = 0$ for $\alpha < 0$ and $\phi = \pi/2$ for $\alpha > 0$. The former case corresponds to Fig. 5.3(b) where $\mathbf{m}$ is perpendicular to the edges; the latter corresponds to Fig. 5.3(a) where $\mathbf{m}$ is parallel to the edges.

If $\phi = 0$ or $\pi/2$, then $\nabla\phi = 0$, and

$$\mathbf{v} \equiv \nabla\phi - \mathbf{S} = [u/(u^2 + \alpha^2)]\mathbf{e}_2, \tag{5.228}$$

$$\mathbf{v} \cdot d\mathbf{e}_3 = v_2\omega_{32} = [u\alpha/(u^2 + \alpha^2)^2]du, \tag{5.229}$$

where v_2 is the component of $\mathbf{v}$ in the direction of $\mathbf{e}_2$. Thus, from Eq. (5.120), we can derive

$$\nabla\mathbf{v} : \nabla\mathbf{e}_3 = 0. \tag{5.230}$$

Considering $\sin 2\phi = 0$ and $\cos 2\phi = \pm 1$, from Eqs. (5.121), (5.122), (5.222) and (5.228), we can obtain

$$(\kappa_{\mathbf{v}} - H)\mathbf{v}^2 = 0, \tag{5.231}$$

$$\nabla \cdot (\mathbf{m}\nabla \times \mathbf{m}) + \nabla \times (\mathbf{m}\nabla \cdot \mathbf{m}) = 0. \tag{5.232}$$

Substituting the above Eqs. (5.223) and (5.230)–(5.232) into Eq. (5.149), we can derive

$$k_c c_0 \alpha^2 / (u^2 + \alpha^2)^2 = 0, \qquad (5.233)$$

which requires $c_0 = 0$ for non-vanishing α.

Now let us turn to the boundary conditions. If $\phi = 0$ for $\alpha < 0$, we have $\bar{\phi} = \pm \pi/2$. At the boundary, $\mathbf{t} // \mathbf{e}_2$ and $u \equiv W/2 = R$, so

$$v_t = R/(R^2 + \alpha^2), \quad v_b = 0. \qquad (5.234)$$

From Eqs. (5.223)–(5.225), we can obtain

$$K = -\alpha^2 / (R^2 + \alpha^2)^2, \qquad (5.235)$$

$$\kappa_m = 0, \quad \tau_m = -\alpha/(R^2 + \alpha^2). \qquad (5.236)$$

Additionally, we have

$$\kappa_n = a\cos^2(\pi/2) + b\sin\pi + c^2\sin(\pi/2) = 0, \qquad (5.237)$$

$$\kappa_g = \omega_{12}/\omega_2 = R/(R^2 + \alpha^2), \qquad (5.238)$$

$$\tau_g = b\cos\pi + (c - a)(\sin\pi)/2 = \alpha/(R^2 + \alpha^2), \qquad (5.239)$$

in terms of the definition of normal curvature, geodesic curvature and geodesic torsion. From Eqs. (5.234)–(5.239), we find that among the boundary conditions (5.144)–(5.151), only Eq. (5.146) is nontrivial, which reduces to

$$\tilde{\lambda}(1 + x^2)\alpha^2 - (\tilde{h} - \tilde{\gamma}x)|\alpha| + \frac{\tilde{k}_f x^2 - 2\tilde{k}}{2(1 + x^2)} = 0 \qquad (5.240)$$

with $x \equiv R/|\alpha|$. Similarly, we obtain the same equation as (5.240) when $\phi = \pi/2$ for $\alpha > 0$.

Guided by the experimental data,[36] we may assume $\tilde{h}$ to be proportional to r_d, the relative concentration difference of the left- and right-handed enantiomers in the experiment, i.e., $\tilde{h} = h_0 r_d$ with a constant h_0. In terms of the experimental data, $|\alpha| \to \infty$ for $r_d \to 0$. Thus, Eq. (5.240) requires $\lambda = 0$ and then

$$|\alpha| = (2\tilde{k} + \tilde{k}_f x^2)/[2(h_0 r_d - \tilde{\gamma}x)(1 + x^2)]. \qquad (5.241)$$

To determine the relation between x and r_d, we need to minimize the average energy per area with respect to $|\alpha|$ for a given width $W = 2R$. The average energy can be obtained from F/A, where A is the area of the twisted ribbon and F is the free energy (5.141). The calculation is straightforward with $c_0 = 0$, Eqs. (5.223), (5.225), (5.228), and

$$dA = \sqrt{u^2 + \alpha^2}\,du d\varphi, \quad ds = \sqrt{R^2 + \alpha^2}\,d\varphi, \quad (5.242)$$

which leads to

$$\bar{F} = \frac{(2\bar{k} - k_f)x + (k_f - 2h_0 R_d |\alpha|)\sqrt{1 + x^2}\,\text{arcsinh}x + 2\gamma|\alpha|(1 + x^2)}{\alpha^2\sqrt{1 + x^2}(\text{arcsinh}x + x\sqrt{1 + x^2})}. \quad (5.243)$$

Minimizing it with respect to $|\alpha|$ and using Eq. (5.241), we obtain

$$x = \beta r_d + O(r_d^3), \quad (5.244)$$

with $\beta \equiv 3h_0/(4\tilde{\gamma})$. This relation reveals that the ratio between the width and pitch of the ribbons is proportional to the relative concentration difference of left- and right-handed enantiomers in the low relative concentration difference region. Equation (5.244) fits well with the experimental data with parameter $\beta = 0.37$ as shown in Fig. 5.12.

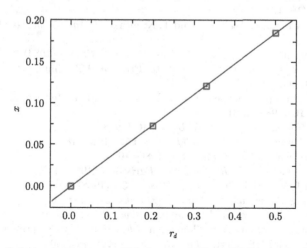

Fig. 5.12.   Relation between $x = R/|\alpha|$ and r_d. The solid line is the fitting curve $x = 0.37 r_d$ and the dots are the experimental data in Ref. [36].

5.4 Challenge

The Gauss–Bonnet theorem tells us that singular points cannot be avoided for a closed vesicle different from toroidal topology when the lipid molecules are tilted from the normal of the membrane surface. Generally speaking, introducing a singular point will cost a certain amount of energy, which has been ignored in our discussions for CLMs. The features of singular points in achiral spherical, torus vesicles and other manifolds at Smectic-C phase has been fully investigated in the previous literature.[46, 47] It is also found that the existence of singular points influences configurations of archiral vesicles at the Smectic-C phase.[48] However, it still remains a challenge to study the effect of singular points on configurations of CLMs at the Smectic-C* phase. In this case, the right-hand term of Eq. (5.133) or (5.143) should be replaced with the sum of δ-function, $\sum_i \sigma_i \delta(\mathbf{r} - \mathbf{r}_i)$, where $\mathbf{r}$ and $\mathbf{r}_i$ represent any point and singular point in the surface of CLM, while σ_i represents the strength of the source or vortex at the singular point $\mathbf{r}_i$.

References

[1] N. Nakashima, S. Asakuma, J. M. Kim, and T. Kunitake, *Chem. Lett.* **1984** (1984) 1709.

[2] K. Yamada, H. Ihara, T. Ide, T. Fukumoto, and C. Hirayama, *Chem. Lett.* **1984** (1984) 1713.

[3] N. Nakashima, S. Asakuma, and T. Kunitake, *J. Am. Chem. Soc.* **107** (1985) 509.

[4] P. Yager and P. Schoen, *Mol. Cryst. Liq. Cryst.*, **106** (1984) 371.

[5] P. Yager, P. Schoen, C. Davies, R. Price, and A. Singh, *Biophys. J.* **48** (1985) 899.

[6] J. H. Fuhrhop, R. Schneider, E. J. Boekema, and W. Helfrich, *J. Am. Chem. Soc.* **110** (1988) 2861.

[7] R. M. Servuss, *Chem. Phys. Lipids* **46** (1988) 37.

[8] J. M. Schnur *et al.*, *Thin Solid Films* **152** (1987) 181.

[9] W. Helfrich, *J. Chem. Phys.* **85** (1986) 1085.

[10] P. G. de Gennes, *C. R. Acad. Sci. Paris* **304** (1987) 259.

[11] W. Helfrich and J. Prost, *Phys. Rev. A* **38** (1988) 3065.

[12] Ou-Yang Zhong-can and Liu Jixing, *Phys. Rev. Lett.* **65** (1990) 1679.

[13] Ou-Yang Zhong-can and Liu Jixing, *Phys. Rev. A* **43** (1991) 6826.

[14] S. Komura and Ou-Yang Zhong-can, *Phys. Rev. Lett.* **81** (1998) 473.

[15] Z. C. Tu and U. Seifert, *Phys. Rev. E* **76** (2007) 031603.

[16] L. P. Eisenhart, *A Treatise on the Differential Geometry of Curves and Surfaces* (Ginn and Co., New York, 1909), p. 146.

[17] A. R. Forsyth, *Lectures on Differential Geometry of Curves and Surface* (Cambridge University Press, 1912).

[18] P. G. de Gennes, *The Physics of Liquid Crystals* (Clarenden, Oxford, 1974).
[19] M. Spivak, *A Comprehensive Introduction to Differential Geometry*, Vol. 3, Ch. 4 (Publisl or Perish, Berkeley, 1979).
[20] W. C. Graustein, *Differential Geometry* (MacMillan and Co., New York, 1934) p. 149.
[21] P. M. Morse and H. Feshbach, *Methods of Theoretical Physics* (McGraw-Hill New York, 1953), p. 43.
[22] M. J. Stephen and J. P. Straley, *Rev. Mod. Phys.* **46** (1974) 618.
[23] J. M. Schnur, *Science* **262** (1993) 1669.
[24] J. V. Selinger and J. M. Schnur, *Phys. Rev. Lett.* **71** (1993) 4091.
[25] B. N. Thomas, C. R. Safinya, R. J. Piano, and N. A. Clark, *Science* **267** (1995) 1635.
[26] J. S. Chappell and P. Yager, *Chem. Phys. Lipids* **58** (1991) 253.
[27] P. Nelson and T. Powers, *Phys. Rev. Lett.* **69** (1992) 3409.
[28] F. M. Konikoff, D. S. Chung, J. M. Donovan, D. M. Small, and M. C. Carey, *J. Clin. Invest.* **90** (1992) 1155.
[29] D. S. Chung, G. B. Benedek, F. M. Konikoff, and J. M. Donovan, *Proc. Natl Acad. Sci. USA* **90** (1993) 11341.
[30] I. Dahl and S. T. Lagerwall, *Ferroelectrics* **58** (1984) 215.
[31] J. V. Selinger, F. C. MacKintosh, and J. M. Schnur, *Phys. Rev. E* **53** (1996) 3804.
[32] M. Caffrey, J. Hogan, and A. S. Rudolph, *Biochem.* **30** (1991) 2134.
[33] J. M. Schnur, B. R. Ratna, J. V. Selinger, A. Singh, G. Jyothi, and K. R. K. Easwaran, *Science* **264** (1994) 945.
[34] Y. Zhao, N. Mahajan, R. Lu, and J. Fang, *Proc. Natl. Acad. Sci. USA* **102** (2005) 7438.
[35] N. Mahajan, Y. Zhao, T. Du, and J. Fang, *Langmuir* **22** (2006) 1973.
[36] R. Oda, I. Huc, M. Schmutz, S. J. Candau, and F. C. MacKintosh, *Nature* **399** (1999) 566.
[37] D. Berthier, T. Buffeteau, J. Léger, R. Oda, and I. Huc, *J. Am. Chem. Soc.* **124** (2002) 13486.
[38] Z. C. Tu and Z. C. Ou-Yang, *J. Phys. A* **37** (2004) 11407.
[39] S. S. Chern and W. H. Chern, *Lecture on Differential Geometry* (Beijing University Press, Beijing, 1983).
[40] C. V. Westenholz, *Differential Forms in Mathematical Physics* (North-Holland, Amsterdam, 1981).
[41] T. C. Lubensky and J. Prost, *J. Phys. (France)* **2** (1992) 371.
[42] D. R. Nelson and L. Peliti, *J. Phys. (France)* **48** (1987) 1085.
[43] Z. C. Tu and Z. C. Ou-Yang, *Phys. Rev. E* **68** (2003) 061915.
[44] Z. C. Ou-Yang, *Phys. Rev. A* **41** (1990) 4517.
[45] M. Mutz and D. Bensimon, *Phys. Rev. A* **43** (1991) 4525.
[46] D. Pettey and T. C. Lubensky, *Phys. Rev. E* **59** (1999) 1834.
[47] V. Vitelli and D. R. Nelson, *Phys. Rev. E* **74** (2006) 021711.
[48] X. Xing, H. Shin, M. J. Bowick, Z. Yao, L. Jia, and M.-H. Li, *Proc. Natl. Acad. Sci. USA* **109** (2012) 5202.

6

Some Untouched Topics

6.1 Nonlocal Theory of Membrane Elasticity

There are two kinds of nonlocal theories of membrane elasticity. One is the area-difference elasticity, the other is the elasticity of membrane with nonlocal interactions between different points.

6.1.1 Area-Difference Elasticity

Since it is very difficult for lipid molecules to flip from one leaflet to another,[1] when the membrane is bent from the planar configuration, the area per lipid molecule in one leaflet should be larger than the equilibrium value while the area per lipid molecule in another leaflet should be smaller than the equilibrium value. Considering the in-plane stretching or compression in each leaf, a nonlocal term $(k_r/2)(\int 2H\mathrm{d}A)^2$ might be added to the free energy of membranes.[2,3] Here, $k_r = k_a t^2/(2A_0)$ with k_a and t being the compression modulus and thickness of the monolayer, respectively, while A_0 is the prescribed area of the membrane. Considering this term, one might express the free energy of a vesicle as

$$F_{AD} = \int \left[\frac{k_c}{2}(2H + c_0)^2 + \bar{k}K \right] \mathrm{d}A + \lambda A + pV + \frac{k_r}{2} \left(\int 2H\mathrm{d}A \right)^2.$$

$$(6.1)$$

Similarly, if the membrane is initially curved with (spontaneous) relative area difference a_0, the nonlocal term $(k_r/2)(\int 2H\mathrm{d}A + a_0)^2$ might be included in the free energy after the membrane is deformed.[4] Thus, the energy of a vesicle can be expressed as

$$F_{ADE} = \int \left[\frac{k_c}{2}(2H + c_0)^2 + \bar{k}K \right] dA + \lambda A + pV + \frac{k_r}{2} \left(\int 2H dA + a_0 \right)^2.$$

(6.2)

In fact, if we make a transformation $C_0 = c_0 + a_0 k_r/k_c$ and $\Lambda = \lambda + a_0^2 k_r/(2A_0) - k_r a_c c_0 - k_r^2 a_0^2/2k_c$, the above free energy is transformed into the form of Eq. (6.1). Thus, it is sufficient for us to consider the free energy (6.1). The budding transitions of axisymmetric fluid-bilayer vesicles have been fully investigated on the basis of area difference elasticity.[4] It is still necessary to discuss the general cases without presumption of axisymmetry.

According to the variational method developed in our previous work,[5-7] the shape equation of vesicles which corresponds to the Euler–Lagrange equation of free energy (6.1) can be derived as

$$\tilde{p} - 2\tilde{\lambda}H + (2H + c_0)(2H^2 - c_0 H - 2K) + \nabla^2(2H) - 4\tilde{k}_r K \int H dA = 0$$

(6.3)

with reduced parameters $\tilde{p} = p/k_c$, $\tilde{\lambda} = \lambda/k_c$ and $\tilde{k}_r = k_r/k_c$. This is a fourth-order nonlinear integro-differential equation, so it is hard for us to find some exact solutions to this equation.

Obviously, a sphere is a solution to Eq. (6.3) which requires the radius R of sphere satisfying

$$\tilde{p}R^2 - (2\tilde{\lambda} + 16\pi\tilde{k}_r)R - c_0(2 - c_0 R) = 0.$$

(6.4)

Comparing this equation with Eq. (3.59), we find that the nonlocal term has an effect on the surface tension.

To check the other axisymmetric solutions, we adopt the representation shown in Fig. 3.16. In this representation, $2H = -h = -[\sin\psi/\rho + (\sin\psi)']$, $K = \sin\psi(\sin\psi)'/\rho$, $\nabla^2(2H) = -(\rho\cos\psi h')'\cos\psi/\rho$ and $dA = 2\pi|\sec\psi|\rho d\rho$. Thus, Eq. (6.3) is transformed into

$$\tilde{p} + \tilde{\lambda}h + (c_0 - h)\left(\frac{h^2}{2} + \frac{c_0 h}{2} - 2K \right)$$

$$-\frac{\cos\psi}{\rho}(\rho\cos\psi h')' + 4\pi\tilde{k}_r \left(\int h|\sec\psi|\rho d\rho \right) K = 0.$$

(6.5)

It is necessary to note that the integral in the above equation is done on the minimal generation curve for the axisymmetric surface.

A torus can be generated by a planar curve expressed by $\sin \psi = \rho/r - R/r$. Substituting it into Eq. (6.5), we still derive $R/r = \sqrt{2}$, while $2\tilde{\lambda}r = c_0(4 - c_0 r) - 16\sqrt{2}\pi^2 \tilde{k}_r r$ and $\tilde{p}r^2 = 8\sqrt{2}\pi^2 \tilde{k}_r r - 2c_0$. That is, the torus with ratio of two generation radii being $\sqrt{2}$ is also the solution to the shape equation of vesicles within the framework of area difference elasticity.

Now, we will check the biconcave surface generated by planar curve expressed by $\sin \psi = \alpha \rho \ln(\rho/\rho_B)$. We find that Eq. (6.5) is satisfied when $\tilde{p} = 0$, $\tilde{\lambda} = (\alpha^2 - c_0^2)/2$, $\alpha = c_0 - \tilde{k}_r(4\pi|z_0| + \alpha A_0)$, where z_0 is the coordinate of the pole shown in Fig. 3.17 while A_0 represents the total area of the membrane. That is, the biconcave surface generated by planar curve expressed by $\sin \psi = \alpha \rho \ln(\rho/\rho_B)$ is also the solution to the shape equation of vesicles within the framework of area difference elasticity. Here, the only difference is that $\alpha \neq c_0$ when we consider the area difference elasticity.

The above three examples imply that the shape equations of vesicles with and without consideration of the area difference elasticity seem to share the same form of solutions. Now, we will verify this proposition is indeed true. Let us assume $c_0 = \bar{c}_0 - 2\tilde{k}_r \int H \, dA$, then Eq. (3.46) is transformed into

$$\tilde{p} - 2\bar{\lambda}H + (2H + \bar{c}_0)(2H^2 - \bar{c}_0 H - 2K) + \nabla^2(2H) = 0, \qquad (6.6)$$

where $\bar{\lambda} = \tilde{\lambda} + (c_0^2 - \bar{c}_0^2)/2$. The above equation has the same form as Eq. (6.3), so the solutions to both equations have the same forms.

6.1.2   Membrane with Nonlocal Interactions

Some lipid molecules contain charged head groups, thus molecules in different regions of membrane can interact with each other when two regions get close to each other. Intuitively, the free energy can be expressed as

$$F_{nint} = \int \left[\frac{k_c}{2}(2H + c_0)^2 + \bar{k}K \right] dA + \lambda A + pV + \varepsilon \int dA \int dA' U(|\mathbf{r} - \mathbf{r}'|), \tag{6.7}$$

where $\mathbf{r}$ and $\mathbf{r}'$ represent the position vectors of different points in the membrane surface while dA and dA' are the area elements corresponding to the points $\mathbf{r}$ and $\mathbf{r}'$, respectively. H and K are the local mean curvature and Gaussian curvature at point $\mathbf{r}$, respectively. ε and $U(\cdot)$ represent the energy scale and the function form of nonlocal interactions, respectively.

Interestingly, we have proved that the Helfrich bending energy (2.97) with vanishing c_0 can be applicable to the bending of graphene.[7–9] If we consider that $U(|\mathbf{r} - \mathbf{r}'|)$ is the van der Waals-like interaction, the relative large camber arch[10, 11] in the edges of bilayer graphene might be understood on the basis of free energy (6.7) without osmotic pressure.

According to the variational method developed in our previous work,[5-7] the shape equation of vesicles which corresponds to the Euler–Lagrange equation of free energy (6.7) can be derived as

$$\tilde{p} - 2\tilde{\lambda}H + (2H + c_0)(2H^2 - c_0H - 2K)$$

$$+ \nabla^2(2H) + 2\tilde{\varepsilon} \int (U_n - 2HU)\mathrm{d}A' = 0, \qquad (6.8)$$

where $\tilde{\varepsilon} \equiv \varepsilon/k_c$, $U_n = (\partial U/\partial R)\hat{\mathbf{R}} \cdot \mathbf{n}$, $\mathbf{R} = \mathbf{r}' - \mathbf{r}$, $R = |\mathbf{R}|$, $\hat{\mathbf{R}} = \mathbf{R}/R$, $U = U(R)$. $\mathrm{d}A'$ represents the area element at point $\mathbf{r}'$. H, K and $\mathbf{n}$ represent the mean curvature, the Gaussian curvature and normal vector of membrane at point $\mathbf{r}$, respectively.

Since the nonlocal term $\int (U_n - 2HU)\mathrm{d}A'$ depends on the vector $\mathbf{r}$ for given function form of U, this term is equivalent to a nonuniform pressure applied on the membrane. A sphere is an obvious solution to Eq. (6.8) because the nonlocal term $\int (U_n - 2HU)\mathrm{d}A'$ gives a constant quantity which corresponds to a uniform pressure. Thus, Eq. (6.8) still reduces to the same form of Eq. (3.59) which determines the radius of the sphere. It is quite complicated to find the solutions corresponding to the shapes rather than spheres because the nonlocal term depends not only on the position of point in the membrane surface, but also on the function form of U. In particular, presuming U to be the van der Waals-like form, can we find some solutions rather than the spherical shape?

6.2 Numerical Simulations

Although some analytical stable membrane configurations have been obtained in the axisymmetric case, on the experimental side, various nonaxisymmetric shapes of spherical topology have long been observed. They may take very complex configurations, and many of them even have no intrinsic geometric symmetry. There are a lot of clear figures obtained by scanning electron microscope of red blood cells (RBCs) in the book *Living Blood Cells and their Ultra-structure*[12] including very complex vesicle shapes such as the *echinocyte type* cells (Ref. [12, Fig. 98]) which have a characteristic shape with crenations or spicules (almost) evenly distributed on the surface, the *acanthocyte type* cells (Ref. [12, Figs. 157 and 159]) which bear a superficial resemblance to *echinocytes* but with much fewer spicules irregularly arranged and bent back at their tips, the *Knizocytes* (Ref. [12, Figs. 106 and 107]) which are tri- and quadri-concave shapes, the *Sickle type* cells (Ref. [12, Fig. 198]) which show a sickle-like shape, and so on. In addition, many other complex shapes were found in the experimental study of transformation pathways of liposomes,[13] in which the shape transformation is induced by the osmotic pressure. A circular biconcave form was used as the initial shape in this study. Many thin stable flexible tube forms were also found. Before the full development of these tubes, certain transient forms appear, which can be described as filaments with small heads. These tubes are curved, so they are also nonaxisymmetric.

These complex shapes have not yet been understood theoretically in the context of bending energy models. Some researchers[4] believe that such exotic shapes may involve other energy contributions such as higher-order-curvature terms and van der Waals attraction of the membrane. However, the conjecture is not so obvious as it seems. We would like to explore if it is possible to describe these complex shapes by a simple curvature model, such as the Helfrich model. The purpose of this section is to search numerically for nonaxisymmetric shapes of spherical topology within the framework of the Helfrich spontaneous curvature (SC) model.

Inspired by its success in finding the nonaxisymmetric ellipsoidal and starfish shaped vesicles,[14, 15] Yan *et al.* have employed the algorithm of brute force energy minimization over a triangulated surface.[16] The method directly minimizes the total energy. The resulting shape has a local energy minimum which depends, in principle, on the initial shape chosen. The following is the detailed algorithm and main results of Yan *et al.*[16]

6.2.1   Surface Evolver

In order to find the locally stable nonaxisymmetric configurations of vesicles, we evaluate the bending energy numerically with the constraint of the constant volume and/or constant area within the SC model. Under the constraint of constant volume V, the parameter λ is understood as the tensile coefficient, while under the constraint of constant area A, the parameter $\triangle P$ is understood as the osmotic pressure.

The software we used to search for the surfaces is the "Surface Evolver" package of computer programs[17] which is based on the discretization of the curvature energy, the area, and the volume on a triangulated surface. The energy in the Evolver can be a combination of surface tension, gravitational energy, squared mean curvature, etc. The constraints can be on vertex positions, or on integrated quantities such as body volume, surface area, etc. The constraints are incorporated in the bending energy. The resulting total energy is minimized by a gradient descent procedure, and the resulting shape is a local energy minimum. These characteristics of the Evolver make it a useful tool for studying nonaxisymmetric shapes in the SC model. In the Evolver, the osmotic pressure is denoted by an internal pressure P and the software can deal with the following energy functional conveniently

$$F = m_1 \int (H - H_0)^2 dA + \lambda \int dA - P \int dV, \qquad (6.9)$$

where m_1 is called the "weight" of the bending energy. Under the definition of $H = (1/2)(C_1 + C_2)$, the model is identical to the SC model under the transformations: $m_1 = 2\kappa_c$, $P = -\triangle P$, and $H_0 = C_0/2$. No particular units of measurement are used in the Evolver. However, in order to relate the program values to a real situation, all the values ought to be within one consistent unit system.

The software has been employed to deal with many geometric problems such as constant mean curvature surfaces, equilibrium foam structures, etc. for several years. It has also been utilized to deal with a wide range of physical problems involving surfaces shaped by surface tension and bending and other energies for a long time. To study Kelvin's conjecture on minimal surfaces, the authors of Ref. [18] used the Surface Evolver to produce the minimal structure of flat-sided polyhedral cells. The Surface Evolver was also used to study the elasticity of dry foams[19,20] and compressed emulsions.[21,22] Just as an exercise, we tested it for the equilibrium condition of a perfect sphere with a given target volume evolved from a cube in the SC model. The equilibrium condition for the energy functional Eq. (6.9) is

$$-Pr^3 + 2\lambda r^2 + 2m_1 H_0 r(-1 + H_0 r) = 0, \qquad (6.10)$$

where r is the radius of the sphere (see Eq. (3.59)). With the parameters $m_1 = 1, H_0 = 1, \lambda = 2$, and the target volume $V = 4.189$, we obtained a stable unit sphere from the Surface Evolver with the area $A = 12.5774$, and the Lagrange multiplier $P = 4.0023$, which do satisfy the equilibrium condition.

Though the surfaces found by such an algorithm correspond to the energy minima, the Surface Evolver has a provision to test the stability by subjecting the resulting shapes to a perturbation of finite amplitude. Each vertex of the triangulated surface is moved by $\mathbf{A} \sin(\nu \omega + \psi)$, where

A is the amplitude vector, ν is the position vector of the vertexes, ω is the wavevector, and ψ is the phase. The parameters **A**, ω and ψ can be set by hand or generated randomly. In the random cases, a random amplitude **A** and a random wavelength L are chosen from a sphere whose radius is the size of the object. We used this feature of the Surface Evolver to test all the shapes reported in the paper.

One should keep in mind two important points of this algorithm. (1) A data file describing the initial shape must be provided in order to initiate the Surface Evolver. It is quite difficult to write the data file for a complex shape. Polyhedra are often used as initial shapes from which a target shape satisfying the constraints, such as target volume, target area, etc., can be obtained by refining and evolving the commands of the software. (2) The final shape satisfying the constrains strongly depends on the choice of the initial shape. One may expect that all the possible shapes in a parameter range $v_1 \leq v \leq v_2$ and $c_{0_1} \leq c_0 \leq c_{0_2}$ can be found by scanning the region step by step from an initial shape. However, this does not happen because (I) for any pair of values of (v, c_0), there is in general a set of coexisting stable shapes (the coexisting shapes mean the shapes with the same parameter values) and (II) the shape found by the algorithm depends strongly on the initial shape. Thus, the shapes generated by the scanning method form just a subset of all the possible shapes. Since normally one can write only simple data file for the starting shape, it is difficult to obtain complex shapes by the scanning procedure. Consider the following example: from a starting shape (which can be stable or unstable); with C_0 and A constant, by gradually changing the value of the target volume V, one can get a sequence of stable shapes (the sequence does not include the initial shape). Since the $(i + 1)$th shape is determined by the ith shape, the whole sequence of generated shapes is determined by the starting shape. A very important point is that the sequence of the generated shapes is insensitive to the initial shape, in that the same sequence is obtained even if the initial shape is slightly distorted. The above scanning process of the example is equivalent to scanning step by step along the reduced volume while the reduced SC is kept constant. Since for any given pair of values of (v, c_0), there exists a set of coexisting shapes S_{v,c_0}, it is obvious that the shapes included in the sequence form just a subset of all the shapes existing in the scanning region. Many (complex) shapes will not show up since in general one can only provide simple data file for the initial shape. To get complex shapes from the simple initial shapes, the sequence's insensitiveness to the initial shape must be broken.

We suggest the following alternative procedure: if some geometric quantities, such as the volume, the area, the reduced volume, etc., of the target shape are chosen to be far from the initial shape, one can imagine that the initial shape will evolve continuously through a long and complex pathway before it finally reaches a stable configuration satisfying all the geometric constraints. Any configuration in the pathway is unstable. One should expect the final shape (as well as the pathway) to be sensitive to the initial shape due to its long and complex pathway. Two shapes only slightly different from each other may lead to two very different final shapes with the same target constraints since the small difference will be enlarged in the course of evolution.

Hence, from an initial shape with parameters (v_i, c_0), there exist different ways to obtain a final shape with target parameters (v_f, c_0). (1) The shape transition procedure involves scanning from v_i to v_f gradually. The shape at each scanning step is stable and the final shape is insensitive to the initial shape. (2) The "jump" procedure: an initial shape evolves directly into a final shape with the target parameters, which is sensitive to the initial shape. The two procedures may produce different final shapes coexisting at the same parameters from the same initial shape. Obviously, the "jump" procedure provides us with the ability to obtain complex shapes from simple initial shapes. Using shapes found by the "jump" procedure as the initial shapes, more interesting shapes can be generated using the shape transition procedure.

However, there is a technical difficulty with the "jump" procedure described above: changing the reduced volume abruptly imposes constraints on both the area A as well as the volume V, which often lead to a singular behavior of the software. The Surface Evolver may not converge within finite number of iterations and singularities might occur. To avoid this problem, we free the constraints on area and abruptly change the volume V to a value far from the initial one. The Surface Evolver thus gains much more freedom to deform the shape in the process of evolving and works with much less singularities.

We calculated the reduced SC and the reduced volume for the shapes and located them in the parameter space spanned by (v, c_0) in order to know in which region of the parameter space these shapes exist. An outline of the procedure is as follows (supposing we start from a sphere of volume 1).

(1) Given certain values of C_0 and λ, let the Surface Evolver evolve the sphere to a target shape with volume V, where V is far from 1. Here, λ is the tensile coefficient, and C_0 is the SC. It is hoped that such a sudden and

big change in the volume will trigger a "random walk" in the configuration space of the surface shapes before it finally settles in a nontrivial locally stable configuration with volume V. This procedure can be applied to any shape to generate more stable shapes.

(2) We choose the complex shapes obtained in the above process and study their shape transition sequences, which can generate more stable and interesting shapes.

Using the procedure described in the preceding sub-section, many striking vesicle shapes were found. Some of the shapes resemble the RBC shapes observed experimentally while some resemble the experimental results of liposomes. To describe these exotic shapes, we adopt the nomenclature used in the RBCs literature for those resembling the observations in RBCs. We report six types of shapes in this paper, the corniculate shape, the knizocyte type shape, the sickle type shape, the acanthocyte type shape, and two tube-like shapes shown in Figs. 6.1(a), 6.2, 6.3, 6.4 and 6.5(a), 6.5(b), respectively. Two thousand to 3000 grid points and an accuracy of at least 1% in the total energy were used.

Though the shapes searched by the algorithm are locally stable, the stability was further tested by subjecting them to perturbations of finite amplitude. All the six shapes are stable under the perturbations. Each of the shapes is mapped into the 2D phase diagram following Seifert *et al.*[36] in reduced volume and the scaled SC.

6.2.1.1 *Corniculate shape*

Figure 6.1(a) shows a corniculate surface with six corns, whose location in the phase diagram is $(v = 0.95, c_0 = 1.35)$. This shape apparently

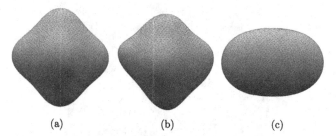

(a)   (b)   (c)

Fig. 6.1. (a) A corniculate shape at reduced volume and SC $v = 0.95$, $c_0 = 1.35$. (b) A heavily distorted shape after subjecting the corniculate shape to perturbations of large amplitude. (c) An axisymmetric ellipsoid evolved from a distorted shape with the same values of v and c_0 as in (a).

has rotational symmetry which is identical to the octahedron and is thus isomorphic to S_4, where S_X denotes the group of permutations of the set X.[23] Though we did not find a similar shape in Ref. [12] this shape may indicate the way of formation of echinocyte III vesicle shapes (Ref. [12, Figs. 98–100]) which have 10–50 corns evenly distributed on a nearly spherical surface. To show that there are coexisting shapes in the parameter space spanned by (v, c_0), keeping (v, c_0) constant, we subjected the shape shown in Fig. 6.1(a) to long wavelength perturbation with large amplitude. It is heavily deformed into a distorted one, as shown by Fig. 6.1(b). Running the Surface Evolver on the deformed shape, we finally got a new stable shape shown by Fig. 6.1(c), an axisymmetric ellipsoid-like shape which has the same parameters as those of Fig. 6.1(a). Since Figs. 6.1(a) and 6.1(c) are at the same location in the parameter space, we use the mark $*1$ in Fig. 6.6 to denote both the shapes. It is obvious that the two coexisting stable shapes will lead to two different shape transition sequences under the same transition procedure.

6.2.1.2 Knizocyte type shape

Knizocytes (Ref. [12, Figs. 106 and 107]) are triconcave and quadriconcave shapes found in the experiments of RBCs. The shape denoted by Fig. 6.2 is a quadriconcave shape and bears a resemblance to the experimentally observed shape (Ref. [12, Fig. 106]). Its location in the phase diagram is $(v = 0.84, c_0 = -1.41)$, denoted by $*2$ in Fig. 6.6. The shape has rotational

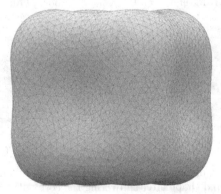

Fig. 6.2. A quadriconcave shape at reduced volume $v = 0.84$ and reduced SC $c_0 = -1.41$.

symmetry which is identical to a cube and is also isomorphic to S_4.[23] This shape may be seen under different circumstances. In fresh blood, it may be observed in certain hemolytic anemias. In addition, if a suspension of cells is examined between a slide and coverslip and an erythrocyte is permitted to adhere to the slide, gentle deformation of the cell by a current of liquid in the preparation may produce this appearance.[12]

6.2.1.3 *Sickle type shape*

Figure 6.3 bears the resemblance of the sickle cells in echinocytic forms (Ref. [12, Fig. 198]). The location in the phase diagram is ($v = 0.74$, $c_0 = -1.48$) and is denoted by ∗3 in Fig. 6.6. The sickle cell is related to sickle cell disease, a hereditary abnormality. Sickle cells appear when the affected blood is exposed to a sufficiently low oxygen tension. The phenomenon can also be seen by sealing a preparation between slide and coverslip and waiting a few hours or leaving the blood for 24–48 hours in a vessel without oxygen.[24]

6.2.1.4 *Acanthocyte type shape*

Figure 6.4 shows a strikingly complex shape without any intrinsic geometric symmetry. The characteristic of this shape is its irregular shape and several irregularly distributed crenations, which are the same as the so-called acanthocyte type cell shapes (Ref. [12, Figs. 157 and 159]) observed experimentally in RBCs. The location of this shape in Fig. 6.6 is ($v = 0.39$,

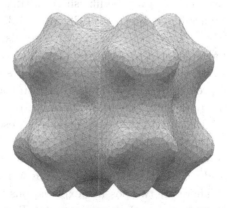

Fig. 6.3. A sickle type shape at reduced volume $v = 0.74$ and reduced SC $c_0 = -1.48$.

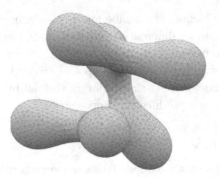

Fig. 6.4.   An acanthocyte type shape obtained by gradually reducing the reduced volume, while the reduced curvature is kept constant, as in Fig. 6.1(a). The reduced volume is $v = 0.38$ and reduced SC $c_0 = 1.35$.

$c_0 = 1.35$) denoted by $*4$. It is obtained by a shape transition procedure starting from Fig. 6.1(a) by gradually increasing the reduced volume v. The term acanthocyte was given by Singer *et al.*[25] to crenated red cells found in hereditary illnesses now characterized by the absence of beta-lipo-protein and serious nervous system alterations. The abnormality appears to develop during the lifespan of the cells within the circulation and to be absent or minimal in the youngest cells.[26] Such complex irregular acanthocyte type shapes are found to be abundant in our study.

6.2.1.5 *Tube-like shapes*

Figures 6.5(a) and 6.5(b) show two interesting shapes: a curved tube-like tail with a biconcave head and a tube-like shape without a distinct head, respectively. Their parameters are ($v = 0.38$, $c_0 = 1.35$) and ($v = 0.46$, $c_0 = 1.35$), respectively. Obviously, Fig. 6.5(a) is coexisting with Fig. 6.4 and its location is also denoted by $*4$ in the phase diagram. Figure 6.5(b) is denoted by $*5$ in the phase diagram. The two shapes were obtained by a shape transition procedure starting from Fig. 6.1(c) and by increasing the reduced volume v gradually. In the study of the transformation pathway of liposomes,[13] Fig. 6.5(b) was experimentally observed as a stable shape. Figure 6.5(a) was also obtained in the same experiment but only as a transient shape before the final state (Fig. 6.5(b)) was achieved. However, according to our study, both shapes of Fig. 6.5(a) as well as Fig. 6.5(b) are found to be stable. There are two possible explanations to account for the discrepancy with the experimental observations: (1) There may exist other energy contributions in such a tiny size which prevent Fig. 6.5(a) from being

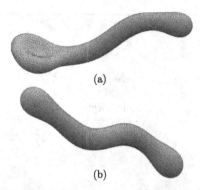

(a)

(b)

Fig. 6.5.   (a) A tube-like shape with a biconcave head at reduced volume $v = 0.38$ and reduced SC $c_0 = 1.35$. (b) A tube-like shape without a distinct head at reduced volume $v = 0.46$ and reduced SC $c_0 = 1.35$. The two shapes belong to the shape transition sequence starting from Fig. 6.1(c) and gradually reducing the reduced volume, while the reduced curvature is kept constant. It is not surprising to see that (a) is coexisting with Fig. 6.4 which belongs to the shape transition sequence starting from Fig. 6.1(a), since Figs. 6.1(a) and 6.5(a) are coexisting shapes.

stable in experiments; (2) By changing the experimental condition, stable configurations like Fig. 6.5(a) may become stable.

Similar observations were also made in the experiments in RBCs. The book[12] provides many examples of vesicles with tube-like tails and various kinds of head shapes. We find that Fig. 182 of Ref. [12] which denotes a kind of poikilocytes in the discocytic form, resembles Fig. 6.5(a). Poikilocytes are related to the thalassemia disorders or Mediterranean anemias. They take on a variety of bizarre erythrocyte shapes. Though the example is not "identical" to Fig. 6.5(a) from the appearance, it indicates that the latter assumption may be correct, i.e., Fig. 6.1(a) may be stable under certain conditions.

The algorithm used in this study has the ability to find complex vesicle shapes starting from simple initial ones under the "jump" procedure described above. With these complex shapes as the starting shapes, more complex shapes are generated by the shape transition procedure. Many strikingly complex shapes have been found within the framework of the SC model. Some of the shapes searched bear resemblance with experimental observations. Apparently, the procedure is also useful in searching for new shapes of high genus and can be used in other curvature models. Among the shapes provided above, we have a strong impression of the existence of

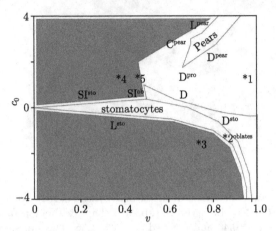

Fig. 6.6.   The schematic copy of the phase diagram in SC model reproduced by permission of the authors of Ref. [36]. The phase diagram shows the shape of lowest curvature energy as a function of the reduced volume v and the reduced SC c_0. The regions where the prolate/dumb-bell, pear-shaped, oblate/discocyte and stomatocytes are stable are separated by transition lines. The shapes included in the initial phase diagram are all axisymmetric, and the shaded regions are those that have not been explored by previous studies. We plot in the phase diagram the locations of the shapes found in the study in order to know to which region these shapes belong. They are denoted by $*n$ where n is an integer. $*1$ corresponds to the two coexisting shapes Figs. 6.1(a) and 6.1(c); $*2$ corresponds to Fig. 6.2; $*3$ corresponds to Fig. 6.3; $*4$ corresponds to two existing shapes Figs. 6.4 and 6.5(a); $*5$ corresponds to Fig. 6.5(b).

complex irregular shapes such as the acanthocyte type shapes and curved tube-like shapes, in the SC model, because they are the first reported irregular shapes of spherical topology in a simple curvature energy model. Our study shows that adding new energy contributions, such as higher-order-curvature terms and van der Waals attraction of the membrane, is not necessary to account for such abnormal shapes. In fact, we have also obtained several other irregular shapes which are not included here.

6.2.2   Phase Field

The other numerical method is the phase field or diffuse interface approach which was developed by Cahn and Hillard for phase transition problems.[27] Du used this method to investigate the shapes of lipid vesicles.[28, 29]

Introduce a phase field function ϕ in a 3D box Ω such that $\phi = -1$ in some region enclosed by a smooth surface M, $\phi = 0$ on that surface, and

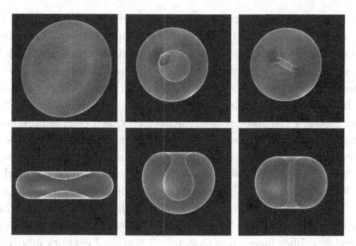

Fig. 6.7. Series of shapes of lipid vesicles obtained from phase field approach (reprint from Ref. [28]).

$\phi = 1$ in the region outside of that surface. In practice, we need to introduce a thin diffuse transition layer containing the surface. The thickness of the transition layer is typically characterized by a small positive parameter ϵ. The key procedure is to construct the geometric quantities such as the area, mean curvature and so on. For the closed vesicle, we need not consider the Gaussian curvature because its integral is a topological invariant.

The following corresponding relation can be derived:

$$\int_{\Omega} \left[\frac{\epsilon}{2} |\nabla \phi|^2 + \frac{1}{4\epsilon} (|\phi|^2 - 1)^2 \right] d\Omega \xrightarrow{\epsilon \to 0} \oint_M dA. \qquad (6.11)$$

$$\frac{1}{\epsilon} \int_{\Omega} \left[\epsilon \nabla^2 \phi + \frac{1}{\epsilon} \phi (1 - \phi^2) \right]^2 d\Omega \xrightarrow{\epsilon \to 0} \oint_M H^2 dA. \qquad (6.12)$$

$$V = \int_{\Omega} \phi d\Omega. \qquad (6.13)$$

For the prescribed surface and volume, minimizing the bending energy $(k_c/2) \int_M H^2 dA$, Du et $al.$ achieved a series of shapes of lipid vesicles as shown in Fig. 6.7.

6.3   Effects of Membrane Skeleton

The cell membrane consists of lipids, proteins, a small quantity of carbohydrates and so on. A simple but widely accepted model for cell membranes is the fluid mosaic model[30] proposed by Singer and Nicolson in 1972. In this model, the cell membrane is considered as a lipid bilayer where the lipid molecules can move freely in the membrane surface like fluid, while the proteins are embedded in the lipid bilayer. Some proteins, so-called integral membrane proteins, traverse entirely in the lipid bilayer and play the role of information and matter communications between the interior of the cell and its outer environment. The others, so-called peripheral membrane proteins, are partially embedded in the bilayer and accomplish the other biological functions. Beneath the lipid membrane, the membrane skeleton, a network of proteins, links with the proteins embedded in the lipid membrane. Mature mammalian and human erythrocytes (i.e., RBCs) lack cell nuclei. Thus, they provide a good experimental model for studying the mechanical properties of cell membranes.[31–34] On the theoretical side, the SC model,[35] rubber membrane model,[36–38] and dual network model[39] have been employed to investigate the mechanical and thermal fluctuation properties of erythrocyte membranes. We will address the elasticity and stability of the composite shell model for cell membranes in this section.

6.3.1 Composite Shell Model of Cell Membranes

A cell membrane can be simplified as a composite shell[40] of lipid bilayer and membrane skeleton. The membrane skeleton, inside the cell membrane, is a network of protein filaments as shown in Fig. 6.8. The joint points of the network are bulk proteins embedded in the lipid bilayer. The whole membrane skeleton seems to float on the sea of the lipid bilayer. It can have a global movement along the surface of the bilayer but the movement of the joints along the normal direction totally couples with the bilayer. From the mechanical point of view, the lipid bilayer can endure the bending

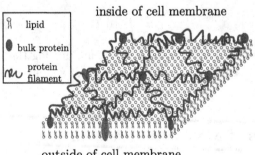

Fig. 6.8. Local schematic picture of the composite shell model for a cell membrane.

deformation but can hardly bear the in-plane shear strain. On the contrary, the membrane skeleton can endure the in-plane shear strain but not the bending deformation. The composite shell overcomes the shortage of the lipid bilayer and the membrane skeleton. It can sustain itself through both bending deformation and in-plane shear strain.

The contour length of protein chain between joints in the membrane skeleton is about 100 nm which is much smaller than the size ($\sim$10 μm) of cell membranes. The lipid bilayer is 2D homogenous. The membrane skeleton is roughly a 2D locally hexagonal lattice. As is well known, the mechanical property of a 2D hexagonal lattice is 2D isotropic.[41] Thus, the composite shell of the lipid bilayer plus the membrane skeleton can still be regarded as a 2D isotropic continuum. Its free energy density should be invariant under the in-plane coordinate transformation and can be written as $G_{\mathrm{cm}} = G_{\mathrm{cm}}(2H, K; 2J, Q)$. We can expand G_{cm} up to the second-order terms of curvatures and strains as

$$G_{\mathrm{cm}} = G_B + (k_b/2)(2J_b)^2 + G_{\mathrm{sk}}, \tag{6.14}$$

where G_B results mainly from the bending energy of the lipid bilayer, which has the form as Eq. (D.2). $(k_b/2)(2J_b)^2$ is the contribution of in-plane compression of the lipid bilayer, where k_b and $2J_b$ are the compression modulus and relative area compression of the lipid bilayer. $G_{\mathrm{sk}} = (k_d/2)(2J)^2 - \tilde{k}Q$ is the in-plane compression and shear energy density which comes from the entropic elasticity of the membrane skeleton. k_d and $\tilde{k}$ are the compression and shear moduli of the membrane skeleton, respectively, their values are experimentally determined as $k_d = \tilde{k} = 4.8\,\mu$N/m.[34] $2J$ and Q are the trace and determinant of the stain tensor of the membrane skeleton. Because there is no in-plane coupling between the lipid bilayer and the membrane skeleton in the composite shell model, thus J_b for the lipid bilayer and J for the membrane skeleton have no local correlation. In Appendix D, we verify that the effect of $(k_b/2)(2J_b)^2$ can be replaced with the surface tension $\lambda = 2k_b J_b$. Considering a closed cell membrane under osmotic pressure p, the free energy can be written as

$$\mathcal{F} = \int G_{\mathrm{cm}}\mathrm{d}A + \Delta p \int \mathrm{d}V. \tag{6.15}$$

Introducing a displacement vector $\mathbf{u} \equiv u_1\mathbf{e}_1 + u_2\mathbf{e}_2 + u_3\mathbf{e}_3$ satisfying

$$2J = \mathrm{div}\,\mathbf{u} - 2Hu_3, \tag{6.16}$$

$$2Q = (\mathrm{div}\,\mathbf{u} - 2Hu_3)^2 + (1/2)(\mathrm{curl}\,\mathbf{u})^2 - (\Diamond\mathbf{u})^2, \tag{6.17}$$

with $\Diamond\mathbf{u}$ being the in-plane part of $\nabla\mathbf{u}$, we can derive the Euler–Lagrange equations corresponding to the free energy (6.15) as

$$(\tilde{k} - 2k_d)\nabla(2J) - \tilde{k}(\Diamond^2\mathbf{u} + K\bar{\mathbf{u}} + \tilde{\nabla}u_3) = 0, \tag{6.18}$$

$$\Delta p + 2k_c[(2H + c_0)(2H^2 - c_0H - 2K) + 2\nabla^2 H] - 2\lambda H$$
$$+ 2H(\tilde{k} - k_d)(2J) - \tilde{k}\Re : \nabla\mathbf{u} = 0, \tag{6.19}$$

where $\bar{\mathbf{u}}$ and $\Diamond^2\mathbf{u}$ are the in-plane components of $\mathbf{u}$ and div $(\Diamond\mathbf{u})$, respectively, and $\Re$ is the curvature tensor related to Eq. (4.5). $\tilde{\nabla}$ is called the gradient operator of the second class, which is shown in our previous work.[6]

Generally speaking, it is difficult to find the analytical solutions to Eqs. (6.18) and (6.19). But we can verify that a spherical membrane with homogenous in-plane strains satisfies these equations. The radius R and the homogenous in-plane strain ε should obey the following relation:

$$\Delta p R^2 + 2(\lambda + 2k_d\varepsilon - \tilde{k}\varepsilon)R + k_c c_0(c_0 R - 2) = 0. \tag{6.20}$$

6.3.2 Stability of Cell Membranes and the Function of Membrane Skeleton

When the osmotic pressure is beyond some threshold, a closed cell membrane will lose its stability and change its shape abruptly. The threshold is called the critical pressure. To obtain it, one should calculate the second-order variation of the free energy (6.15), which reads

$$\delta^2\mathcal{F} = \int k_c[(\nabla^2\Omega_3)^2 + (2H + c_0)\nabla(2H\Omega_3) \cdot \nabla\Omega_3]\mathrm{d}A$$

$$+ \int [4k_c(2H^2 - K)^2 + k_c K(c_0^2 - 4H^2) + 2\lambda K - 2H\Delta p]\Omega_3^2\mathrm{d}A$$

$$+ \int \left[k_c\left(14H^2 + 2c_0 H - 4K - \frac{c_0^2}{2}\right) - \lambda \right] \Omega_3\nabla^2\Omega_3\mathrm{d}A$$

$$- 2k_c \int (2H + c_0)[\nabla\Omega_3 \cdot \tilde{\nabla}\Omega_3 + 2\Omega_3\nabla \cdot \tilde{\nabla}\Omega_3]\mathrm{d}A$$

$$- k_d \int [(\mathbf{v} \cdot \nabla + 2H\Omega_3)(\mathrm{div}\,\mathbf{v} - 2H\Omega_3)]\mathrm{d}A$$

$$+ \frac{\tilde{k}}{2} \int (\mathrm{curl}\,\mathbf{v})^2\mathrm{d}A - \tilde{k}\int K\bar{\mathbf{v}}^2\mathrm{d}A + \tilde{k}\int \Omega_3\tilde{\nabla} \cdot \mathbf{v}\mathrm{d}A$$

$$+ \tilde{k}\int 2H\Omega_3(\mathrm{div}\,\mathbf{v} - 2H\Omega_3)\mathrm{d}A - \tilde{k}\int \Omega_3\Re : \nabla\mathbf{v}\mathrm{d}A, \tag{6.21}$$

where $\mathbf{v} = \Omega_1\mathbf{e}_1+\Omega_2\mathbf{e}_2+\Omega_3\mathbf{e}_3$ is the infinitesimal displacement vector of the cell membrane whose in-plane component is denoted as $\bar{\mathbf{v}} = \Omega_1\mathbf{e}_1 + \Omega_2\mathbf{e}_2$.

In terms of the Hodge decomposed theorem,[42] $\mathbf{v}$ can be expressed by two scalar functions Ω and χ as

$$\mathbf{v}\cdot d\mathbf{r} = d\Omega + *d\chi, \qquad (6.22)$$

where $*$ is the Hodge star.[6, 42] Then we have div $\mathbf{v} = \nabla^2\Omega$ and curl $\mathbf{v} = \nabla^2\chi$. For the spherical cell membrane satisfying Eq. (6.20), Eq. (6.21) can be divided into two parts: one is

$$\delta^2\mathcal{F}_1 = \frac{\tilde{k}}{2}\int\left[(\nabla^2\chi)^2 + \frac{2}{R^2}\chi\nabla^2\chi\right]dA; \qquad (6.23)$$

another is

$$\delta^2\mathcal{F}_2 = \int\Omega_3^2\left(\frac{2c_0 k_c}{R^3} + \frac{\Delta p}{R} + \frac{4k_d - 2\tilde{k}}{R^2}\right)dA$$

$$+ \int\Omega_3\nabla^2\Omega_3\left(\frac{k_c c_0}{R} + \frac{2k_c}{R^2} + \frac{\Delta p R}{2}\right)dA$$

$$+ \int k_c(\nabla^2\Omega_3)^2 dA + \frac{4k_d - 2\tilde{k}}{R}\int\Omega_3\nabla^2\Omega dA$$

$$+ k_d\int(\nabla^2\Omega)^2 dA + \frac{\tilde{k}}{R^2}\int\Omega\nabla^2\Omega dA. \qquad (6.24)$$

It is easy to verify that $\delta^2\mathcal{F}_1$ is always positive on a spherical surface. Then the stability of the spherical cell membrane is merely determined by $\delta^2\mathcal{F}_2$. By analogy with our previous work,[43] we can prove that $\delta^2\mathcal{F}_2$ is also positive if

$$\Delta p < p_l \equiv \frac{2\tilde{k}(2k_d - \tilde{k})}{[k_d l(l+1) - \tilde{k}]R} + \frac{2k_c}{R^3}[l(l+1) - c_0 R], \qquad (6.25)$$

for any integer $l \geq 2$. Thus, the critical pressure is

$$p_c \equiv \min\{p_l(l = 2, 3, 4, \ldots)\}. \qquad (6.26)$$

Obviously, if $\tilde{k} = 0$, i.e., the effect of membrane skeleton vanishes in the cell membrane, p_c degenerates into the critical pressure (3.71) of a spherical lipid vesicle.

When $\tilde{k}k_d(2k_d - \tilde{k})R^2/[k_c(6k_d - \tilde{k})^2] > 1$, the critical pressure is derived from Eqs. (6.25) and (6.26) as

$$p_c = (4/R^2)\sqrt{(\tilde{k}/k_d)(2k_d - \tilde{k})k_c}. \qquad (6.27)$$

As an example, let us consider a cell membrane with typical values of $\tilde{k} = k_d = 4.8\,\mu\text{N/m},$[34] $k_c = 10^{-19}\,\text{J}$, and $R \approx 10\,\mu m$. Through a simple manipulation, we find that $\tilde{k}k_d(2k_d - \tilde{k})R^2/[k_c(6k_d - \tilde{k})^2] \gg 1$, and so Eq. (6.27) holds, from which we obtain the critical pressure $p_c = 0.03$ Pa. However, if the membrane skeleton vanishes, $\tilde{k} = 0$, we calculate $p_c = 0.001$ Pa from Eqs. (6.25) and (6.26). This example reveals a mechanical function of the membrane skeleton: it highly enhances the stability of cell membranes.

As a byproduct, Eq. (6.27) also gives the critical pressure

$$p_c = \sqrt{4/[3(1 - \nu^2)]}\; Y(h/R)^2 \tag{6.28}$$

for a spherical thin solid shell of 3D isotropic materials if we take $k_c = Yh^3/[12(1-\nu^2)]$, $k_d = Yh/(1-\nu^2)$, and $\bar{k}/k_c = \tilde{k}/k_d = 1-\nu$. This formula is the same as the classic strict result obtained by Pogorelov from another method.[44]

6.4   Application in Elasticity of Low-Dimensional Carbon Materials

In this section, we will discuss whether and to what extent the membrane theory can be applied to nanostructures, especially the graphitic structures, such as graphene and carbon nanotubes. This topic has been investigated in Refs. [7, 45].

6.4.1   Revised Lenosky Model

Graphene is a single layer of carbon atoms with a 2D honeycomb lattice as shown in Fig. 6.9(a). We will construct a theory to describe the elasticity of graphene-like structure.

We start from the concise formula proposed by Lenosky et al. in 1992 to describe the deformation energy of a single layer of curved graphite[46]

$$E_g = \frac{\epsilon_0}{2}\sum_{(ij)}(r_{ij} - r_0)^2 + \epsilon_1\sum_i\left(\sum_{(j)}\mathbf{u}_{ij}\right)^2$$

$$+ \epsilon_2\sum_{(ij)}(1 - \mathbf{n}_i \cdot \mathbf{n}_j) + \epsilon_3\sum_{(ij)}(\mathbf{n}_i \cdot \mathbf{u}_{ij})(\mathbf{n}_j \cdot \mathbf{u}_{ji}). \tag{6.29}$$

The first two terms are the contributions of bond length and bond angle changes to the energy. The last two terms are the contributions from the π-electron resonance. In the first term, r_0 is the initial bond length of planar graphite, and r_{ij} is the bond length between atoms i and j after the

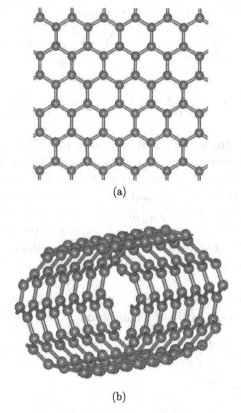

(a)

(b)

Fig. 6.9.   (a) Graphene. (b) single-walled carbon nanotube.

deformations. In the remaining terms, $\mathbf{u}_{ij}$ is a unit vector pointing from atom i to its neighbor j, and $\mathbf{n}_i$ is the unit vector normal to the plane determined by the three neighbors of atom i. The summation $\sum_{(j)}$ is taken over the three nearest neighbor atoms j to atom i, and $\sum_{(ij)}$ is taken over all the nearest neighbor atoms. The parameters $(\epsilon_1, \epsilon_2, \epsilon_3) = (0.96, 1.29, 0.05)\,\mathrm{eV}$ were determined by Lenosky $et\ al.$[46] through local density approximation. The value of ϵ_0 was given by Zhou $et\ al.$ as $\epsilon_0 = 57\,\mathrm{eV/\AA^2}$ through the force-constant method.[47]

In the above energy form, the second term requires that the energy cost due to in-plane bond angle changes is the same as that due to out-of-plane bond angle changes. However, the experiment by inelastic neutron scattering techniques reveals that the energy costs due to in-plane and out-of plane bond angle changes are quite different from each other. To describe this effect, we revise the Lenosky lattice model as[7]

$$E_g = \frac{\epsilon_0}{2} \sum_{(ij)} (r_{ij} - r_0)^2 + \epsilon_{1t} \sum_i \sum_{(j<k)} (\mathbf{u}_{ij}^t \cdot \mathbf{u}_{ik}^t + 1/2)^2$$

$$+ \epsilon_{1n} \sum_i \left(\sum_{(j)} \mathbf{u}_{ij}^n \right)^2 + \epsilon_2 \sum_{(ij)} (1 - \mathbf{n}_i \cdot \mathbf{n}_j), \qquad (6.30)$$

where $\mathbf{u}_{ij}^t = \mathbf{u}_{ij} - (\mathbf{n}_i \cdot \mathbf{u}_{ij})\mathbf{n}_i$ and $\mathbf{u}_{ij}^n = \mathbf{n}_i \cdot \mathbf{u}_{ij}$. If the three nearest neighbor atoms to atom i are labeled as 1,2,3, the summation $\sum_{(j<k)}$ is understood as $\sum_{1 \le j < k \le 3}$. The second and third terms of Eq. (6.30) represent the energy costs due to in-plane and out-of-plane bond angle changes, respectively. We have omitted the term $\epsilon_3 \sum_{(ij)} (\mathbf{n}_i \cdot \mathbf{u}_{ij})(\mathbf{n}_j \cdot \mathbf{u}_{ji})$ relative to the original Lenosky model (6.29) because its contribution is very small in terms of the results by Lenosky *et al.* Using the first-principles method, we fit parameters as $r_0 = 1.41$ Å, $\epsilon_0 = 46.34\,\mathrm{eV/Å}^2$, $\epsilon_{1t} = 4.48\,\mathrm{eV}$, $\epsilon_{1n} = 1.04\,\mathrm{eV}$, and $\epsilon_2 = 1.24\,\mathrm{eV}$.[7]

The continuum limit form of the revised Lenosky lattice model (6.30) is derived up to the second-order magnitudes, which reads

$$E_g = \int \left[\frac{k_c}{2}(2H)^2 - \bar{k}K + \frac{k_d}{2}(2J)^2 - \tilde{k}Q \right] dA, \qquad (6.31)$$

with four parameters

$$k_c = (9\epsilon_{1n} + 6\epsilon_2)r_0^2/(8\Omega_0), \qquad (6.32)$$

$$\bar{k} = 3\epsilon_2 r_0^2/(4\Omega_0), \qquad (6.33)$$

$$k_d = 9(\epsilon_0 r_0^2 + 3\epsilon_{1t})/(16\Omega_0), \qquad (6.34)$$

$$\tilde{k} = 3(\epsilon_0 r_0^2 + 9\epsilon_{1t})/(8\Omega_0), \qquad (6.35)$$

where $\Omega_0 = 3\sqrt{3}r_0^2/4$ is the occupied area per atom. The continuum form (6.31) was first derived in our previous work[48] which is, in fact, the natural conclusion of the symmetry of graphene: the curved graphene comprises a lot of hexagons, which has approximately local hexagonal symmetry. In fact, 2D structures with hexagonal symmetry are 2D isotropic.[41] Thus, the elasticity of the graphene can be reasonably described by the shell theory of 2D isotropic materials.

Using the values of r_0, ϵ_{1t}, ϵ_{1n}, and ϵ_2 obtained from the first-principles calculations, we have $k_c = 1.62\,\mathrm{eV}$, $\bar{k} = 0.72\,\mathrm{eV}$, $k_d = 22.97\,\mathrm{eV/Å}^2$, and

$\tilde{k} = 19.19\,\mathrm{eV/\text{Å}^2}$. Because the results of first-principles calculation are applicable for zero temperature, only the results derived from the experiments at low temperature can be used as reference values to compare with them. The value of k_c is close to the value $1.77\,\mathrm{eV}$ estimated by Komatsu[49-51] at low temperature (less than $60\,\mathrm{K}$). The value $\tilde{k}/k_d = 0.83$ is quite close to the experimental value 0.8 derived from the in-plane elastic constants of graphite.[52] The elastic properties of graphene can be described by Eq. (6.31) with four parameters k_c, $\bar{k}$, k_d, and $\tilde{k}$, where the energy density is the same as the free energy density of solid shell with 2D isotropic materials.

6.4.2 Carbon Nanotubes

A single-walled carbon nanotube (SWNT) can be regarded as a seamless cylinder wrapped up from a graphitic sheet, as shown in Fig. 6.9(b), whose diameter is in nanometer scale and length is from tens of nanometers to several micrometers if we ignore its two end caps.

The early researches on the elasticity of carbon nanotubes focused on their Young's modulus Y and Poisson ratio ν. An SWNT is a single layer of carbon atoms. What is the thickness h of the atomic layer? It is a widely controversial question. Three typical values of the thickness are adopted. The first one is about 0.7 Å obtained from fitting the atomic scale model with the elastic shell theory of 3D isotropic materials.[53-59] The second one is about 1.4 Å derived from molecular dynamics or finite element method.[60,61] The third one is about 3.4 Å adopting the layer distance of bulk graphite.[62-66] Recently, Huang $et\ al.$ have investigated the effective thickness of SWNTs and found that it depends on the type of loadings.[67]

As mentioned above, different thickness leads to different Young's modulus, which implies that the Young's modulus and thickness of SWNTs are not well-defined physical quantities. Here we may ask: What are the fundamental quantities for SWNTs?

An SWNT is also a single layer of graphite, whose deformation energy can also be described as the revised Lenosky model (6.30). The corresponding continuum limit is Eq. (6.31) which contains four elastic constants k_c, $\bar{k}$, k_d, and $\tilde{k}$. These four quantities avoid the controversial thickness of SWNTs. We suggest to use them as the fundamental quantities for SWNTs from which we can obtain some reduced quantities as follows.

Let us consider a cylinder under an axial loading with line density f along the circumference. The corresponding axial and circumferential strains are denoted as ε_{11} and ε_{22}. With Eq. (6.31), the free energy of this system is written as

$$\mathcal{F} \approx 2\pi\rho L[(k_d/2)(\varepsilon_{11} + \varepsilon_{22})^2 - \tilde{k}\varepsilon_{11}\varepsilon_{22} - f\varepsilon_{11}], \qquad (6.36)$$

where L and ρ are the length and radius of the SWNT, respectively. The in-plane Young's modulus and Poisson ratio can be defined as $Y_s = f/\varepsilon_{11}$ and $\nu_s = -\varepsilon_{22}/\varepsilon_{11}$. From $\partial\mathcal{F}/\partial\varepsilon_{11} = 0$ and $\partial\mathcal{F}/\partial\varepsilon_{22} = 0$, we derive

$$Y_s = \tilde{k}(2 - \tilde{k}/k_d) = 22.35 \text{ eV/Å}^2, \qquad (6.37)$$

$$\nu_s = 1 - \tilde{k}/k_d = 0.165, \qquad (6.38)$$

where the value of Y_s is close to the in-plane Young's modulus derived from Table 6.1. It is between 20 and 23 eV/Å^2 obtained by Sánchez-Portal et al.[68] It is much larger than the value 15 eV/Å^2 obtained by Arroyo et al.[69] and Zhang et al.,[70] and 17 eV/Å^2 by Caillerie et al.,[71] but smaller than 34.6 eV/Å^2 for armchair tube by Wang.[72] The value of ν_s is close to the value 0.16–0.19 obtained by Yakobson et al.,[73] Kudin et al.,[53] Pantano et al.,[56,57] Wang et al.,[59] Hernandez et al.,[62] and Shen et al.[63]

The other quantity, the bending rigidity D, is also widely discussed in literature. In terms of Eq. (6.31), the energy per area of a SWNT without the in-plane strains can be expressed as

$$G_g = k_c/(2\rho^2) \equiv D/(2\rho^2). \qquad (6.39)$$

Thus, the bending rigidity

$$D = k_c = 1.62 \text{ eV}, \qquad (6.40)$$

which is quite close to the value 1.49–1.72 eV obtained by Kudin et al.[53] and Sánchez-Portal et al.[68] through ab initio calculations. It is a little larger than the values 0.85–1.22 eV obtained by Yakobson et al.,[73] Pantano et al.,[56,57] and Wang.[72]

In terms of Eqs. (6.37)–(6.40), we can infer the values of k_d, $\tilde{k}$, k_c from the previous literature, which are listed in Table 6.1. There is still a lack of literature on $\tilde{k}$ except for our previous work[7,48,75] and recent work by Zelisko and his coworkers.[76] Because the term related to $\tilde{k}$ in the free energy (6.31) can be transformed into the boundary term with the aid of the Gauss–Bonnet formula, $\tilde{k}$ should be implicitly interplayed with the boundary energy which need further investigations in the future.

Table 6.1. The values of Y_s, ν_s, k_d, $\tilde{k}$, k_c and $\bar{k}$.

Authors	Y_s (eV/Å²)	ν_s	k_d (eV/Å²)	$\tilde{k}$ (eV/Å²)	k_c (eV)	$\bar{k}$ (eV)	Method	Refs.
Yakobson et al.	22.69	0.19	23.54	19.06	0.85	—	MD	[73]
Tu and Ou-Yang	22.03	0.34	24.88	16.44	1.17	0.75	LDA	[48]
Kudin et al.	21.69	0.15	22.19	18.86	1.49–1.53	—	Ab initio	[53]
Zhou et al.	23.59	0.24	25.03	19.02	1.14	—	TB	[54]
Pantano et al.	22.5	0.19	23.34	18.91	1.09	—	SM & FEM	[56, 57]
Chen and Cao	34.38	—	—	—	—	—	SM	[58]
Wang et al.	21.36	0.16	21.92	18.41	0.82	—	Ab initio	[59]
Sears and Batra	20.94	0.21	21.90	17.30	3.28	—	MD	[60]
Tserpes et al.	22.05	—	—	—	—	—	FEM	[61]
Lu	21.25	0.28	23.06	16.60	—	—	MD	[74]
Hernandez et al.	25.50	0.18	26.35	21.61	—	—	TB	[62]
Shen and Li	23.38	0.16	23.99	20.16	—	—	Force-field	[63]
Li and Chou	21.25	—	—	—	—	—	SM	[64]
Bao et al.	19.13	—	—	—	—	—	MD	[65]
Zhou et al.	17.00	0.32	18.94	12.88	—	—	LDA	[66]
Sánchez-Portal	19.41–22.40	0.12–0.19	19.92–23.00	16.73–19.31	1.49–1.72	—	Ab initio	[68]
Arroyo et al.	15.19	0.40	18.08	10.85	0.69	—	FEM	[69]
Zhang et al.	14.75	—	—	—	—	—	CTIP	[70]
Caillerie et al.	17.31	0.26	18.57	13.74	—	—	CTIP	[71]
Wang	34.63 or 17.31	—	—	—	1.12 or 1.21	—	CTIP	[72]
Tu and Ou-Yang	22.35	0.16	22.97	19.19	1.62	0.72	LDA	[7]

Notes: MD = molecular dynamics; TB = tight-binding; SM = structure mechanics; FEM = finite element method; LDA = local density approach; CTIP = continuum theory of interatomic potential.

References

[1]   M. Sheetz and S. Singer, *Proc. Natl. Acad. Sci.* **71** (1974) 4457.
[2]   E. Evans, *Biophys. J.* **30** (1980) 265.
[3]   S. Svetina, M. Brumen, and B. Žekš, *Stud. Biophys.* **110** (1985) 177.
[4]   L. Miao, U. Seifert, M. Wortis, and H. Döbereiner *Phys. Rev. E* **49** (1994) 5389.
[5]   Z. C. Tu and Z. C. Ou-Yang, *Phys. Rev. E* **68** (2003) 061915.
[6]   Z. C. Tu and Z. C. Ou-Yang, *J. Phys. A: Math. Gen.* **37** (2004) 11407.
[7]   Z. C. Tu and Z. C. Ou-Yang, *J. Comput. Theor. Nanosci.* **5** (2008) 422.
[8]   Z. C. Ou-Yang, Z. B. Su, and C. L. Wang, *Phys. Rev. Lett.* **78** (1997) 4055.
[9]   Z. C. Tu and Z. C. Ou-Yang, *Phys. Rev. B* **65** (2002) 233407.
[10] Z. Liu, K. Suenaga, P. Harris, and S. Iijima *Phys. Rev. Lett.* **102** (2009) 015501.
[11] J. Huang, F. Ding, B. Yakobson, P. Lu, L. Qi, and J. Li *Proc. Natl. Acad. Sci.* **106** (2009) 10103.
[12] M. Bessis, *Living Blood Cells and their Ultrastructure* (Springer-Verlag, Berlin 1973).
[13] H. Hotani, *J. Mol. Biol.* **178** (1984) 113.
[14] M. Jaric, U. Seifert, W. Wintz, and M. Wortis, *Phys. Rev. E* **52** (1995) 6623.
[15] M. Wintz, H. G. Dobereiner, and U. Seifert, *Europhys. Lett.* **23** (1996) 404.
[16] Y. Jie, L. Quanhui, L. Jixing, and O.-Y. Zhong-can, *Phys. Rev. E* **58** (1998) 4730.
[17] K. Brakke, *Exp. Math.* **1** (1992) 141.
[18] D. Weaire and R. Phelan, *Phil. Mag. Lett.* **69** (1994) 107.
[19] D. A. Reinelt and A. M. Kraynik, *J. Fluid. Mech.* **311** (1996) 327.
[20] A. M. Kraynik and D. A. Reinelt, *J. Colloid Interface Sci.* **181** (1996) 511.
[21] M.-D. Lacasse, G. S. Grest, D. Levine, T. G. Mason, and D. A. Weitz, *Phys. Rev. Lett.* **76** (1996) 3448.
[22] M.-D. Lacasse, G. S. Grest, and D. Levine, *Phys. Rev. E* **54** (1996) 5436.
[23] M. A. Armstrong, *Groups and Symmetry* (Springer-Verlag, Berlin, 1988).
[24] W. N. Jensen, D. L. Rucknagel, and W. J. Taylor, *J. Laborat. Clin. Invest* **56** (1960) 854.
[25] K. Singer, B. Fisher, and M. A. Perlstein, *Blood* **7** (1952) 577.
[26] C. Reed, P. Ways, and E. Simon, *Nouv. Rev. jr. Hemat.* **3** (1963) 59.
[27] J. W. Cahn and J. E. Hillard, *J. Chem. Phys.* **28** (1958) 258.
[28] Q. Du, C. Liu, and X. Wang, *J. Comput. Phys.* **198** (2004) 450.
[29] Q. Du, *Phil. Mag.* **91** (2011) 165.
[30] S. J. Singer and G. L. Nicolson, *Science* **175** (1972) 720.
[31] E. A. Evans, R. Waugh, and L. Melnik, *Biophys. J.* **16** (1976) 585.
[32] E. A. Evans, *Biophys. J.* **43** (1983) 27.
[33] H. Engelhardt and E. Sackmann, *Biophys. J.* **54** (1988) 495.
[34] G. Lenormand, S. Hénon, A. Richert, J. Siméon, and F. Gallet *Biophys. J.* **81** (2001) 43.
[35] W. Helfrich, *Z. Naturforsch. C* **28** (1973) 693.

[36] E. A. Evans, *Biophys. J.* **13** (1973) 941.

[37] Y. C. Fung and P. Tong, *Biophys. J.* **8** (1968) 175.

[38] E. A. Evans, *Biophys. J.* **13** (1973) 926.

[39] D. H. Boal, U. Seifert, and A. Zilker, *Phys. Rev. Lett.* **69** (1992) 3405.

[40] E. Sackmann, A. R. Bausch, and L. Vonna, *Physics of Composite Cell Membrane and Actin Based Cytoskeleton*, in *Physics of Bio-Molecules and Cells*, eds. H. Flyvbjerg, F. Jülicher, P. Ormos and F. David (Springer, Berlin, 2002).

[41] J. F. Nye, *Physical Properties of Crystals* (Clarendon Press, Oxford, 1985).

[42] C. V. Westenholz, *Differential Forms in Mathematical Physics* (North-Holland, Amsterdam, 1981).

[43] Z. C. Tu, L. Q. Ge, J. B. Li, and Z. C. Ou-Yang, *Phys. Rev. E* **72** (2005) 021806.

[44] A. V. Pogorelov, *Bendings of Surfaces and Stability of Shells* (American Mathematical Society, Providence, 1980).

[45] I. M. Mladenov, P. A. Djondjorov, M. Ts. Hadzhilazova, and V. M. Vassilev, *Commun. Theor. Phys.* **59** (2013) 213.

[46] T. Lenosky, X. Gonze, M. Teter, and V. Elser, *Nature* **355** (1992) 333.

[47] X. Zhou, H. Chen, J. J. Zhou, and Z. C. Ou-Yang, *Physica B* **304** (2001) 86.

[48] Z. C. Tu and Z. C. Ou-Yang, *Phys. Rev. B* **65** (2002) 233407.

[49] K. Komatsu, *J. Phys. Soc. Jpn.* **10** (1955) 346.

[50] K. Komatsu, *J. Phys. Chem. Solids* **6** (1958) 380.

[51] T. Nihira and T. Iwata, *Phys. Rev. B* **68** (2003) 134305.

[52] O. L. Blakeslee, D. G. Proctor, E. J. Seldin, G. B. Spence, and T. Weng, *J. Appl. Phys.* **41** (1970) 3373.

[53] K. N. Kudin, G. E. Scuseria, and B. I. Yakobson, *Phys. Rev. B* **64** (2001) 235406.

[54] X. Zhou, J. Zhou, and Z. C. Ou-Yang, *Phys. Rev. B* **62** (2000) 13692.

[55] T. Vodenitcharova and L. C. Zhang, *Phys. Rev. B* **68** (2003) 165401.

[56] A. Pantano, M. C. Boyce, and D. M. Parks, *Phys. Rev. Lett.* **91** (2003) 145504.

[57] A. Pantano, D. M. Parks, and M. C. Boyce, *J. Mech. Phys. Solids* **52** (2004) 789.

[58] X. Chen and G. Cao, *Nanotechnology* **17** (2006) 1004.

[59] L. Wang, Q. Zheng, J. Z. Liu, and Q. Jiang, *Phys. Rev. Lett.* **95** (2005) 105501.

[60] A. Sears and R. C. Batra, *Phys. Rev. B* **69** (2004) 235406.

[61] K. I. Tserpes and P. Papanikos, *Composites: Part B* **36** (2005) 468.

[62] E. Hernandez, C. Goze, P. Bernier, and A. Rubio, *Appl. Phys. A: Solids Surf.* **68** (1999) 287.

[63] L. Shen and J. Li, *Phys. Rev. B* **71** (2005) 165427.

[64] C. Li and T. Chou, *Int. J. Solids Struct.* **40** (2003) 2487.

[65] W. X. Bao, C. C. Zhu, and W. Z. Cui, *Physica B* **352** (2004) 156.

[66] G. Zhou, W. Duan, and B. Gu, *Chem. Phys. Lett.* **333** (2001) 344.

[67] Y. Huang, J. Wu, and K. C. Hwang, *Phys. Rev. B* **74** (2006) 245413.

[68] D. Sánchez-Portal, E. Artacho, J. M. Soler, A. Rubio, and P. Ordejón, *Phys. Rev. B* **59** (1999) 12678.

[69] M. Arroyo and T. Belytschko, *Phys. Rev. B* **69** (2004) 115415.

[70] P. Zhang, Y. Huang, P. H. Geubelle, P. A. Klein, and K. C. Hwang, *Int. J. Solids Struct.* **39** (2002) 3893.

[71] D. Caillerie, A. Mourad, and A. Raoult, *J. Elasticity* **84** (2006) 33.

[72] Q. Wang, *Int. J. Solids Struct.* **41** (2004) 5451.

[73] B. I. Yakobson, C. J. Brabec, and J. Bernholc, *Phys. Rev. Lett.* **76** (1996) 2511.

[74] J. P. Lu, *Phys. Rev. Lett.* **79** (1997) 1297.

[75] O. Y. Zhongcan, Z. B. Su, and C. L. Wang, *Phys. Rev. Lett.* **78** (1997) 4055.

[76] M. Zelisko, F. Ahmadpoor, H. Gao, and P. Sharma, *Phys. Rev. Lett.* **119** (2017) 068002.

Appendix A

Tensor Calculus

Let $(x^1, x^2, \ldots, x^N)$ be a coordinate system describing an N dimensional space and $\phi^\alpha(x^1, x^2, \ldots, x^N)$, $\alpha = 1, 2, \ldots, N$ be N independent, single valued, continuous and differentiable functions, then the functions

$$\bar{x}^\alpha = \phi^\alpha(x^1, x^2, \ldots, x^N) \quad (\alpha = 1, 2, \ldots, N) \tag{A.1}$$

can be used to define a new coordinate system $(\bar{x}^1, \bar{x}^2, \ldots, \bar{x}^N)$ for the same N dimensional space. Since ϕ^α are independent of each other, the Jacobian determinant $|\partial \bar{x}^\alpha / \partial x^j|$ does not vanish and Eq. (A.1) can be solved to get its inverse transformation

$$x^i = \psi^i(\bar{x}^1, \bar{x}^2, \ldots, \bar{x}^N) \quad (i = 1, 2, \ldots, N). \tag{A.2}$$

Obviously,

$$d\bar{x}^\alpha = \frac{\partial \bar{x}^\alpha}{\partial x^i} dx^i, \quad dx^i = \frac{\partial x^i}{\partial \bar{x}^\alpha} d\bar{x}^\alpha, \tag{A.3}$$

and

$$\frac{\partial \bar{x}^\alpha}{\partial x^j} \frac{\partial x^j}{\partial \bar{x}^\beta} = \delta^\alpha_\beta, \quad \frac{\partial x^i}{\partial \bar{x}^\alpha} \frac{\partial \bar{x}^\alpha}{\partial x^j} = \delta^i_j, \tag{A.4}$$

where the summation convention is used throughout the whole discussion. Under the transformation (A.2), the transformed function $\bar{A}^\alpha$ of any N functions A^i of x^j and A^i are called first-order contravariant tensors (or contravariant vectors) if they satisfy the condition

$$\bar{A}^\alpha = \frac{\partial \bar{x}^\alpha}{\partial x^i} A^i \quad (\alpha, i = 1, 2, \ldots, N). \tag{A.5}$$

259

Obviously, the products of contravariant vectors $A^i, B^i, C^i, \ldots (i = 1, 2, \ldots, N)$

$$A^{ij} = B^i C^j \quad (i, j = 1, 2, \ldots, N),$$
$$A^{ijk} = B^i C^j D^k \quad (i, j, k = 1, 2, \ldots, N),$$
$$\cdots$$

satisfy the relations

$$\bar{A}^{\alpha\beta} = (\partial \bar{x}^\alpha / \partial x^i)(\partial \bar{x}^\beta / \partial x^j) A^{ij},$$
$$\bar{A}^{\alpha\beta\gamma} = (\partial \bar{x}^\alpha / \partial x^i)(\partial \bar{x}^\beta / \partial x^j)(\partial \bar{x}^\gamma / \partial x^k) A^{ijk}, \qquad (A.6)$$
$$\cdots$$

They are called contravariant tensors of second order, third order, etc.

Contrary to the transformation of contravariant tensor, the transformation of any scalar function $f(x^1, x^2, \ldots, x^N)$ from the set of independent variables x^i $(i = 1, 2, \ldots, N)$ into the set of variables $\bar{x}^\alpha$ $(\alpha = 1, 2, \ldots, N)$ satisfies the relations

$$\frac{\partial f}{\partial \bar{x}^\alpha} = \frac{\partial x^j}{\partial \bar{x}^\alpha} \frac{\partial f}{\partial x^j}, \quad \frac{\partial f}{\partial x^i} = \frac{\partial \bar{x}^\alpha}{\partial x^i} \frac{\partial f}{\partial \bar{x}^\alpha}. \qquad (A.7)$$

If in the transformation of a set of functions of independent variables, $(x^1, x^2, \ldots, x^N)$, $A_i(x^1, x^2, \ldots, x^N)$ $(i = 1, 2, \ldots, N)$, into a set of functions $\bar{A}_\alpha(\bar{x}^1, \bar{x}^2, \ldots, x^N)$, $(\alpha = 1, 2, \ldots, N)$ of independent variables $(\bar{x}^1, \bar{x}^2, \ldots, x^N)$ it follows the law of transformation

$$\bar{A}_\alpha = \frac{\partial x^i}{\partial \bar{x}^\alpha} A_i, \quad A_i = \frac{\partial \bar{x}^\alpha}{\partial x^i} \bar{A}_\alpha, \qquad (A.8)$$

then A_i and $\bar{A}_\alpha$ are called covariant vectors (covariant tensors of the first order). Higher order covariant tensors $\bar{A}_{\alpha\beta}, \bar{A}_{\alpha\beta\gamma}, \ldots$ are defined by

$$\bar{A}_{\alpha\beta} = (\partial x^i / \partial \bar{x}^\alpha)(\partial x^j / \partial \bar{x}^\beta) A_{ij},$$
$$\bar{A}_{\alpha\beta\gamma} = (\partial x^i / \partial \bar{x}^\alpha)(\partial x^j / \partial \bar{x}^\beta)(\partial x^k / \partial \bar{x}^\gamma) A_{ijk},$$
$$\cdots$$
$$A_{ij} = (\partial \bar{x}^\alpha / \partial x^i)(\partial \bar{x}^\beta / \partial x^j) \bar{A}_{\alpha\beta} \qquad (A.9)$$
$$A_{ijk} = (\partial \bar{x}^\alpha / \partial x^i)(\partial \bar{x}^\beta / \partial x^j)(\partial \bar{x}^\gamma / \partial x^k) \bar{A}_{\alpha\beta\gamma},$$
$$\cdots$$

The products of the components of a tensor of contravariant order s and the components of a tensor of covariant order p form a mixed tensor of contravariant order s and covariant order p. The products of the components of a mixed tensor of contravariant order s and covariant order p and a mixed tensor of contravariant order t and covariant order q form a mixed tensor of contravariant order $s + t$ and covariant order $p + q$. This tensor is called the outer product of the two tensors. Under coordinate transformation, the mixed tensor $A^{t_1 t_2 \cdots t_s}_{q_1 q_2 \cdots q_p}$ satisfies the relations

$$\bar{A}^{\alpha_1 \alpha_2 \cdots \alpha_s}_{\beta_1 \beta_2 \cdots \beta_p} = \frac{\partial \bar{x}^{\alpha_1}}{\partial x^{t_1}} \frac{\partial \bar{x}^{\alpha_2}}{\partial x^{t_2}} \cdots \frac{\partial \bar{x}^{\alpha_s}}{\partial x^{t_s}} \frac{\partial x^{q_1}}{\partial \bar{x}^{\beta_1}} \frac{\partial x^{q_2}}{\partial \bar{x}^{\beta_2}} \cdots \frac{\partial x^{q_p}}{\partial \bar{x}^{\beta_p}} A^{t_1 t_2 \cdots t_s}_{q_1 q_2 \cdots q_p},$$

$$A^{t_1 t_2 \cdots t_s}_{q_1 q_2 \cdots q_p} = \frac{\partial x^{t_1}}{\partial \bar{x}^{\alpha_1}} \frac{\partial x^{t_2}}{\partial \bar{x}^{\alpha_2}} \cdots \frac{\partial x^{t_s}}{\partial \bar{x}^{\alpha_s}} \frac{\partial \bar{x}^{\beta_1}}{\partial x^{q_1}} \frac{\partial \bar{x}^{\beta_2}}{\partial x^{q_2}} \cdots \frac{\partial \bar{x}^{\beta_p}}{\partial x^{q_p}} \bar{A}^{\alpha_1 \alpha_2 \cdots \alpha_s}_{\beta_1 \beta_2 \cdots \beta_p}.$$

(A.10)

It is important to note that the order of the superscripts and the order of the subscripts cannot be changed at will, different order implies different tensors. Only tensors of the same contravariant order and the same covariant order can be added together. In a tensor equation, all the individual terms must have the same contravariant order and the same covariant order. The product of a contravariant vector $A^i(x^1, x^2, \ldots, x^N)$ and a covariant vector $B_i(x^1, x^2, \ldots, x^N)$ satisfies the relation

$$A^i B_i = \frac{\partial x^i}{\partial \bar{x}^\alpha} \bar{A}^\alpha \frac{\partial \bar{x}^\beta}{\partial x^i} \bar{B}_\beta.$$

This means $A^i B_i$ is a scalar or a tensor of the zeroth order. For any mixed tensor, e.g., A^{ij}_{lmn}, one has that

$$\bar{A}^{\alpha \beta}_{\gamma \delta \beta} = \frac{\partial \bar{x}^\alpha}{\partial x^i} \frac{\partial \bar{x}^\beta}{\partial x^j} \frac{\partial x^l}{\partial \bar{x}^\gamma} \frac{\partial x^m}{\partial \bar{x}^\delta} \frac{\partial x^n}{\partial \bar{x}^\beta} A^{ij}_{lmn}$$

$$= \frac{\partial \bar{x}^\alpha}{\partial x^i} \frac{\partial x^l}{\partial \bar{x}^\gamma} \frac{\partial x^m}{\partial \bar{x}^\delta} \delta^n_j A^{ij}_{lmn}$$

$$= \frac{\partial \bar{x}^\alpha}{\partial x^i} \frac{\partial x^l}{\partial \bar{x}^\gamma} \frac{\partial x^m}{\partial \bar{x}^\delta} A^{ij}_{lmj}.$$

This shows that $\bar{A}^{\alpha \beta}_{\gamma \delta \beta}$ is a mixed tensor of the first order in the contravariant tensor and of the second order in the covariant tensor. This process is called contraction of the tensor. It reduces the order of the tensor by two when one pair of the superscript and subscript is equal. The inner product of two tensors is simply an outer product of two tensors with contraction.

Direct method of testing whether a quantity satisfies the law of transformation Eq. (A.10) is rather troublesome. A simple way of testing whether a quantity is a tensor is provided by the quotient law. The quotient law states that N^p functions of x^i form the components of a tensor of order p provided an inner product of these functions with an arbitrary tensor is itself a tensor. A proof for the following particular case ensures the validity of the quotient law. The set of functions A_{ij} forms the components of a tensor of the type indicated by its indices if

$$A_{ij}B^i = C_j,$$

provided that B^i is an arbitrary tensor and C_j is a covariant vector. To transform these quantities to the $\bar{x}^\alpha$ coordinate system, it becomes

$$\bar{A}_{ij}\bar{B}^i = \bar{C}_j.$$

According to Eq. (A.10), it takes the form

$$\bar{A}_{ij}\frac{\partial \bar{x}^i}{\partial x^\alpha}B^\alpha = \frac{\partial x^\beta}{\partial \bar{x}^j}C_\beta = \frac{\partial x^\beta}{\partial \bar{x}^j}A_{\alpha\beta}B^\alpha$$

or

$$\left(\bar{A}_{ij}\frac{\partial \bar{x}^i}{\partial x^\alpha} - \frac{\partial x^\beta}{\partial \bar{x}^j}A_{\alpha\beta}\right)B^\alpha = 0$$

The inner multiplication of this equation with $\partial \bar{x}^j/\partial x^\beta$ gives

$$\left(\bar{A}_{ij}\frac{\partial \bar{x}^i}{\partial x^\alpha}\frac{\partial \bar{x}^j}{\partial x^\beta} - A_{\alpha\beta}\right)B^\alpha = 0.$$

Since B^α is an arbitrary contravariant tensor, it follows that

$$\bar{A}_{ij}\frac{\partial \bar{x}^i}{\partial x^\alpha}\frac{\partial \bar{x}^j}{\partial x^\beta} = A_{\alpha\beta}.$$

This proves that A_{ij} is a covariant tensor of the second order.

Equation (A.9) indicates that in a rectangular coordinate system, the differentiation of any quantity with respect to one of the coordinates increases the covariant order of that quantity by one. However, in other curvilinear coordinate systems, the differentiation of a quantity by one of the coordinates does not necessarily satisfy the law of transformation for covariant tensor, Eq. (A.8), unless the quantity is a scalar. In order to raise the covariant order by one, it is necessary to use the so-called covariant differentiation instead of ordinary differentiation. For a mixed tensor $A_{r_1 r_2 \cdots r_p}^{u_1 u_2 \cdots u_s}$, its covariant derivative $A_{r_1 r_2 \cdots r_p, n}^{u_1 u_2 \cdots u_s}$ with respect to x^n is defined by

$$A_{r_1 r_2 \cdots r_p, n}^{u_1 u_2 \cdots u_s} = \frac{\partial}{\partial x^n}A_{r_1 r_2 \cdots r_p}^{u_1 u_2 \cdots u_s} + \sum_{\alpha=1}^{s}\Gamma_{kn}^{u_\alpha}A_{r_1 r_2 \cdots r_p}^{u_1 u_2 \cdots u_{\alpha-1}ku_{\alpha+1}\cdots u}$$

$$- \sum_{\beta=1}^{p}\Gamma_{r_\beta n}^{l}A_{r_1 r_2 \cdots r_{\beta-1}lr_{\beta+1}\cdots r_p}^{u_1 u_2 \cdots u_s}, \tag{A.11}$$

where $\Gamma^i_{jk} = (1/2)g^{il}(g_{jl,k}+g_{kl,j}-g_{gk,l})$ is the Christoffel symbol introduced in Eqs. (2.33) and (3.16). It should be noted that the symbol ",n" used in covariant differentiation has different meaning from the ",i" used in the main text, even in the symbol $\Gamma^l_{jk,i}$, where it means simple differentiation with respect to the coordinate i.

The covariant derivative of the sum of two tensors is equal to the sum of the covariant derivative of each tensor. The covariant derivative of the outer product (or inner product) of two tensors is equal to the sum of the outer product (or inner product) of one tensor and the covariant derivative of the other tensor, e.g.,

$$(A_{ij}B^l)_{,m} = A_{ij,m}B^l + A_{ij}B^l_{,m},$$
$$(A_{ij}B^j)_{,m} = A_{ij,m}b^j + A_{ij}B^j_{,m}. \qquad (A.12)$$

These can be proved directly by the definition of covariant differentiation. If all the components of the tensor $A^{u_1 u_2 \cdots u_s}_{r_1 r_2 \cdots r_p}$ are functions of the parameter t along the curve $x^i = x^i(t)$ then the derivative of $A^{u_1 u_2 \cdots u_s}_{r_1 r_2 \cdots r_p}$ with respect to t should be replaced by the intrinsic derivative (absolute derivative) $(\delta/\delta t)A^{u_1 u_2 \cdots u_s}_{r_1 r_2 \cdots r_p}$ order to keep the order of the original tensor. Intrinsic derivative is defined by

$$\frac{\delta}{\delta t}A^{u_1 u_2 \cdots u_s}_{r_1 r_2 \cdots r_p} \equiv A^{u_1 u_2 \cdots u_s}_{r_1 r_2 \cdots r_p, k}\frac{dx^k}{dt}. \qquad (A.13)$$

Corresponding to the permutation symbols e_{lmn} and e^{lmn} introduced in Eq. (3.6), there are the permutation tensors ϵ_{lmn} and ϵ^{lmn} satisfying the relations

$$\bar{\epsilon}_{lmn} = \epsilon_{ijk}\frac{\partial x^i}{\partial \bar{x}^l}\frac{\partial x^j}{\partial \bar{x}^m}\frac{\partial x^k}{\partial \bar{x}^n}, \quad \epsilon^{lmn} = \bar{\epsilon}_{ijk}\frac{\partial x^l}{\partial \bar{x}^i}\frac{\partial x^m}{\partial \bar{x}^j}\frac{\partial x^n}{\partial \bar{x}^k}, \qquad (A.14)$$

where

$$\epsilon_{ijk} = \sqrt{g}e^{ijk}, \quad \epsilon^{ijk} = \epsilon_{ijk}/\sqrt{g}.$$

In terms of the permutation tensor, the curl of the covariant vector, $\mathbf{A}$, $\mathbf{B} = \nabla \times \mathbf{A}$, may be written as

$$B^j = \epsilon^{jkl}A_{l,k} \qquad (A.15)$$

The cross product of two covariant vectors $\mathbf{A}$ and $\mathbf{B}$, $\mathbf{C} = \mathbf{A} \times \mathbf{B}$, may be written as

$$C^i = \epsilon^{ijk}A_jB_k. \qquad (A.16)$$

Appendix B

The Gradient Operator

Chapter 2 has shown that the necessary and sufficient condition for the parametric curves $u = $ const. and $v = $ const. on the surface $\mathbf{Y}(u, v)$ to be orthogonal is that the fundamental magnitude of the first order $g_{uv} = 0$. Consider displacements $d\mathbf{Y}(u, v) = \mathbf{Y}_{,u}du + \mathbf{Y}_{,v}dv$ and $\delta\mathbf{Y}(u, v) = \mathbf{Y}_{,u}\delta u + \mathbf{Y}_{,v}\delta v$ on the surface $\mathbf{Y}(u, v)$. The inclination angle ψ between $d\mathbf{Y}$ and $\delta\mathbf{Y}$ satisfies the relations:

$$ds\delta s \cos\psi = d\mathbf{Y} \cdot \delta\mathbf{Y},$$

$$ds\delta s \sin\psi = |d\mathbf{Y} \times \delta\mathbf{Y}|$$

$$= |du\delta v - dv\delta u| \cdot |\mathbf{Y}_{,u} \times \mathbf{Y}_{,v}| = \sqrt{g}|du\delta v - dv\delta u|.$$

The necessary and sufficient conditions for these two displacements are perpendicular to each other such that

$$g_{uu}du\delta u + g_{uv}(du\delta v + \delta u dv) + g_{vv}dv\delta v = 0. \tag{B.1}$$

The gradient of the scalar function $\phi(u, v)$ at any point P on the curve is defined as a vector whose direction is that direction on the surface at P along which the curve lies, which gives the maximum arc-rate of increase of ϕ, and whose magnitude is this maximum rate of increase. In space, besides the two components on the surface, the gradient operator has another component along the normal direction $\mathbf{n}$ of the surface, that is, $\mathbf{n}\partial/\partial n$. Thus, we deduce[1]

$$\nabla = \nabla' + \mathbf{n}\frac{\partial}{\partial n}. \tag{B.2}$$

A curve $\phi = $ const is called a level curve of the function. Let C and C' be two neighboring-level curves corresponding to ϕ and $\phi + d\phi$ (Fig. B.1). Let PQ be an element of the orthogonal trajectory of the level curves intercepted

265

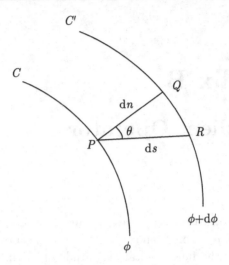

Fig. B.1.   Two neighboring level curves corresponding to ϕ and $\phi + \mathrm{d}\phi$.

between C and C' and $\mathrm{d}n$ be the length of the element. Let PR be an element of arc of another curve through P cutting C at R with length $\mathrm{d}s = PR$. Clearly, PQ is the shortest distance between C and C' at P and its direction gives the maximum rate of increase of ϕ at P. Thus, the gradient of ϕ at P, $\nabla'\phi$ is equal to $\mathrm{d}\phi/\mathrm{d}n$ and its direction is along PQ. The rate of increase of ϕ in the direction PR is given by $\mathrm{d}\phi/\mathrm{d}s = (\mathrm{d}\phi/\mathrm{d}n)/(\mathrm{d}n/\mathrm{d}s) = (\mathrm{d}\phi/\mathrm{d}n)\cos\theta$, where θ is the inclination of PR to PQ. The relation may be written as $\mathrm{d}\phi = \mathrm{d}\mathbf{s} \cdot \nabla'\phi$. The curves $\phi = \mathrm{const}$ will be parallel provided $\nabla'\phi$ is the same for all points on the same curve. The curve $\phi = \mathrm{const}$. will be orthogonal to the curve $\psi = \mathrm{const}$. if

$$\nabla'\phi \cdot \nabla'\psi = 0. \tag{B.3}$$

Let $(\delta u, \delta v)$ be an infinitesimal displacement along the curve $\phi(u, v) = \mathrm{const}$, then

$$\phi_{,u}\delta u + \phi_{,v}\delta v = 0.$$

According to Eq. (B.1), a displacement $(\mathrm{d}u, \mathrm{d}v)$ perpendicular to $(\delta u, \delta v)$ satisfies the relation

$$\frac{\mathrm{d}u}{\mathrm{d}v} = \frac{g_{vv}\phi_{,u} - g_{uv}\phi_{,v}}{g_{uu}\phi_{,v} - g_{uv}\phi_{,u}}.$$

Hence, the vector $(g_{vv}\phi_{,u} - g_{uv}\phi_{,v})\mathbf{Y}_{,u} + (g_{uu}\phi_{,v} - g_{uv}\phi_{,u})\mathbf{Y}_{,v}$ is perpendicular to the curve $\phi = \mathrm{const}$., or in other words it is in the direction of

$\nabla'\psi$ and

$$\nabla'\phi = (1/g)(g_{vv}\phi_{,u} - g_{uv}\phi_{,v})\mathbf{Y}_{,u} + (1/g)(g_{uu}\phi_{,v} - g_{uv}\phi_{,u})\mathbf{Y}_{,v}$$
$$= g^{ij}\mathbf{Y}_{,i}\partial\phi/\partial j.$$

Therefore, the gradient operator ∇' is given by

$$\nabla' = g^{ij}\mathbf{Y}_{,i}\frac{\partial}{\partial j},$$

and the gradient operator ∇ is given by

$$\nabla = \nabla' + \mathbf{n}\frac{\partial}{\partial n}. \tag{B.4}$$

Reference

[1] C. E. Weatherburn, *Differential Geometry of Three Dimensions* (Cambridge University Press, London, 1955), Vol. 1, p. 220.

Appendix C

Elastic Theory of Membranes Viewed from Force Balance

A 2D continuum can be simplified as a surface as shown in Fig. C.1. At each point, we can select a frame $\{\mathbf{e}_1, \mathbf{e}_2, \mathbf{e}_3\}$. A pressure p is loaded on the surface in the inverse direction of the normal vector $\mathbf{e}_3$. Let us cut a region enclosed in any curve C from the surface. $\mathbf{t}$ is the tangent vector at point of curve C. $\mathbf{b}$ is normal to $\mathbf{t}$ and in the tangent plane. The force and moment per length performed by the other region on curve C are denoted as $\mathbf{f}$ and $\mathbf{m}$, respectively. Through Newton's laws, the force and moment balance conditions are obtained as

$$\oint_C \mathbf{f} \, \mathrm{d}s - \int p\mathbf{e}_3 \, \mathrm{d}A = 0, \tag{C.1}$$

$$\oint_C \mathbf{m} \, \mathrm{d}s + \oint_C \mathbf{r} \times \mathbf{f} \, \mathrm{d}s - \int \mathbf{r} \times p\mathbf{e}_3 \, \mathrm{d}A = 0, \tag{C.2}$$

where $\mathrm{d}s$ and $\mathrm{d}A$ are the arc length element of curve C and area element of the region enclosed in curve C, respectively.

Define two second-order tensors $\mathfrak{S}$ and $\mathfrak{M}$ such that

$$\mathfrak{S} \cdot \mathbf{b} = \mathbf{f}, \quad \mathfrak{M} \cdot \mathbf{b} = \mathbf{m}. \tag{C.3}$$

These two tensors can be called the stress tensor and bending moment tensor, respectively. Using the Stokes' theorem, we can derive

$$\int (\mathrm{div}\,\mathfrak{S} - p\mathbf{e}_3) \, \mathrm{d}A = 0, \tag{C.4}$$

$$\int (\mathrm{div}\,\mathfrak{M} + \mathbf{e}_1 \times \mathfrak{S}_1 + \mathbf{e}_2 \times \mathfrak{S}_2) \, \mathrm{d}A = 0. \tag{C.5}$$

where $\mathfrak{S}_1 = \mathfrak{S} \cdot \mathbf{e}_1$ and $\mathfrak{S}_2 = \mathfrak{S} \cdot \mathbf{e}_2$. Since the integral is performed on the region enclosed in an arbitrary curve C, from the above two equations, we

269

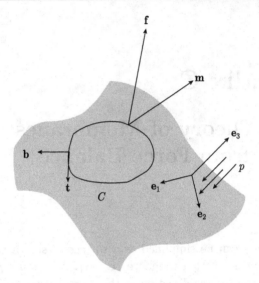

Fig. C.1.   Force and moment in a 2D continuum.

obtain the force and moment balance conditions of 2D continua as:

$$\operatorname{div} \mathfrak{S} = p\mathbf{e}_3, \qquad (C.6)$$

$$\operatorname{div} \mathfrak{M} = \mathfrak{S}_1 \times \mathbf{e}_1 + \mathfrak{S}_2 \times \mathbf{e}_2. \qquad (C.7)$$

The shape equation (3.46) can also be derived from the above two equations.

Appendix D

A Different Viewpoint of Surface Tension

Although the lipid bilayer cannot withstand the in-plane shear strain, it can still endure the in-plane compression strain. The in-plane compression modulus, k_b, of lipid bilayers is about $0.24\,\text{N/m}$.[1] Considering this point, we may write the free energy of a closed lipid vesicle as

$$\mathcal{F} = \Delta p \int \mathrm{d}V + \int G_B \mathrm{d}A + \int \frac{k_b}{2}(2J_b)^2 \mathrm{d}A, \qquad (D.1)$$

where

$$G_B = (k_c/2)(2H + c_0)^2 - \bar{k}K, \qquad (D.2)$$

and J_b is the in-plane compression or stretching strain. We emphasize that the contribution of chemical potential is omitted when we write the above free energy.

The first-order variation of the free energy (D.1) reveals that $2J_b$ is a constant and then

$$\Delta p - 2(2k_b J_b)H + 2k_c \nabla^2 H + k_c(2H + c_0)(2H^2 - c_0 H - 2K) = 0. \tag*{(D.3)}$$

Comparing the above equation with the shape equation (3.46) of lipid vesicles, we deduce that

$$\lambda = 2k_b J_b. \qquad (D.4)$$

In the discussion on the stability of closed lipid vesicles, we have seen that the surface tensor λ has no effect on the critical pressure. The second-order variation of the free energy (D.1) can give the same conclusion. $\delta^2[\Delta p \int \mathrm{d}V + \int G_B \mathrm{d}A]$ has been shown in Eq. (3.54) with vanishing λ.

The additional term is

$$\delta^2 \int \frac{k_b}{2}(2J_b)^2 \mathrm{d}A = \int k_b(\mathrm{div}\,\mathbf{v} - 2H\Omega_3)^2 \mathrm{d}A, \qquad (\mathrm{D.5})$$

where $\mathbf{v} = \Omega_1\mathbf{e}_1 + \Omega_2\mathbf{e}_2 + \Omega_3\mathbf{e}_3$ represents the infinitesimal displacement vector of the vesicle surface. We can always select the proper deformation modes such that $\mathrm{div}\,\mathbf{v} - 2H\Omega_3 = 0$ and then $\delta^2 \int (k_b/2)(2J_b)^2 \mathrm{d}A$ vanishes, but $\delta^2[\Delta p \int \mathrm{d}V + \int G_B \mathrm{d}A]$ is not affected. That is, the critical pressure is determined merely by $\delta^2[\Delta p \int \mathrm{d}V + \int G_B \mathrm{d}A]$, which is independent of the compression modulus of lipid bilayer k_b.

Reference

[1] W. Rawicz, K. C. Olbrich, T. McIntosh, D. Needham and E. Evans, *Biophys. J.* **79** (2000) 328.

Index

Printed in the United States
By Bookmasters